CORE
Mathematics
for Cambridge IGCSE

Audrey Simpson

CAMBRIDGE
UNIVERSITY PRESS

CAMBRIDGE UNIVERSITY PRESS

Cambridge, New York, Melbourne, Madrid, Cape Town, Singapore,
São Paulo, Delhi, Dubai, Tokyo, Mexico City

Cambridge University Press
4381/4 Ansari Road, Daryaganj, Delhi 110002, India

www.cambridge.org
Information on this title: www.cambridge.org/9780521727921

First Published 2011

Printed in India at Replika Press Pvt. Ltd.

A catalogue for this publication is available from the British Library

ISBN 978-0-521-72792-1 Paperback

Contents

Introduction

Core Mathematics for Cambridge IGCSE can be used in three different ways.

- As a stand-alone textbook it forms the entire two-year course for the Cambridge IGCSE Mathematics Core Level examination from University of Cambridge International Examinations. It is designed to be worked through sequentially so that students studying alone and those in the classroom are taken step by step through the course. All elements of the Core course are covered. Each topic is carefully explained so that the book can be used for self-study or as a back-up for classroom work.
- The book can also be used as the first year of a two-year course leading to the Cambridge IGCSE Mathematics Extended Level examination. Students would then have the option of taking the Core examination at the end of their first year before going on to study the second book (*Extended Mathematics for Cambridge IGCSE*) in their second year. This would give them valuable examination practice. They would then take the Examination at the Extended level at the end of their second year.
- Alternatively, students studying for the Extended Level examination can work through each chapter in the Core Book followed by the corresponding chapter in the Extended Book, taking two years to work through both books.

Students will find that the structure of the book enables them to proceed at their own pace by reading the explanatory text, following the worked examples and then working through the exercises with frequent checking of the answers at the back of the book. They are encouraged to give sufficient working in their answers to show that they understand the methods required to obtain the correct answer.

A final section gives suggestions for establishing a good revision programme, and hints to enable students to gain the best possible marks in their examination.

Audrey Simpson

Acknowledgements

I would like to thank Professor Gordon Kirby for his invaluable advice and encouragement. I am also grateful for his efforts to check my work patiently for errors, both mathematical and stylistic.

I am also indebted to my sister, Pat Victor, for the times she sorted out frustrating problems with both my computer and the software needed for the production of the manuscript.

Audrey Simpson

Note

A note about the currencies used in this book:

1 dollar ($) is worth 100 cents.
1 rupee [Re (singular) or Rs (plural)] is worth 100 paise (singular paisa).
1 pound (£) is worth 100 pence.
1 euro (€) is worth 100 cents.

Unless greater accuracy is required, answers to money questions which are not whole numbers, should be given to 2 decimal places.

For example, $12.7 would be given as $12.70, which is read 'twelve dollars and seventy cents'.

Chapter 1

Understanding Number

By the end of this chapter, you should know more about the different types of numbers that you need to study for the rest of the course. You may feel that you know most of it already, but please work through it as there are plenty of things in it that will help you gain valuable marks in your examination. Treat it as revision if you like.

Read the text and try to do the worked examples before you look at the answers. Cover up the answers, write down your answers to the examples, then compare them with the given working to see if you definitely understood the concept. Try all the exercises and check your answers with those at the back as you go along. It is best not to do the whole exercise (possibly incorrectly) before you check the answers. You want to practise getting the answers right, not wrong!

If you were an athlete, and wanted to win competitions, you would train your body to do your workouts correctly and efficiently from the beginning. For mathematics, if you want to get the best possible marks in your examinations, you should try to train your mind to do the work correctly *and to set out your answers neatly and legibly* from the outset. If your answers are not clear and your examiner cannot read them, you are in danger of losing valuable marks. Every mark counts, so start now!

Essential Skills

To get the most from this course you should know the multiplication tables from 2 to 10 and be able to recall them without hesitation. It is also important to know the facts about addition and subtraction.

Try the following mini-test and see how quickly you can answer the questions without a calculator.

(a) 4×6	(b) 3×7	(c) 8×5	(d) 9×8
(e) 2×7	(f) 6×9	(g) 8×8	(h) 9×5
(i) 7×7	(j) 3×6	(k) $6 + 7$	(l) $5 + 8$
(m) $9 + 7$	(n) $3 + 5 + 9$	(o) $8 + 9$	(p) $11 + 9$
(q) $13 + 6$	(r) $3 + 4 + 5$	(s) $16 + 5$	(t) $4 + 17$
(u) $9 - 4$	(v) $11 - 7$	(w) $15 - 9$	(x) $7 - 4$
(y) $16 \div 8$	(z) $24 \div 6$		

Here is a rough guide, just for you, so that you can get an idea of how much you might need to practise.

Time: 1 minute, 25 correct   Excellent!
         3 minutes, 20 correct   Very good!
         5 minutes, 18 correct   Not bad!
         More than 5 minutes   Keep practising!

Answers

(a) 24	(b) 21	(c) 40	(d) 72	(e) 14	(f) 54
(g) 64	(h) 45	(i) 49	(j) 18	(k) 13	(l) 13
(m) 16	(n) 17	(o) 17	(p) 20	(q) 19	(r) 12
(s) 21	(t) 21	(u) 5	(v) 4	(w) 6	(x) 3
(y) 2	(z) 4				

Sets of Numbers

The numbers that we use today have developed over a period of time as the need arose. At first humans needed numbers just to count things, so the simplest set of numbers was the set of **natural** or **counting** numbers. We use the symbol N to represent the counting numbers, and we use curly brackets to list some of these numbers.

$$N = \{1, 2, 3, 4 \ldots\}$$

The dots at the end mean 'and so on' because the list goes on forever. (Lists like these are often shown in curly brackets, however, this is not essential.)

When addition and subtraction were introduced, a new set of numbers was needed.

For example, I had three goats. Three were stolen. How many goats do I have now?

We know that the answer is none or zero, which does not appear in the counting numbers. Subtraction also meant that negative numbers were needed, as we will see later in this chapter. Our next set of numbers is the set of **integers**, which have the symbol Z, and include negative whole numbers, zero and the natural numbers.

$$Z = \{\ldots -3, -2, -1, 0, 1, 2, 3 \ldots\}$$

After addition and subtraction came division and multiplication. What happens when we divide two by three?

The answer is that we get the fraction $\frac{2}{3}$. But where does that fit in with our latest set of numbers? We need another set which includes all the fractions or **rational numbers**. This is the set Q.

Practical Work

- Make yourself an integer number line on a long strip of paper.

- Mark on it the integers from −20 through zero to +20. Make sure they are evenly spaced.

- Fold the strip and stick it on the inside cover of your exercise book so that you can unfold it whenever you need it later in the course.

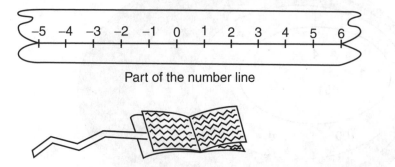

Part of the number line

Rational numbers can all be expressed as fractions or ratios made up of one integer over another. Remember, for example, that 5 can be written as $\frac{5}{1}$, so integers themselves are included in the set of rational numbers. We can only list some examples of this set because there is an infinite number of members belonging to **Q**.
Some examples of rational numbers are:

$$\frac{2}{3}, \quad \frac{5}{2}, \quad -2\frac{1}{2}, \quad \frac{3}{100}, \quad 5, \quad 0, \quad 29, \quad -500, \quad \text{etc.}$$

The last set we need for our number sets is the set of **real numbers, R**. This includes all the previous sets and also the irrational numbers. **Irrational numbers** are numbers which *cannot* be written as fractions (or ratios) made up of one integer over another.
The Greek letter π (which is spelled and pronounced as pi) is used to represent what is perhaps the most famous irrational number. Pi is the number you get when you divide the length of the circumference of a circle by its diameter. *You can never find the value of π exactly.* We will do some experiments later in the course to see how close we can get to the calculated value of π.
Irrational numbers include square roots of numbers that are not perfect squares themselves, and as we find in the case of π, irrational numbers are decimals that go on and on forever, and never repeat any pattern.
The number $\pi\,(= \mathbf{3.14159265358979323846264\ldots})$ has been calculated to billions of places of decimal by high-powered computers, using a more advanced method than measuring the circumference and diameter of a circle. However, no recurring pattern has been found.
Recurring decimals are not irrational numbers because they can always be written as fractions.

For example, $0.66666666666\ldots = \dfrac{2}{3}$, and $0.285714285714285714\ldots = \dfrac{2}{7}$.

Recurring decimals do, of course, have a repeating pattern unlike irrational numbers.
Write down the sequence of numbers that recur in the decimal equivalent of $\dfrac{2}{7}$.
The diagram will help you to see how these sets of numbers build up:

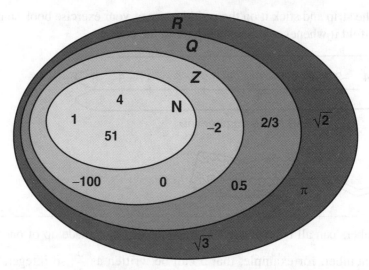

Each number type has been drawn with two or three examples in it.
Another way to show these sets is on number lines. Some examples of each set are shown below. The arrows show that the sets go on forever in that direction.

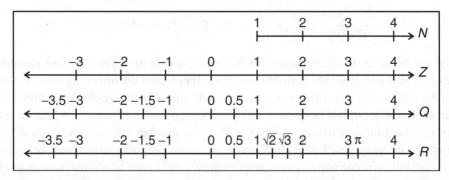

Example 1

$$2 \quad \sqrt{3} \quad \dfrac{1}{1000} \quad -99 \quad 2\dfrac{1}{2} \quad -\dfrac{1}{4} \quad \pi \quad 0.3 \quad 0 \quad 2005$$

From the list given above select:
(a) the natural numbers   (b) the integers   (c) the rational numbers
(d) the irrational numbers   (e) the real numbers

Answer 1

(a) The natural numbers (N) are: 2 and 2005

(b) The integers (Z) are: $-99, 0, 2$ and 2005 (because each larger set includes the set before it)

(c) The rational numbers (Q) are: $-99, -\dfrac{1}{4}, 0, \dfrac{1}{1000}, 0.3, 2\dfrac{1}{2}, 2, 2005$

(d) The irrational numbers are: $\sqrt{3}$ and π (because these are decimals that go on forever with no repeating pattern)

(e) The real numbers (R) are: $2, \sqrt{3}, \dfrac{1}{1000}, -99, 2\dfrac{1}{2}, -\dfrac{1}{4}, \pi, 0.3, 2005, 0$

Within the above sets of numbers there are other, smaller sets. Some of these sets are discussed below.

Prime Numbers, Factors and Multiples

In this section we will use natural numbers only.

Prime numbers are natural numbers that can only be divided by themselves or by 1. Some examples of prime numbers are:

$$2, 3, 5, 7, 11, 13, 17 \dots$$

Notice that 1 is *not* counted as a prime number, and 2 is the only *even* prime number.

Example 2

Write a list of all the prime numbers between 20 and 35.

Answer 2

23, 29, 31

(all the other numbers between 20 and 35 can be divided by numbers other than just themselves or 1)

The **factors** of a number are the natural numbers that can be multiplied together to make the number.

For example, 2 and 3 are factors of 6 because $2 \times 3 = 6$.

The **multiples** of a number are obtained by multiplying the number by other natural numbers.

For example, the multiples of 12 would include 24, 36, 48 and so on.

NOTE: You may need to find a way of remembering which are factors of a number, and which are multiples of the number.

Perhaps you can remember that multiples are bigger than the original number, or that they are in the times (multiply) table for that number.

So the multiples of 2 are 2, 4, 6, 8, 10 ...

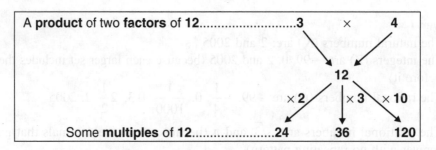

The factors of 12 in the diagram above are shown multiplied together. This is called a **product of factors**. So numbers that are multiplied together are called **factors**, and the result of multiplying them together is called the **product**. There are other factors of 12. Altogether the factors of 12 are: 1, 2, 3, 4, 6 and 12 (all the numbers that will divide 12 without leaving a remainder).

Of particular interest are the prime factors. The prime numbers among the factors of 12 are 2 and 3. We can write 12 as a **product** of its **prime factors**:

$$12 = 2 \times 2 \times 3$$

or we can **list** the prime factors of 12: {2, 3}.

A factor tree is a neat method for finding prime factors of larger numbers. The following example will show you how to make a factor tree.

Example 3

Write 200 as a product of its prime factors.

Answer 3

First make a list of the smaller prime numbers: 2, 3, 5, 7 …
Start by dividing by 2, and repeat until the number will no longer divide by 2. Then work through your list in order, trying 3, then 5 and so on.

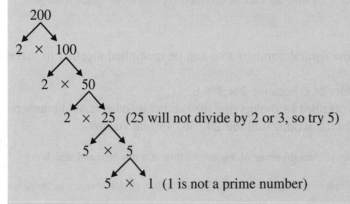

The answer is: $200 = 2 \times 2 \times 2 \times 5 \times 5$.
(Check this with your calculator)

Example 4

(a) List all the factors of 18.

(b) List the prime factors of 18.

(c) Write 18 as a product of its prime factors.

(d) List three multiples of 18.

Answer 4

(a) {1, 2, 3, 6, 9, 18}

(b) {2, 3}

(c) $18 = 2 \times 3 \times 3$

(d) For example, 36 (18 × 2), 54 (18 × 3), 90 (18 × 5)

Exercise 1.1

1. 5,   −100,   −3.67,   π,   0,   1507, $\dfrac{99}{7}$, $\dfrac{6}{1}$

   From the list above:
   (a) write down all the real numbers,
   (b) write down all the rational numbers,
   (c) write down all the integers,
   (d) write down all the natural numbers.
   (e) One of the numbers is irrational. Which is it?

2. (a) List all the factors of 30.
   (b) List the prime factors of 30.
   (c) Write 30 as a product of its prime factors.
       (Use your calculator to check your answer.)
   (d) Write down three multiples of 30.

3. 1,   4,   30,   45,   5,   15,   9,   1500,   3,   10
   From this list choose:
   (a) the multiples of 15,
   (b) the factors of 15.

4. Use a factor tree to find the prime factors of 240.
   Write your answer:
   (a) as a list of prime factors,
   (b) as a product of prime factors.

5. Write down all the prime numbers between 20 and 40.

6. Use your calculator if necessary to find out which of the following numbers are prime numbers.
   37,   49,   53,   81,   87,   93,   101

7. Write down a list of numbers between 80 and 90, including 80 and 90.

   From your list find:

   (a) two prime numbers,

   (b) three multiples of 5,

   (c) a factor of 348.

Highest Common Factor (HCF) and Lowest Common Multiple (LCM)

Common means 'belonging to all'.

We often need to find the factors or multiples of two (or more) numbers that belong to both (or all) the numbers. One way to do this is to list all the factors or multiples of both numbers and see which factors or multiples occur in both lists.

The following example shows how this is done.

Example 5

(a) (i) List all the factors of 30.

   (ii) List all the factors of 20.

   (iii) From your two lists find the common factors of 20 and 30 (not including 1).

(b) (i) List the first four multiples of 30 (not including 30 itself).

   (ii) List the first five multiples of 20 (not including 20 itself).

   (iii) From your two lists, find any common multiples.

(c) Find the HCF of 30 and 20.

(d) Find the LCM of 30 and 20.

Answer 5

(a) (i) {1, 2, 3, 5, 6, 10, 15, 30}   (b) (i) {60, 90, 120, 150}

   (ii) {1, 2, 4, 5, 10, 20}   (ii) {40, 60, 80, 100, 120}

   (iii) {2, 5, 10}   (iii) {60, 120}

(c) 10   (d) 60

Using the above example you should see that finding the highest common factor (HCF) of 20 and 30 is simple. It is the highest number that appears in both lists of factors of both the numbers. The HCF of 20 and 30 is 10.

Similarly, the lowest common multiple of 20 and 30 is the smallest number that appears in both lists of multiples. The LCM of 20 and 30 is 60.

Tests of Divisibility

Before you go any further you might like to try some tests of divisibility which can help you save time in these questions. These tests show what will divide into a number without leaving a remainder.

- **Divisibility by 2:** All even numbers divide by 2. (All even numbers end in 2, 4, 6, 8 or 0.)

- **Divisibility by 3:** This is a rather surprising test, but it does work!

Add all the digits (individual numbers) of the entire number together. If the result is 3, 6 or 9 then the number will divide by 3. If the result is 10 or more, keep adding the digits until you get to a single digit. This is called finding the **digital root** of the number. If the digital root is 3, 6 or 9 then the number will divide by 3.

For example, the digital root of 2115 is $2 + 1 + 1 + 5 = 9$, so 2115 will divide by 3 (check it on your calculator).

Of course it does not matter what order the digits of the number appear or if any zeroes appear in the number so 5121, 2511, 12510, 105120 (and so on), will all divide by 3.

To find the digital root of 3672:

$$3 + 6 + 7 + 2 = 18$$
$$1 + 8 = 9$$

So the digital root of 3672 is 9. Hence, 3672 will divide by 3.

- **Divisibility by 5:** All numbers ending in 5 or 0 will divide by 5. Therefore, 3672 will not divide by 5 whereas 3670 will.

- **Divisibility by 6:** All *even* numbers with a digital root of 3, 6 or 9 will divide by 6. 3672 will divide by 6.

- **Divisibility by 9:** All numbers with a digital root of 9 will divide by 9. 3672 will divide by 9.

Example 6

(a) Test 552 for divisibility by 2, 3, 5, 6 and 9.

(b) Test 6165 for divisibility by 2, 3, 5, 6 and 9.

Answer 6

(a) 552 is even so it will divide by 2.

$5 + 5 + 2 = 12 \rightarrow 1 + 2 = 3$, so it will divide by 3.

552 does not end in 5 or 0, so it will not divide by 5.

552 is even *and* it will divide by 3, so it will also divide by 6.

The digital root of 552 is 3, not 9, so it will not divide by 9.

(b) 6165 is not even, so it will not divide by 2.

$6 + 1 + 6 + 5 = 18 \rightarrow 1 + 8 = 9$, so it will divide by 3.

6165 ends in 5, so it will divide by 5.

Although 6165 will divide by 3 it is not even, so it will not divide by 6.

The digital root of 6165 is 9, so it will divide by 9.

Exercise 1.2

1. (a) List all the factors of 8. Then list all the factors of 12.
   (b) Find the highest common factor of 8 and 12.

2. Find the highest common factor of 21 and 42.

3. (a) List all the factors of:
   (i) 15   (ii) 35   (iii) 20
   (b) Write down the highest common factor of 15, 35 and 20.

4. (a) List the first six multiples of 12 and of 8.
   (b) Write down the lowest common multiple of 12 and 8.

5. Find the lowest common multiple of 3, 5 and 12.

6. Test 21603 for divisibility by 2, 3, 5 and 9. Explain your reasoning
   (see Example 6).

7. Test 515196 for divisibility by 2, 3, 5, 6 and 9. Explain your reasoning.

Operations and Inverses

Mathematical operations like addition, or division, have inverses which 'undo' the operation.

For example, $2 \times 3 = 6$, and $6 \div 3 = 2$.

Division is the inverse of multiplication because it 'undoes' multiplication.

Also multiplication is the inverse of division, as you can see below.

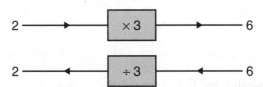

What do you think is the inverse of addition? Look at the next diagram.

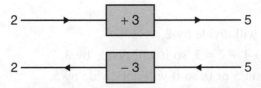

Squares and Square Roots, Cubes and Cube Roots

The square of a number is the result of multiplying a number by itself.

For example, the square of 9 is $9 \times 9 = 81$, the square of 11 is $11 \times 11 = 121$, and the square of 35 is $35 \times 35 = 1225$.

The compact way of showing that the number is to be squared is to write it to the power of 2.

So the square of 9 is written as $9^2 = 81$. Similarly $11^2 = 121$, and $35^2 = 1225$.

Finding the square root of a number *undoes* the squaring, so for example, the square root of 81 is 9, the square root of 121 is 11, and the square root of 1225 is 35.

The compact way of showing that the square root of a number is to be found is to use the square root sign: $\sqrt{}$.

So the square root of 81 is written as $\sqrt{81} = 9$, also $\sqrt{121} = 11$ and $\sqrt{1225} = 35$.

You should be able to see that squaring and finding the square root undo each other. As we have seen above, operations which 'undo' each other are inverses of each other. Hence, squaring and finding the square root are **inverse operations**.

We will come across more inverse operations later in the course.

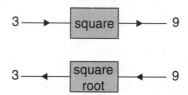

We can find the square of any number. My calculator tells me that the square of 2.41 is 5.8081. It also tells me that $\sqrt{468.2896} = 21.64$.

However, not all numbers have exact square roots. For example, $\sqrt{2}$, $\sqrt{3}$ or $\sqrt{5}$ are numbers with decimals that 'go on forever' without any repeating pattern, and are irrational numbers.

$\sqrt{49} = 7$ exactly, so $\sqrt{49}$ is a rational number. As we see above, $\sqrt{468.2896} = 21.64$, so $\sqrt{468.2896}$ is a rational number.

My calculator tells me that $\sqrt{7.2} = 2.683281573$, so $\sqrt{7.2}$ *looks* as if it could be an irrational number, although we cannot tell for certain by this method alone.

NOTE: It would be helpful to learn to recognise some of the irrational numbers, such as: π, $\sqrt{2}$, $\sqrt{3}$, $\sqrt{5}$, $\sqrt{7}$ and $\sqrt{10}$, so that you can give examples when required.

The numbers that you get when you square the natural numbers are called **perfect squares**. They are called perfect squares because on being square rooted they give whole numbers. The first four perfect squares are 1, 4, 9, 16. Write down the next three square numbers.

The **cube** of a number is the result of multiplying that number by itself twice.

The cube of $4 = 4 \times 4 \times 4 = 16 \times 4 = 64$.

The compact way of writing the cube of a number is to write it to the power 3, so the cube of 4 is: $4^3 = 4 \times 4 \times 4 = 64$.

The cubes of first four numbers are 1, 8, 27, 64. What is the cube of the next number? Cube numbers are also often called **cubic numbers**.

NOTE: It would be useful to make a list of the first five square numbers and the first five cube numbers and learn to recognise them. You will come across them quite often.

The inverse of cubing a number is finding the cube root, and you may find that your calculator has a cube root button if you look carefully.

The cube root sign is $\sqrt[3]{}$, so $\sqrt[3]{27} = 3$, (because $3 \times 3 \times 3 = 27$).

Example 7

(a) 12, 6, 7, 36, 125, 5, 15, 4

From the list of numbers choose:

(i) a perfect square   (ii) the square root of 49

(iii) 6^2 (iv) $\sqrt{25}$ (v) 5^3

(b) Use your calculator to find:

(i) 42.3^2 (ii) $\sqrt{9.61}$ (iii) 1.6^3 (iv) $\sqrt[3]{64}$

Answer 7

(a) (i) 36 or 4   (ii) 7   (iii) 36   (iv) 5   (v) 125

(b) (i) 1789.29   (ii) 3.1   (iii) 4.096   (iv) 4

Exercise 1.3

1. For each of the operations below state the inverse.

   (a) multiply   (b) subtract   (c) square   (d) cube root

2. Write down:

   (a) the square of 6   (b) the square root of 9

   (c) 2^3 (d) $\sqrt{25}$ (e) 10^2 (f) 10^3

3. Use your calculator to find:

   (a) 5.2^2 (b) $\sqrt{82.81}$ (c) $\sqrt{100}$ (d) $\sqrt[3]{1000}$

4. $\sqrt{256}$ $\sqrt{6.1}$ $\sqrt{841}$ $\sqrt{7}$ $\sqrt{449.44}$

   Use your calculator to choose from the above list:

   (a) three numbers that you think are rational,

   (b) two numbers that you think are irrational.

   In each case write down all the figures on your calculator display.

5. Write a list of the first seven square numbers.

6. Fill in the gaps in this list of cube numbers.

   1, 8, ….., 64, ……, 216.

7. Using your answers to questions 5 and 6, write down a number which is both a perfect square and a perfect cube.

8. Using your calculator, find another number which is both a perfect square and a perfect cube.

9. 1   2   3   4   5   6   7   8   9   10   11

   Copy the table on the next page. Enter each of the numbers in the list above in the correct rows in your table. (Some numbers may fit in more than one row.)

Natural numbers	
Prime numbers	
Even numbers	
Multiples of 3	
Square numbers	
Cube numbers	
Factors of 20	

Directed Numbers

We have looked at integers, which are positive (with a plus sign) or negative (with a minus sign) whole numbers, or zero, which has no sign.

Directed numbers are also positive or negative but include the whole set of real numbers, hence they also include rational and irrational numbers, as well as integers.

They are called **directed numbers** because they indicate a direction along a number line.

Think of a thermometer that measures temperatures above and below zero.

If the temperature starts at 4° C and *falls* by 5° C it will end at −1° C. This can be written as 4 − 5 = −1.

The minus sign in front of the 5 shows the direction in which the temperature has moved from 4.

The minus sign in front of the 1 shows that it is 1 degree *below* zero. If the temperature starts at 4° C and *rises* by 5° C it will end at +9° C. This can be written as 4 + 5 = +9.

The plus sign shows that the temperature is 9 degrees *above* zero. In practice we do not usually write in the plus sign. If a number is written without a sign it is assumed that it is positive. We are not restricted to whole numbers, so 4 − 5.5 = −1.5.

Example 8

(a) Use the thermometer shown above to find the new temperature in each case below.
  (i) The temperature starts at −6° C and rises by 4° C.
  (ii) The temperature starts at −1° C and falls by 2° C.
  (iii) The temperature starts at −2.5° C and rises by 5.5° C.

(b) Use the thermometer to work out the following:
  (i) 3 − 6   (ii) −5 + 9   (iii) −1 − 3.5   (iv) 3 − 5 + 6
  (v) the difference between 4° C and 7° C
  (vi) the difference between −2° C and −4° C
  (vii) the difference between −2° C and 4° C

(c) Which is warmer, 2° C or −5° C?

Answer 8

(a) (i) $-6 + 4 = -2$, so the new temperature is $-2°$ C

(ii) $-1 - 2 = -3$, so the new temperature is $-3°$ C

(iii) $-2.5 + 5.5 = +3$, so the new temperature is $+3°$ C, (or just $3°$ C)

(b) (i) $3 - 6 = -3$ (ii) $-5 + 9 = +4$ (or just 4)

(iii) $-1 - 3.5 = -4.5$ (iv) $3 - 5 + 6 = -2 + 6 = 4$

(v) $3°$ C (look at the thermometer)   (vi) $2°$ C   (vii) $6°$ C

(c) $2°$ C is warmer than $-5°$ C

Exercise 1.4

1. Draw a thermometer, with temperatures between $-10°$ C and $+10°$ C.
   Use it to complete the following statements.
   (a) $-10 + 5 =$ (b) $-2 - 3 =$ (c) $5 - 8 =$ (d) $0 - 7 =$ (e) $6 + 2 - 3 =$

2.

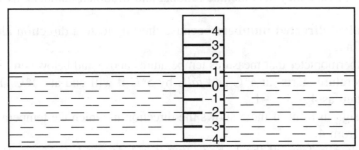

The diagram shows a marker in a reservoir which is used to show the level (in metres) of the water. Copy the diagram and use it to answer the following questions.

(a) Overnight the water level sinks from the level shown in the diagram to -1.5 metres. By how many metres has the water level in the reservoir fallen?

(b) The water level falls another 2.1 metres. What is the new level?

(c) By how much does the water have to rise to bring the level up to 2 metres?

3.

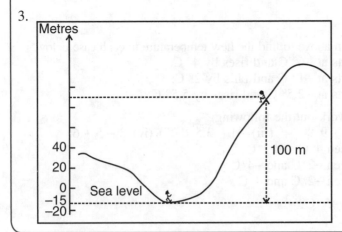

The diagram shows the cross-section of a mountain region. Sea level is 0 metres. A climber starts at 15 metres *below* sea level and climbs 100 metres. How high is he above sea level now?

4.

Bank Account			
Start	Money in	Money out	Balance
−$216			−$216
	$503		(a)
		$290	(b)
	(c)		$0.00

My bank account is overdrawn by $216. The balance (the amount of money I have in the bank) is shown in the first line in the diagram as −$216. This means that I owe the bank $216.

(a) I pay in $503. What should my account balance show now?

(b) I write a cheque for $290 to pay for my electricity. Am I still overdrawn?

(c) If so, how much would I need to pay in to clear my debt?

We will learn about the directed numbers in Chapter 3.

Important Mathematical Symbols

You are already familiar with some mathematical symbols.

For example, $+$, $-$, $\times$, $\div$, π, $\sqrt{}$ and $=$.

Another symbol which is sometimes used is $\neq$ which means 'is not equal to'.

For example, $4 \neq 7$, or 'four is not equal to seven'.

We also need to be able to use symbols to mean 'is greater (or larger) than', or 'is less (or smaller) than'.

For example, we need a mathematical way of writing 'four is less (or smaller) than seven'. This is written as $4 < 7$.

We can also write $7 > 4$. This means that 'seven is greater than four'.

Suppose we wanted to say that the number of days in February is greater than or equal to 28? This would be written as: Days in February $\geqslant 28$.

So $\geqslant$ means greater than *or equal to* and $>$ means *strictly* greater than. What do you think $\leqslant$ means?

The signs $>$ and $<$ are called **inequality signs**.

NOTE: If you have difficulty remembering the signs, you might be able to remember that the inequality sign points to the smaller number, or even that the smaller end of the sign is on the side of the smaller number.

Ordering Integers

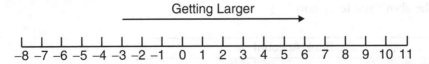

Getting Larger

$$-8 \ -7 \ -6 \ -5 \ -4 \ -3 \ -2 \ -1 \ \ 0 \ \ 1 \ \ 2 \ \ 3 \ \ 4 \ \ 5 \ \ 6 \ \ 7 \ \ 8 \ \ 9 \ \ 10 \ \ 11$$

The number line above shows the integers from −8 to 11. The rest of the real numbers fit in their correct places along the line, so −2.5 would be halfway between −3 and −2.
The numbers get *larger* as you go from *left* to *right*.
For example, 8 > 3 (as we know).
Also 1 > −2, and −4 < 0 and so on.
This is also true for all the positive and negative real numbers, so −6.25 < 3.5.

Example 9
Use the number line you made earlier to insert the correct symbol between the following pairs of numbers:

(a) 7 20   (b) −5 10   (c) 2 ..... −1

(d) −8 ..... −19   (e) 4.5 ..... −6.5   (f) $-\dfrac{1}{2}$ ..... −4

Answer 9

(a) 7 < 20   (b) −5 < 10   (c) 2 > −1

(d) −8 > −19   (e) 4.5 > −6.5   (f) $-\dfrac{1}{2}$ > −4

Exercise 1.5

1. Write down the symbol for:

    (a) pi   (b) square root   (c) cube root

    (d) is not equal to   (e) is less than   (f) is greater than or equal to

2. Fill in the correct inequality sign between each of the following pairs of numbers.

    (a) 2 4   (b) −2 ..... −5   (c) −10 4   (d) −1 0

3. Arrange the following integers in the correct order, starting with the smallest.

    100,   −1,   −100,   0,   −89,   −76,   75,   101,   61,   −62

Standard Form

Sometimes we have to work with very large numbers (the distance from earth to the moon is approximately 384400000 metres), or very small numbers (the thickness of a page in one of my books is approximately 0.0000213 metres).
There is a neater way of writing these numbers without having to use so many zeroes.

It is called **standard form**. Using standard form we write numbers in the form, $a \times 10^n$ where a is a number greater than or equal to 1 and less than 10 ($1 \leqslant a < 10$) and n is an integer.

To write 230000 (two hundred and thirty thousand) in standard form:

- First identify the place where the decimal point belongs (since it is not shown). We know that if the decimal point is not shown, it actually comes after the last digit. So 230000 could be written as 230000.0.
- Next count how many places the decimal point would have to be moved back until it is between the 2 and the 3. You will see that it is 5 places.
- So $230000 = 2.3 \times 10^5$. This is read as 'two hundred and thirty thousand is equal to two point three times ten to the power five'.

You will learn more about powers and working with standard form in a later chapter. For the moment, you just have to understand how to write numbers in standard form.

To write 0.000003546 in standard form:

- Count how many places, the decimal point would have to be moved forward, to lie between 3 and 5. It is 6 places.
- So $0.000003546 = 3.546 \times 10^{-6}$. (This is ten to the power of **negative** six.)

NOTE: If it is a problem to remember which power to use, you should notice that numbers less than one have a negative power and numbers greater than 10 have a positive power in standard form.

Example 10

(a) Write in standard form:

    (i) 20015   (ii) 175   (iii) 3200000

    (iv) 0.127   (v) 0.00506

(b) Write in the normal way:

    (i) 9.013×10^{-3} (ii) 1.0007×10^7

Answer 10

(a) (i) 2.0015×10^4 (ii) 1.75×10^2 (iii) 3.2×10^6

    (iv) 1.27×10^{-1} (v) 5.06×10^{-3}

(b) (i) 0.009013   (ii) 10007000

Order of Working in Calculations

Ram was asked to calculate $5 + 2 \times 3$, without using a calculator. His answer was 21. He checked his answer with a calculator. The calculator answer was 11.

What has happened?

Both Ram and the calculator were correct in different ways.

Ram first added 5 and 2 and then multiplied by 3, ($5 + 2 = 7$ then $7 \times 3 = 21$).

The calculator multiplied 2 and 3 first and then added 5, ($2 \times 3 = 6$ then $6 + 5 = 11$).

It is clearly not satisfactory to get two different answers to the same question, so an order of working had to be decided to ensure that all calculations yield the same answer. The accepted order is:

- First **B**rackets.
- Next **D**ivision and **M**ultiplication (in either order).
- Lastly **A**ddition and **S**ubtraction (in either order).

NOTE: There are different ways of remembering this order. For example, the made up word BoDMAS is often used.

Try to follow this example.
To calculate $7 + 3 \times 2 - (6 - 2) \div 2$

B......brackets......	$= 7 + 3 \times 2 - \mathbf{4} \div 2$	$[\,(6 - 2) = \mathbf{4}\,]$
D......division......	$= 7 + 3 \times 2 - \mathbf{2}$	$[\,4 \div 2 = \mathbf{2}\,]$
M......multiplication......	$= 7 + \mathbf{6} - 2$	$[3 \times 2 = \mathbf{6}]$
A......addition.......	$= \mathbf{13} - 2$	$[7 + 6 = \mathbf{13}]$
S......subtraction......	$= \mathbf{11}$	$[13 - 2 = \mathbf{11}]$

Answer: $7 + 3 \times 2 - (6 - 2) \div 2 = 11$

Try putting this in your calculator in exactly the same order as it is written and see if your calculator arrives at the same answer when you press the 'equals' button. Most calculators now use this form of logic (order of working), but you need to be sure about your own. There will be more about this in Chapter 4.

Work out the following by doing the multiplication first.
$$4 \times 6 \div 2$$
$$4 \times 6 \div 2 = 24 \div 2 = 12$$

Now do the same sum but do the division first.
$$4 \times 6 \div 2 = 4 \times 3 = 12$$

You should note that multiplication and division can be done in either order. Can you find a rule for addition and subtraction?

It is very important that you learn this order of working, and know how to use it.

Example 11

Work out the following, showing your working:

(a) $4 + 3 \times 10 - 6 \div 2$ (b) $(4 + 3) \times 10 - 6 \div 2$
(c) $4 + (3 \times 10) - 6 \div 2$ (d) $4 + 3 \times (10 - 6) \div 2$
(e) $(4 + 3) \times (10 - 6) \div 2$ (f) $4 + (3 \times 10 - 6) \div 2$

Answer 11

(a) $4 + 3 \times 10 - 6 \div 2$ (b) $(4 + 3) \times 10 - 6 \div 2$
   $= 4 + 30 - 3 = 31$ $= 7 \times 10 - 3$
                                          $= 70 - 3 = 67$

(c) $4 + (3 \times 10) - 6 \div 2$
   $= 4 + 30 - 3 = 31$

(This is the same as (a) because the multiplication is done first anyway, and so does not need brackets.)

(d) $4 + 3 \times (10 - 6) \div 2$

$= 4 + 3 \times 4 \div 2$ (Notice that $3 \times 4 \div 2 = 12 \div 2 = 6$

$= 4 + 6 = 10$ *or* $3 \times 4 \div 2 = 3 \times 2 = 6$)

(e) $(4 + 3) \times (10 - 6) \div 2$

$= 7 \times 4 \div 2$

$= 14$

(f) $4 + (3 \times 10 - 6) \div 2$ (The working inside the brackets also follows

$= 4 + (30 - 6) \div 2$ BoDMAS, so 3×10 first, then -6)

$= 4 + 24 \div 2$

$= 4 + 12$

$= 16$

Setting out your working

It is important to be able to communicate in mathematics. You have to be able to explain to another person (who may be your examiner!), how you have arrived at your answer in a mathematical and concise way.

If you write an equals sign, the things that come before and after that sign must be equal to each other.

Look at how two students answer the same question, showing their working.

Rita writes: $(10 + 2) \div 4 = 10 + 2 = 12 \div 4 = 3$

Sara writes: $(10 + 2) \div 4 = 12 \div 4 = 3$

Which is the easier to follow?

In the first case, Rita has written $10 + 2 = 12 \div 4$. But this is not true!

Sara has set out her work so that the equals sign means exactly that. She has also used a new line between each bit of working which makes it easier to read.

The examples throughout this book will show you how to set out your work, so do practice this right from the beginning. In general, writing one equals sign per line is good practice. However, please note that in this book, for reasons of economy and space, it has not always been possible to restrict working to one equals sign per line.

Exercise 1.6

1. Write in standard form:
   (a) 12000   (b) 365   (c) 59103   (d) 6000   (e) 7010400
2. Write in standard form:
   (a) 0.0035   (b) 0.156   (c) 0.0005   (d) 0.0000043   (e) 0.0102
3. Write in standard form:
   (a) 0.00345   (b) 520160   (c) 112
   (d) 0.001   (e) 0.1001   (f) 2 million

4. Write as a normal number:
   (a) 5.6×10^3 (b) 2.7×10^{-4} (c) 1.16×10^{-2} (d) 6×10^5 (e) 2×10^{-3}

5. Calculate the following, without using a calculator:
   (a) $4 + 7 \times 2$ (b) $12 \div 3 \times 2 + 6$
   (c) $1 + 2 + 3 - (2 \times 3)$ (d) $(4 + 5) \div (4 - 1)$
   Check your answers with a calculator.

6. Use your calculator to work out the following:
   (a) $(5 + 7 - 2) \div (6 - 4)$ (b) $2 \times 3 + 5 \times 7$ (c) $3 \times (14 - 7) - 2$
   Check your answers by calculating without the calculator.

7. Put brackets in the right places to make each of these sums correct:
   (a) $5 - 3 \times 4 = 8$ (b) $9 + 50 - 24 + 2 = 22$ (c) $31 - 15 \div 10 - 2 = 2$

Exercise 1.7

Mixed Exercise

1. (a) $\{-5, -4, -3, -2, -1, 0, 1, 2, 3, 4, 5, 6, 7, 8, 9, 10, 11, 12, 13, 14, 15, 16, 17\}$
      Using the set of numbers above, answer true or false to the following:
        (i) all the numbers come from the set of real numbers,
        (ii) all the numbers come from the set of rational numbers,
        (iii) all the numbers come from the set of natural numbers,
        (iv) all the numbers come from the set of integers.
   (b) Insert an inequality sign to make the following true:
        (i) -4 ...... 3 (ii) 0 ...... -2 (iii) 5 ...... -5 (iv) 3 ...... -2
   (c) List (in curly brackets):
        (i) the set of prime numbers less than 10,   (ii) the set of factors of 45,
        (iii) the set of multiples of 3 less than 20.

2. Find the LCM of:
   (a) 12 and 20   (b) 5 and 15 and 90

3. Find the HCF of:
   (a) 16 and 12   (b) 20 and 8 and 12

4. Calculate the following, using a calculator if necessary:
   (a) 2.1^2 (b) 3^3 (c) $\sqrt{28.09}$ (d) $\sqrt{81}$ (e) $\sqrt[3]{125}$

5. Write 600 as a product of its prime factors.

6. List all the factors of 160.

Examination Questions

7. In May, the average temperature in Kiev was 12°C. In February, the average temperature was 26°C lower than in May. What was the average temperature in February? (0580/01 May/June 2008 q 6)

8. Work out the value of $\dfrac{9-3\times7}{3\times2}$. (0580/01 May/June 2007 q 1)

9. Work out $4^3 - 5^2$. (0580/01 May/June 2004 q 1)

10. The Dead Sea shore is 395 metres **below** sea level.

    Hebron is 447 metres **above** sea level.

    Find the difference in height. (0580/01 May/June 2004 q 2)

11. Look at the numbers 21, 35, 49, 31, 24.

    From this list write down:

    (a) a square number,

    (b) a prime number. (0580/01 May/June 2004 q 4)

12. At a weather centre the temperature at midnight was −21° C.

    By noon the next day it had risen to −4° C.

    By how many degrees had the temperature risen?

    (0580/01 Oct/Nov 2004 q 1)

13. Place brackets in the following calculation to make it a correct statement.
    $10 - 5 \times 9 + 3 = 60$ (0580/01 Oct/Nov 2004 q 2)

14. (a) List all the factors of 30.

    (b) Write down the prime factors of 30. (1 is not a prime number.)
    (0580/01 Oct/Nov 2004 q 18)

15. Write down a multiple of 4 and 14 which is less than 30.
    (0580/01 Oct/Nov 2008 q 1)

16. Write 0.00362 in standard form.

    (0580/01 May/June 2008 q 7)

Chapter 2

Fractions, Decimals and Percentages

This chapter starts with a reminder about common fractions, using diagrams to help you visualise them. Visualisation can be a help in the understanding of the methods of working with fractions. This section is worth reading to reinforce your understanding.

This chapter should give you the basic skills for working with fractions, decimals and percentages that you will need later in the course, when we start using real life examples. You may already have a good grasp of the basic ideas, but misunderstandings and errors in the handling of fractions are often the cause of difficulties in arithmetic and algebra. Make sure you can complete the Examples and Exercises confidently.

You should **not** use a calculator when working through this chapter. The numbers used in the chapter are simple enough to work on without a calculator. It is important that you first understand the principles without using a calculator so that you will be able to work more easily with algebra. We will go on to more difficult work requiring a calculator in a later chapter.

Remember: no calculator in this chapter!

Essential Skills

Make sure you can calculate the following. Look back to the previous chapter if you need a reminder.

1. Find the LCM of the following numbers.
   (a) 2, 5   (b) 7, 14   (c) 3, 8, 12   (d) 3, 5, 12, 60
2. Find the HCF of the following numbers.
   (a) 12, 36   (b) 18, 24   (c) 50, 150, 200   (d) 40, 24, 56
3. Test these numbers for divisibility by 2, 3, 5, 6 and 9.
   (a) 180   (b) 3006   (c) 1305

Answers

1. (a) 10   (b) 14   (c) 24   (d) 60
2. (a) 12   (b) 6   (c) 50   (d) 8

3. (a) 180 divides by 2, 3, 5, 6 and 9   (b) 3006 divides by 2, 3, 6 and 9
    (c) 1305 divides by 3, 5 and 9

Understanding Common Fractions

When we use the word fraction we normally think of numbers like $\frac{7}{8}$, $\frac{2}{3}$ or $\frac{1}{2}$.

These are actually *common* or *vulgar* fractions.

In your IGCSE course the word fraction will normally mean common fraction, but sometimes it will help you to understand your work if you remember that decimals (decimal fractions) and percentages are also fractions. Percentages are fractions with a denominator of a hundred. For example, 21% is the same as $\frac{21}{100}$.

As you know, common fractions have a number above the fraction line, and another number below the fraction line. These numbers are called the **numerator** and the **denominator** respectively. So in the fraction $\frac{2}{3}$, 2 is the numerator and 3 is the denominator.

You can think of the denominator as the name of the fraction with the numerator showing the number of fractions with this name. The diagram below should help you to see this.

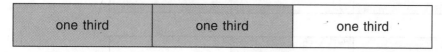

The strip in the diagram has been divided into three equal parts.

Each part is one third $\left(\frac{1}{3}\right)$ of the whole strip.

Three thirds $\left(\frac{3}{3}\right)$ make up the whole strip.

Two thirds $\left(\frac{2}{3}\right)$ are shaded.

The **numerator** is the top number in a common fraction.
The **denominator** is the bottom number in a common fraction.
The denominator shows into how many equal parts the whole strip has been divided. The denominator tells us the name of the fraction, in this case 'thirds'.
The numerator shows the number of these fractions, in the case 2 'thirds' have been shaded. Look at the next diagram to see this drawn out.

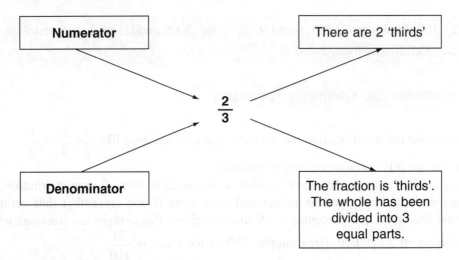

Mixed numbers and improper fractions

Mixed numbers have a whole part and a fraction part. The mixed number $1\frac{2}{3}$ means there is one whole part and 2 thirds. The diagram below shows two strips, each divided into three equal parts.

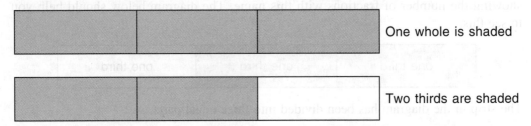

One whole is shaded

Two thirds are shaded

This diagram also shows how we can write a mixed number as an improper fraction.
An **improper** fraction is a mixed number written entirely in fractions, so the numerator is larger than the denominator. The diagram shows the shaded parts of the two strips as either $1\frac{2}{3}$ or $\frac{5}{3}$ (or 5 thirds).

NOTE: An improper fraction is often referred to as a top heavy fraction, which describes it well because the top number is larger than the lower number.

Equivalent fractions

Fractions can be given different names, and if the rules for doing this are followed the resulting fraction is the same size as the original.
Equivalent fractions are fractions of the same size, but different denominators (names) and numerators. Look at the diagram below.

The diagram shows the strip divided into three equal parts with the fraction $\frac{2}{3}$ shaded as before. If we divide *each* third into two equal parts you should see that there are now six equal parts and four of these are equivalent in size to 2 thirds. The next diagram shows this.

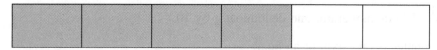

This shows that $\frac{2}{3} = \frac{4}{6}$. These two fractions are called equivalent fractions because they represent the same amount of the whole strip. The rule for finding equivalent fractions is that the denominator *and* numerator have to be multiplied, (or divided) by the same number. In this case the first fraction has had the numerator and denominator multiplied by 2. You will find out more about this later.

More examples of fractions

We can work with things other than strips of paper to understand fractions. Imagine a bag containing 20 sweets. You want to share these sweets equally between four people. The 20 sweets would have to be divided into 4 equal parts. There would be 5 sweets in each part. Each part would be one quarter of the whole.

This could be written as $\frac{1}{4} \times 20 = 5$, as shown below.

How many counters would be there in $\frac{2}{3}$ of 15 counters?

You can use this diagram of 15 counters.

Look at this clock face.

We know that 15 minutes is a quarter of an hour, and that there are 60 minutes in one

hour. The hour is divided into sixty equal parts. So each minute is 1 sixtieth $\left(\frac{1}{60}\right)$ of an hour. Therefore, fifteen minutes is fifteen sixtieths of one hour.

Simplifying shows that $\dfrac{15}{60} = \dfrac{1}{4}$. (Divide numerator and denominator by 15.)

How do we work out what fraction of an hour is ten minutes?

Write ten sixtieths and simplify.

$\dfrac{10}{60} = \dfrac{1}{6}$. (Divide numerator and denominator by 10.)

So ten minutes is one sixth of an hour.

Other shapes can be divided into equal parts.

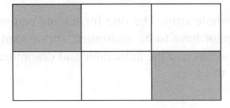 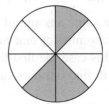

The rectangle has been divided into six equal areas, and so each is $\dfrac{1}{6}$ of the rectangle.

Two of these are shaded. This means that $\dfrac{2}{6}$, which is equivalent to $\dfrac{1}{3}$ of the rectangle, is

shaded. The circle has been divided into 8 equal parts, and 3 are shaded, so $\dfrac{3}{8}$ of the circle

is shaded.

Working with Common Fractions

Changing a mixed number to an improper fraction

For example, the steps to be followed to change $3\dfrac{1}{5}$ to an improper fraction are given
below.

- Multiply: $3 \times 5 = 15$ (there are 15 fifths in three wholes so $3\dfrac{1}{5} = \dfrac{15}{5} + \dfrac{1}{5}$)
- Add: $15 + 1 = 16$ (add the extra 1 fifth)
- Answer: $\dfrac{16}{5}$ (16 fifths)

Changing an improper fraction to a mixed number

For example, the steps to be followed to change $\dfrac{23}{4}$ to a mixed number are given below.

- Divide: $23 \div 4 = 5$ remainder 3 (23 quarters = 5 wholes with 3 quarters left over)
- Answer: $5\dfrac{3}{4}$

Equivalent fractions

For example, to change $\dfrac{4}{10}$ to equivalent fractions,

- either multiply numerator and denominator by the same number: $\dfrac{4\times2}{10\times2}=\dfrac{8}{20}$

- or divide numerator and denominator by the same number: $\dfrac{4\div2}{10\div2}=\dfrac{2}{5}$.

Addition and subtraction of fractions

Before fractions are added or subtracted, we have to make sure they have the same name. For example, look at this diagram, which represents the addition sum $\dfrac{3}{4}+\dfrac{1}{8}$. The only way we can add these two is to write them with the same name (denominator).

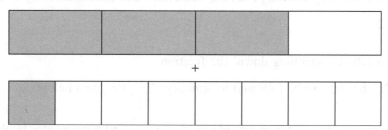

To do this we divide *each* of the quarters into two equal parts, to make eighths. The three quarters has become six eighths and can now be added to the one eighth as in the next diagram.

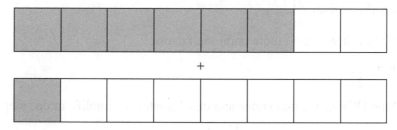

It is now easy to see that $\dfrac{3}{4}+\dfrac{1}{8}=\dfrac{6}{8}+\dfrac{1}{8}=\dfrac{7}{8}$.

The answer is seven eighths.

It can be easier to add or subtract mixed numbers by changing them to top heavy (improper) fractions first as you will see in Example 1, part (f) (iv).

You may have to change both fractions to equivalent fractions with the same denominator.

For example, $\dfrac{2}{3}+\dfrac{4}{5}$.

Follow these steps to see how to work this out.

- Change any mixed numbers to improper fractions.
- Find the lowest common multiple of both denominators (LCM of 3 and 5 is 15).

- Change both fractions to equivalent fractions with the same denominator $\left(\dfrac{2\times5}{3\times5}+\dfrac{4\times3}{5\times3}=\dfrac{10}{15}+\dfrac{12}{15} \right)$.

- Add or subtract the fractions in the usual way $\left(\dfrac{10}{15}+\dfrac{12}{15}=\dfrac{22}{15} \right)$.

- Simplify and change to a mixed number if necessary $\left(\dfrac{22}{15}=1\dfrac{7}{15} \right)$.

- Answer: $\dfrac{2}{3}+\dfrac{4}{5}=1\dfrac{7}{15}$

Simplifying fractions

Simplifying fractions refers to writing them in the simplest equivalent form. For example, $\dfrac{4}{10}$ can be simplified by dividing both the numerator *and* denominator by 2. This means that $\dfrac{4}{10}=\dfrac{2}{5}$.

This is often called 'cancelling down' the fraction.

For example, the steps to be followed to simplify $\dfrac{42}{162}$ are given below.

Either

- Find any common factor and divide the numerator and denominator by this number:

$$\frac{42\div2}{162\div2}=\frac{21}{81}$$

- Repeat if possible: $\dfrac{21\div3}{81\div3}=\dfrac{7}{27}$

- Stop when there are no more common factors.

- Answer: $\dfrac{7}{27}$

or

- Find the HCF of the numerator and denominator to simplify in one step:

$$\frac{42\div6}{162\div6}=\frac{7}{27}$$

Example 1

(a) Change the top heavy (improper) fraction $\dfrac{7}{2}$ to a mixed number.

(b) Change the mixed number $4\dfrac{3}{5}$ to an improper fraction.

(c) Which of these fractions are equivalent?

$$\frac{20}{50}, \frac{2}{5}, \frac{3}{10}, \frac{4}{10}, \frac{3}{5}, \frac{8}{20}$$

(d) Change $\dfrac{4}{5}$ to twentieths.

(e) Fill in the blank spaces to give equivalent fractions.

$$\dfrac{\ }{16} = \dfrac{3}{8} = \dfrac{30}{\ } = \dfrac{15}{\ }$$

(f) Calculate, simplifying and writing the answers as mixed numbers if necessary:

   (i) $3 + \dfrac{5}{6}$ (ii) $\dfrac{5}{8} + \dfrac{1}{2}$ (iii) $\dfrac{3}{4} - \dfrac{2}{3}$ (iv) $2\dfrac{1}{3} + 4\dfrac{5}{6}$ (v) $1 - \dfrac{7}{9}$

(g) Write each of the following fractions in their simplest forms.

   (i) $\dfrac{5}{40}$ (ii) $\dfrac{6}{48}$ (iii) $\dfrac{18}{72}$

(h) How many sheep are there in 3 fifths of a flock of 25 sheep?

(i) How many students are there in $\dfrac{2}{7}$ of a class of 35?

Answer 1

(a) $\dfrac{7}{2} = 3\dfrac{1}{2}$ (b) $4\dfrac{3}{5} = \dfrac{23}{5}$ (c) $\dfrac{20}{50} = \dfrac{2}{5} = \dfrac{4}{10} = \dfrac{8}{20}$

(d) $\dfrac{4}{5} = \dfrac{4 \times 4}{5 \times 4} = \dfrac{16}{20}$ (e) $\dfrac{6}{16} = \dfrac{3}{8} = \dfrac{30}{80} = \dfrac{15}{40}$

(f) (i) $3 + \dfrac{5}{6} = 3\dfrac{5}{6}$ (ii) $\dfrac{5}{8} + \dfrac{1}{2} = \dfrac{5}{8} + \dfrac{4}{8} = \dfrac{9}{8} = 1\dfrac{1}{8}$

   (iii) $\dfrac{3}{4} - \dfrac{2}{3} = \dfrac{3 \times 3}{4 \times 3} - \dfrac{2 \times 4}{3 \times 4} = \dfrac{9}{12} - \dfrac{8}{12} = \dfrac{1}{12}$

   (iv) $2\dfrac{1}{3} + 4\dfrac{5}{6} = \dfrac{7}{3} + \dfrac{29}{6} = \dfrac{14}{6} + \dfrac{29}{6} = \dfrac{43}{6} = 7\dfrac{1}{6}$ (v) $1 - \dfrac{7}{9} = \dfrac{9}{9} - \dfrac{7}{9} = \dfrac{2}{9}$

(g) (i) $\dfrac{5}{40} = \dfrac{5 \div 5}{40 \div 5} = \dfrac{1}{8}$ (ii) $\dfrac{6}{48} = \dfrac{1}{8}$ (iii) $\dfrac{18}{72} = \dfrac{18 \div 2}{72 \div 2} = \dfrac{9}{36} = \dfrac{1}{4}$

(h) One fifth of the flock is 5 sheep, so 3 fifths is 15 sheep.

(i) $\dfrac{1}{7}$ of 35 = 5, so $\dfrac{2}{7}$ of 35 = 10. Answer: 10 students

Exercise 2.1

1. Change to mixed numbers:

   (a) $\dfrac{19}{5}$ (b) $\dfrac{201}{10}$ (c) $\dfrac{33}{2}$

2. Change to improper fractions:

   (a) $3\dfrac{7}{8}$ (b) $100\dfrac{1}{2}$ (c) $3\dfrac{11}{12}$

3. Fill in the blank spaces to give equivalent fractions:

$$\frac{5}{-} = \frac{10}{30} = \frac{7}{3} = \frac{-}{-} = \frac{21}{-}$$

4. Write the following as hundredths (denominator = 100):

   (a) $\frac{7}{10}$ (b) $\frac{4}{25}$ (c) $\frac{19}{20}$ (d) $\frac{52}{200}$ (e) $\frac{81}{900}$

5. Calculate the following, simplifying and writing your answers as mixed numbers if necessary:

   (a) $\frac{3}{7} + \frac{2}{7}$ (b) $\frac{4}{5} - \frac{3}{5}$ (c) $\frac{7}{12} - \frac{1}{6}$ (d) $\frac{2}{9} + \frac{3}{4}$

   (e) $2\frac{1}{5} + 1\frac{3}{4}$ (f) $3\frac{2}{5} - 1\frac{1}{2}$ (g) $1 - \frac{6}{7}$ (h) $1 - \frac{5}{12}$

6. Simplify:

   (a) $\frac{22}{77}$ (b) $\frac{60}{72}$ (c) $\frac{45}{60}$ (d) $\frac{45}{360}$

7. How many sweets would be in a bag of 28 sweets after $\frac{1}{4}$ of them had been eaten?

8. One third of a class of 45 students has gone away on a field trip. How many students have gone on the trip?

Multiplying and dividing fractions

The first part of the diagram below shows a strip divided into thirds, with one third shaded. We can use this figure to work out $\frac{1}{2} \times \frac{1}{3}$, which means $\frac{1}{2}$ **of** $\frac{1}{3}$.

The second part of the figure shows the same strip with the shaded third divided into two equal parts. Each of these is one half of a third of the strip.

You should see that each of these is equal to one sixth of the whole strip.

So $\frac{1}{2} \times \frac{1}{3} = \frac{1}{6}$ which means $\frac{1}{2}$ of $\frac{1}{3} = \frac{1}{6}$.

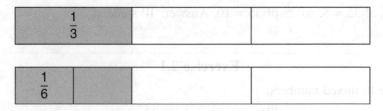

You will probably find multiplying and dividing fractions easier than adding and subtracting. The rules for multiplying fractions are:

- Change any mixed numbers to top heavy (improper) fractions.
- Write any whole numbers over one.

- Multiply the numerators together, and multiply the denominators together.
- Simplify the answer if necessary, and change to a mixed number if necessary.

Applying these rules to our example above:

$$\frac{1}{2} \times \frac{1}{3} = \frac{1 \times 1}{2 \times 3} = \frac{1}{6}$$

Example 2

(a) Multiply the following fractions, simplifying and writing your answers as mixed numbers if necessary.

(i) $3 \times \frac{3}{4}$ (ii) $\frac{5}{6} \times \frac{2}{3}$ (iii) $\frac{6}{7} \times 2\frac{1}{3}$ (iv) $1\frac{2}{3} \times 2\frac{1}{5}$

(b) Calculate the following:

(i) $\left(\frac{3}{5}\right)^2$ (ii) $\left(\frac{2}{3}\right)^3$

Answer 2

(a) (i) $3 \times \frac{3}{4} = \frac{3}{1} \times \frac{3}{4} = \frac{9}{4} = 2\frac{1}{4}$ (ii) $\frac{5}{6} \times \frac{2}{3} = \frac{10}{18} = \frac{5}{9}$

  (iii) $\frac{6}{7} \times 2\frac{1}{3} = \frac{6}{7} \times \frac{7}{3} = \frac{42}{21} = \frac{2}{1} = 2$ (iv) $1\frac{2}{3} \times 2\frac{1}{5} = \frac{5}{3} \times \frac{11}{5} = \frac{55}{15} = \frac{11}{3} = 3\frac{2}{3}$

(b) (i) $\left(\frac{3}{5}\right)^2 = \frac{3}{5} \times \frac{3}{5} = \frac{9}{25}$ (ii) $\left(\frac{2}{3}\right)^3 = \frac{2}{3} \times \frac{2}{3} \times \frac{2}{3} = \frac{8}{27}$

You have probably noticed that in Example 2(a) parts (ii), (iii) and (iv) the working could have been shortened considerably by simplifying earlier. We will look at this now. In Example 2(a) (ii):

$$\frac{5}{6} \times \frac{2}{3} = \frac{5 \times 2}{6 \times 3} = \frac{5 \times 2 \div 2}{6 \times 3 \div 2} = \frac{5}{9}$$

So we could have simplified before doing the multiplication:

$$\frac{5}{6} \times \frac{2}{3} = \frac{5}{3 \times 3} = \frac{5}{9} \text{ (dividing the top and bottom of the fraction by 2)}$$

In Example 2(a) (iii):

$$\frac{6}{7} \times 2\frac{1}{3} = \frac{6}{7} \times \frac{7}{3} = \frac{6}{3} = 2 \text{ (divide top and bottom by 7 first, and then by 3)}$$

Try Example 2(a) (iv) yourself.

Warning: This only works for multiplication, so do not use it in addition or subtraction!

How can we visualise division? Remember that if you do the division $10 \div 2$ you are finding how many twos there are in ten. The answer of course is 5.

Think about $\dfrac{3}{4} \div \dfrac{1}{8}$. This means 'how many eighths are there in three quarters?'

The next diagram shows one strip with $\dfrac{3}{4}$ shaded, and another divided into eight equal

parts and shaded to show that 6 eighths will go exactly into $\dfrac{3}{4}$. So the answer is 6.

i.e. $\dfrac{3}{4} \div \dfrac{1}{8} = 6$

$\frac{1}{4}$	$\frac{1}{4}$	$\frac{1}{4}$	

$\frac{1}{8}$	$\frac{1}{8}$	$\frac{1}{8}$	$\frac{1}{8}$	$\frac{1}{8}$	$\frac{1}{8}$		

The rules for dividing fractions are:

- Change any mixed numbers to top heavy (improper) fractions.
- Write any whole numbers over one.
- Change the division sign to multiplication.
- Turn the second fraction upside down.
- Proceed as for multiplication.

Using these rules for $\dfrac{3}{4} \div \dfrac{1}{8}$, we get:

$$\dfrac{3}{4} \div \dfrac{1}{8} = \dfrac{3}{4} \times \dfrac{8}{1} = \dfrac{24}{4} = 6$$

Example 3

Do the following divisions:

(a) $\dfrac{5}{6} \div 3$ 　　(b) $\dfrac{3}{4} \div \dfrac{1}{2}$ 　　(c) $1\dfrac{2}{5} \div 4\dfrac{3}{5}$ 　　(d) $\dfrac{2}{5} \div \dfrac{5}{8}$

Answer 3

(a) $\dfrac{5}{6} \div 3 = \dfrac{5}{6} \div \dfrac{3}{1} = \dfrac{5}{6} \times \dfrac{1}{3} = \dfrac{5}{18}$

(b) $\dfrac{3}{4} \div \dfrac{1}{2} = \dfrac{3}{4} \times \dfrac{2}{1} = \dfrac{6}{4} = \dfrac{3}{2} = 1\dfrac{1}{2}$ 　(or $\dfrac{3}{4} \times \dfrac{2}{1} = \dfrac{3}{2} = 1\dfrac{1}{2}$ by dividing top and bottom by 2)

(c) $1\dfrac{2}{5} \div 4\dfrac{3}{5} = \dfrac{7}{5} \div \dfrac{23}{5} = \dfrac{7}{5} \times \dfrac{5}{23} = \dfrac{7}{23}$ 　(by dividing top and bottom by 5)

(d) $\dfrac{2}{5} \div \dfrac{5}{8} = \dfrac{2}{5} \times \dfrac{8}{5} = \dfrac{16}{25}$ 　(Be careful! You can not divide top and bottom by 5 here!)

Looking at these examples you should see that you can do the simplifying shortcut *only* *after* the second fraction has been inverted and the sign has changed from division to multiplication.

Exercise 2.2

Calculate the following, simplifying your answers and changing to mixed numbers as necessary.

1. (a) $3 \times \dfrac{1}{5}$ (b) $3 \times \dfrac{2}{5}$ (c) $\dfrac{3}{4} \times 10$

2. (a) $\dfrac{1}{3} \times \dfrac{1}{6}$ (b) $\dfrac{3}{4} \times \dfrac{1}{7}$ (c) $\dfrac{5}{8} \times \dfrac{3}{4}$

3. (a) $\dfrac{5}{7} \times \dfrac{1}{10}$ (b) $\dfrac{7}{8} \times \dfrac{2}{5}$ (c) $\dfrac{4}{9} \times \dfrac{3}{8}$

4. (a) $2\dfrac{1}{2} \times 3\dfrac{1}{4}$ (b) $5\dfrac{3}{5} \times 2\dfrac{1}{8}$ (c) $1\dfrac{1}{3} \times 2\dfrac{1}{3}$

5. (a) $3 \div \dfrac{2}{5}$ (b) $\dfrac{4}{3} \div 3$ (c) $\dfrac{2}{7} \div 4$

6. (a) $\dfrac{1}{2} \div \dfrac{3}{7}$ (b) $\dfrac{3}{7} \div \dfrac{1}{2}$ (c) $\dfrac{5}{9} \div \dfrac{6}{7}$

7. (a) $\dfrac{2}{9} \div \dfrac{2}{3}$ (b) $\dfrac{2}{3} \div \dfrac{2}{9}$ (c) $\dfrac{5}{8} \div \dfrac{3}{8}$

8. (a) $2\dfrac{3}{4} \div 1\dfrac{1}{2}$ (b) $5\dfrac{1}{2} \div \dfrac{3}{4}$ (c) $3\dfrac{3}{5} \div 1\dfrac{2}{3}$

9. (a) $\dfrac{2}{5} \div \dfrac{1}{15}$ (b) $\dfrac{3}{5} \div \dfrac{5}{6}$ (c) $2\dfrac{1}{3} \div \dfrac{3}{7}$

Working with decimals

Decimals (or decimal fractions) are usually easier to work with than (common) fractions, so the rules and a few examples should be sufficient to remind you how to do each operation. We will abbreviate to decimals and fractions because these terms are generally understood to mean decimal fractions and common fractions.

Addition and subtraction of decimals

For example, to simplify 3 + 1.205 + 40.016 follow the steps given below.

- Add '.0' to the whole number to remind you where the decimal point belongs. (3.0 + 1.205 + 40.016)
- Write the numbers in column form, but with the decimal points in a vertical line.

$$
\begin{array}{r}
3.0 \\
1.205 + \\
40.016 + \\
\end{array}
$$

- Starting from the right add (or subtract) using the normal methods of addition (or subtraction).

- Place the decimal point in the answer vertically under the other decimal points.

$$
\begin{array}{r}
3.0 \\
1.205\ + \\
40.016\ + \\
\hline
44.221
\end{array}
$$

Multiplying decimals

For example, to simplify 2.16×0.002 follow the steps given below.

- At first ignore the decimal points.
- Starting from the right multiply using the normal methods.
- Count how many digits (numbers) come after the decimal points.
- Starting from the right count back this number of places and insert the decimal point, inserting zeroes if necessary.

$$
\begin{array}{r}
2.16 \\
0.002\ \times \\
\hline
0.00432
\end{array}
$$

(There are 5 digits after the decimal points, so counting 5 places from the right it is necessary to insert 2 zeroes.)

- Answer: $2.16 \times 0.002 = 0.00432$

To multiply by 10, 100, 1000 and so on:

For example, 0.013×100

- Count the number of zeroes (2 in this example) in the number you are multiplying by.
- Move the decimal point to the *right* by the same number of places, inserting zeroes if necessary. ($0.013 \times 100 = 1.3$)
- Answer: $0.013 \times 100 = 1.3$

NOTE: Remember that *multiplying* by 10, 100 and so on will make the answer *larger*.

Dividing decimals

For example, to divide 63.6 by 0.012, follow the steps given below.

- Write the first number over the second number $\left(\dfrac{63.6}{0.012} \right)$.

- Multiply top and bottom by 10, 100, 1000 or 10000 until the lower number is a whole number (in this case we need to use 1000, i.e., $\dfrac{63.6 \times 1000}{0.012 \times 1000} = \dfrac{63600}{12}$).

- Divide the new lower number into the top number. ($63600 \div 12 = 5300$)
- Answer: 63.6 divided by 0.012 = 5300.

To divide by 10, 100, 1000 and so on:

For example, divide 0.234 by 1000.

- Count the number of zeroes (three in this case) in the number you are to divide by.
- Move the decimal point 3 places *left*, filling in zeroes if necessary. ($0.234 \div 1000 = 0.000234$)
- Answer: $0.234 \div 1000 = 0.000234$

NOTE: Remember that dividing by 10, 1000, 1000 and so on will make the number smaller.

Example 4

Calculate the following.

(a) 2.501 + 12.6   (b) 45.3173 − 1.012   (c) 3.513 × 100
(d) 0.012 × 10   (e) 4.12 × 1000   (f) 2.1 × 1.1
(g) 0.16 ÷ 100   (h) 31.323 ÷ 0.03

Answer 4

(a) 2.501 + 12.6 =   2.501
                     12.6   +
                    ─────────
                    15.101

(b) 45.3173 − 1.012 = 45.3173
                        1.012   −
                      ───────────
                       44.3053

(c) 3.513 × 100 = 351.3   (d) 0.012 × 10 = 0.12   (e) 4.12 × 1000 = 4120

(f) 2.1 × 1.1 =   2.1
                   1.1 ×
                  ────────
                  2.31

(g) 0.16 ÷ 100 = 0.0016

(h) $31.323 ÷ 0.03 = \dfrac{31.323 \times 100}{0.03 \times 100} = \dfrac{3132.3}{3} = 1044.1$

Exercise 2.3

Calculate the following.

1. 3.5 + 0.16 + 10.2   2. 501 + 1.67 + 0.3   3. 17.95 − 1.4
4. 6.119 − 2.01   5. 13.41 × 1000   6. 0.0169 × 1000
7. 6.017 ÷ 100   8. 10.2 × 3.1   9. 18.96 ÷ 1.2

Conversion between Common Fractions, Decimals and Percentages

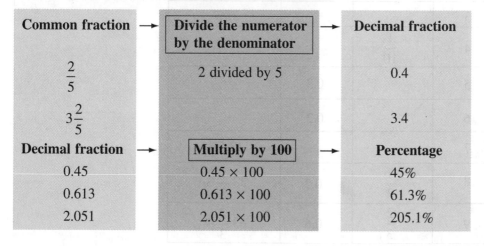

Common fraction	Divide the numerator by the denominator	Decimal fraction
$\dfrac{2}{5}$	2 divided by 5	0.4
$3\dfrac{2}{5}$		3.4

Decimal fraction	Multiply by 100	Percentage
0.45	0.45 × 100	45%
0.613	0.613 × 100	61.3%
2.051	2.051 × 100	205.1%

Divide by 100 to change a percentage to decimal fraction.

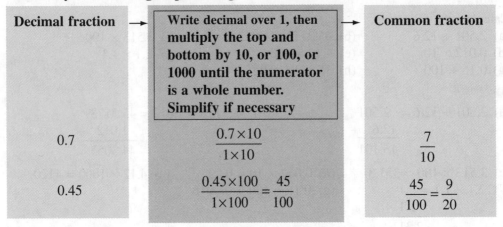

Decimal fraction →	Write decimal over 1, then multiply the top and bottom by 10, or 100, or 1000 until the numerator is a whole number. Simplify if necessary	→ Common fraction
0.7	$\dfrac{0.7 \times 10}{1 \times 10}$	$\dfrac{7}{10}$
0.45	$\dfrac{0.45 \times 100}{1 \times 100} = \dfrac{45}{100}$	$\dfrac{45}{100} = \dfrac{9}{20}$

Exercise 2.4

Copy and complete the table for conversions between common fractions, decimals and percentages. It is a good idea to learn these as they occur quite frequently and you can save time if you know them. The last two have been done for you, and it is *definitely* a good idea to learn these!

The dot above the number means that the number repeats forever. For example $0.\dot{3}$ means 0.33333333333 … . It is called 'zero point three recurring'.

		Fraction	Decimal	Percentage
1		$\dfrac{1}{2}$		
2			0.25	
3				75%
4		$\dfrac{1}{10}$		
5			0.3	
6			0.2	
7				12.5%
8		$\dfrac{1}{3}$	$0.\dot{3}$	$33\dfrac{1}{3}\%$
9		$\dfrac{2}{3}$	$0.\dot{6}$	$63\dfrac{2}{3}\%$

Percentages

It may help you to visualise percentages and compare them with fractions if you imagine a stack of, say 100 counters, as in the diagram. Imagine that the counters are numbered from 1 to 100, with 1 at the bottom of the stack.

Each of the counters is $\dfrac{1}{100}$ of the whole stack, so each counter is 1% of the stack. The whole stack is 100% of the stack or one whole.

Now you can see that half way up is 50%, one quarter of the way up is 25%, $\dfrac{1}{10}$ of the way up is 10% and so on.

Copy the diagram and mark in $\dfrac{3}{4}$ and its corresponding percentage, 20% and any others that you can think of.

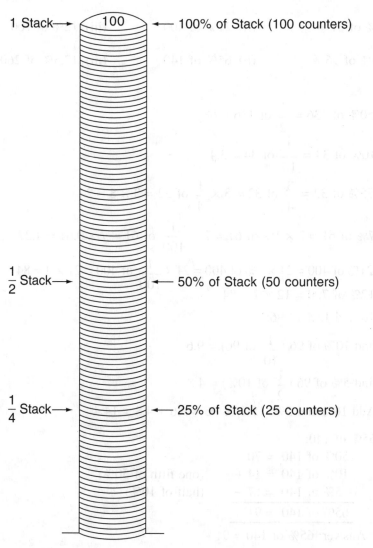

Calculating percentages of an amount

There are several ways to calculate percentages quickly.

The first is to know the common percentages (50%, 10%, 25% and so on) and their corresponding fractions (see part (a) of Example 5).

The second is to find 1% by dividing by 100, and then multiply by whatever percentage is required (see part (b) of Example 5).

Lastly some percentages can be 'built up' from smaller percentages that are easy to find (see part (c) of Example 5).

Example 5

(a) Find:

(i) 50% of 136   (ii) 10% of 34   (iii) 75% of 32

(b) Find:

(i) 7% of 61   (ii) 21% of 400   (iii) 12% of 700

(c) Find:

(i) 15% of 96   (ii) 65% of 140   (iii) 17.5% of 260

Answer 5

(a)   (i) $50\% \text{ of } 136 = \dfrac{1}{2} \text{ of } 136 = 68$

     (ii) $10\% \text{ of } 34 = \dfrac{1}{10} \text{ of } 34 = 3.4$

     (iii) $75\% \text{ of } 32 = \dfrac{3}{4} \text{ of } 32 = 3 \times \dfrac{1}{4} \text{ of } 32 = 3 \times 8 = 24$

(b)   (i) $7\% \text{ of } 61 = 7 \times 1\% \text{ of } 61 = 7 \times \dfrac{1}{100} \text{ of } 61 = 7 \times 0.61 = 4.27$

     (ii) $21\% \text{ of } 400 = 21 \times 1\% \text{ of } 400 = 21 \times \dfrac{1}{100} \text{ of } 400 = 21 \times 4 = 84$

     (iii) $12\% \text{ of } 700 = 12 \times 7 = 84$

(c)   (i)   To find 15% of 96:

       find $10\% \text{ of } 96 \ (\dfrac{1}{10} \text{ of } 96) = 9.6$

       find $5\% \text{ of } 96 \ (\dfrac{1}{2} \text{ of } 10\%) = 4.8$

       Add 10% and 5%: 15% of 96 = 9.6 + 4.8 = 14.4

     (ii)   65% of 140:

         50% of 140 = 70

         10% of 140 = 14 +   (one fifth of 50%)

          5% of 140 = 7 +   (half of 10%)

        ———————————

        65% of 140 = 91

       Answer: 65% of 140 = 91

(iii) 17.5% of 260:

$$
\begin{array}{ll}
10\% \text{ of } 260 = 26 & \\
5\% \text{ of } 260 = 13 + & (\text{half of } 10\%) \\
2.5\% \text{ of } 260 = 6.5 + & (\text{half of } 5\%) \\
\hline
17.5\% \text{ of } 260 = 45.5 &
\end{array}
$$

Answer: 17.5% of 260 = 45.5

Exercise 2.5

Calculate the following, showing your method:

1. 75% of 64
2. 30% of 1550
3. 9% of 3400
4. 55.5% of 680
5. 3% of 73

Finding one number as a percentage of another

We sometimes need to express one number as a percentage of another. For example, you get 6 answers correct out of 20 in a test. What is your percentage mark?

- First make a fraction by writing the first number over the second. $\left(\dfrac{6}{20}\right)$

- Change the fraction to a percentage by multiplying by 100 over 1.

$$\left(\frac{6}{20} \times \frac{100}{1} = \frac{600}{20} = 30\%\right)$$

There is an alternative method that can sometimes be used, if the denominator of the fraction is a factor of 100.

Alternative method

- First make the fraction as before. $\left(\dfrac{6}{20}\right)$

- Change to the equivalent fraction with the denominator as 100.

$$\left(\frac{6 \times 5}{20 \times 5} = \frac{30}{100} = 30\%\right)$$

Example 6

(a) Find 25 as a percentage of 40.   (b) Find 15 as a percentage of 25.

Answer 6

(a) $\dfrac{25}{40} \times \dfrac{100}{1} = \dfrac{250}{4} = 62.5\%$

(b) $\dfrac{15 \times 4}{25 \times 4} = \dfrac{60}{100} = 60\%$

Exercise 2.6

Calculate the first number as a percentage of the second:

1. 35, 140   2. 72, 600   3. 23, 50   4. 40, 125

5. 17, 250   6. 90, 180   7. 12, 6   8. 29, 1000

Ordering Quantities

It is often easiest, when comparing fractions, decimals and percentages to change them all to decimals. Alternatively, compare fractions by finding equivalent fractions with the same denominator.

Example 7

(a) Using the symbols >, < or = insert the correct sign to make the following statements true.

   (i) 22 21   (ii) 0.75 ..... $\dfrac{3}{4}$ (iii) 0.25 25   (iv) $\dfrac{1}{3}$ 0.3

(b) Write the following in order of size, starting with the smallest:

   (i) 0.48, 0.408, 0.390, 0.399

   (ii) $\dfrac{2}{5}, \dfrac{3}{5}, \dfrac{3}{10}, \dfrac{9}{20}$

   (iii) 33%, 0.5, $\dfrac{3}{10}, \dfrac{1}{3}$

(c) Rafi loves eating naan. Do you think he would rather have two thirds or three quarters of a naan? Why?

(d) Find a fraction which is between each of the following pairs:

   (i) $\dfrac{7}{10}$ and $\dfrac{9}{10}$ (ii) $\dfrac{1}{2}$ and $\dfrac{3}{4}$ (iii) $\dfrac{6}{8}$ and $\dfrac{7}{8}$

Answer 7

(a) (i) 22 > 21   (ii) 0.75 = $\dfrac{3}{4}$ (iii) 0.25 < 25   (iv) $\dfrac{1}{3}$ > 0.3

(b) (i) 0.390 < 0.399 < 0.408 < 0.48

   (ii) Changing $\dfrac{2}{5}, \dfrac{3}{5}, \dfrac{3}{10}, \dfrac{9}{20}$ to twentieths

      $\dfrac{8}{20}, \dfrac{12}{20}, \dfrac{6}{20}, \dfrac{9}{20}$

   and putting in order,

      $\dfrac{6}{20}, \dfrac{8}{20}, \dfrac{9}{20}, \dfrac{12}{20}$

simplifying again,

$$\frac{3}{10} < \frac{2}{5} < \frac{9}{20} < \frac{3}{5}$$

(iii) Changing 33%, 0.5, $\frac{3}{10}$, $\frac{1}{3}$ to decimals,

0.33, 0.5, 0.3, 0.3333 ...

putting in order,

0.3, 0.33, 0.3333 ..., 0.5,

and re-writing as before $\frac{3}{10} < 33\% < \frac{1}{3} < 0.5$

(c) $\frac{3}{4} = 75\%$ and $\frac{2}{3} = 66\frac{2}{3}\%$, so Rafi would rather have $\frac{3}{4}$ of the naan.

(d) (i) $\frac{8}{10}$ is between $\frac{7}{10}$ and $\frac{9}{10}$. $\frac{8}{10} = \frac{4}{5}$

Answer: $\frac{4}{5}$

(ii) $\frac{1}{2}$ and $\frac{3}{4}$ can be changed to their equivalent fractions $\frac{4}{8}$ and $\frac{6}{8}$, so $\frac{5}{8}$

is between $\frac{1}{2}$ and $\frac{3}{4}$.

Answer: $\frac{5}{8}$

(iii) $\frac{6}{8}$ and $\frac{7}{8}$ are equivalent to $\frac{12}{16}$ and $\frac{14}{16}$, so

Answer: $\frac{13}{16}$

Exercise 2.7

1. Find a fraction that lies between $\frac{3}{5}$ and $\frac{4}{5}$.

2. Place the following in order of size, starting with the smallest:
   4.51, 4.579, 4.098, 4.105

3. Place these fractions in order of size, starting with the smallest:
   $\frac{4}{5}$, $\frac{2}{3}$, $\frac{3}{4}$, $\frac{17}{20}$

4. Place the following in order of size, starting with the smallest:
   $\frac{33}{100}$, $33\frac{1}{3}\%$, $\frac{3}{25}$, $\frac{3}{50}$, $\frac{67}{200}$

Exercise 2.8

Mixed Exercise

1. Write as a fraction in its lowest terms.
   (a) 75%
   (b) 0.07   (0580/01 May/June 2004 q 3)

2. Without using a calculator, work out $2\frac{1}{4} \div \frac{1}{2}$ as a single fraction.
   Show all your working.
                                                              (0580/01 May/June 2004 q 6)

3. $\frac{3}{5} \div \frac{7}{10} = \frac{6}{7}$
   Show how this calculation is done without using a calculator. Write down the working.
                                                              (0580/01 Oct/Nov 2004 q 9)

4. =   <   >
   Use one of the above to complete each of these statements.
   (a) 23 32
   (b) 9% 0.09
                                                              (0580/01 Oct/Nov 2004 q 11)

5. (a) Work out each of the following as a decimal.
   (i) 28%   (ii) $\frac{275}{1000}$ (iii) $\frac{2}{7}$

   (b) Write 28%, $\frac{275}{1000}$ and $\frac{2}{7}$ in order of size, writing the smallest first.
                                                              (0580/01 Oct/Nov 2003 q 15)

6. (a) Write in order of size, with smallest first.
   0.68, $\frac{33}{50}$, 67%
   (b) Convert 0.68 into a fraction in its lowest terms.
                                                              (0580/01 May/June 2003 q 3)

7. Show all your working for the following calculations. (The answers are given so it is only your working that will be given marks.)
   (a) $\frac{1}{2} + \frac{2}{3} = 1\frac{1}{6}$
   (b) $1\frac{1}{5} \times 1\frac{3}{4} = 2\frac{1}{10}$ (0580/01 May/June 2003 q 14)

8. Write the following in order, with the smallest first.
   $\frac{3}{5}$,   0.58,   62%   (0580/01 May/June 2007 q 2)

9. Rehana pays $284 in tax. This is $\frac{2}{9}$ of the money she earns. How much does she earn?
                                                              (0580/01 May/June 2008 q 9)

Chapter 3

Beginning Algebra

Algebra is a tool for doing arithmetic with unknown or variable quantities. We use letters to represent these quantities. If the quantities are unknown, algebra may enable us to work out what they are. If the quantities are variable, we can replace the letters with numbers when we need to.

Learning to use algebra is like learning a language. We need clear rules for the language so that we can all understand each other. You have met rules like these before, in Chapter 1, when you learned that the same order of working in arithmetic is needed if we are all to get the same answer.

Algebra can also be thought of as a form of mathematical shorthand, as you will see in this chapter. The early part of the chapter may seem too elementary if you are already familiar with variables or unknown quantities as well as pure numbers, but do read through it.

Essential Skills

1. (a) Calculate:
   - (i) $2 \times 6 + 3 \times 5$
   - (ii) $3 \times (6 - 4)$
   - (iii) $1 + 2 \times 3 - 4 \div 2 + 5 \times (6 - 3)$

   (b) What is the difference between temperatures of $-6°C$ and $-2°C$?

   (c) Insert a pair of brackets in this statement to make it true:
   $8 - 6 + 2 + 7 - 3 = 4$

   (d) (i) What is the sum of 5 and 6?
      (ii) What is the product of 5 and 6?

   (e) (i) What is the HCF of 20, 45 and 15?
      (ii) Rewrite 20, 45 and 15 as products of this factor and one other in each case.

   (f) Rewrite as an improper fraction:
   - (i) $4\dfrac{3}{4}$
   - (ii) $3\dfrac{5}{8}$
   - (iii) $12\dfrac{2}{3}$

Answers

1. (a) (i) $2 \times 6 + 3 \times 5$
   $= 12 + 15 = 27$

   (ii) $3 \times (6 - 4)$
   $= 3 \times 2 = 6$

   (iii) $1 + 2 \times 3 - 4 \div 2 + 5 \times (6 - 3)$

       $= 1 + 2 \times 3 - 4 \div 2 + 5 \times 3$

       $= 1 + 6 - 2 + 15$

       $= 20$

(b) The difference between $-6°C$ and $-2°C = 4°C$

(c) $8 - (6 + 2) + 7 - 3 = 4$

(d) (i) $5 + 6 = 11$ (ii) $5 \times 6 = 30$

(e) (i) Factors of 20 = {1, 2, 4, 5, 10, 20}

       Factors of 45 = {1, 3, 5, 9, 15, 45}

       Factors of 15 = { 1, 3, 5, 15}

       HCF of 20, 45 and 15 is 5

   (ii) $20 = 5 \times 4$

       $45 = 5 \times 9$

       $15 = 5 \times 3$

(f) (i) $\dfrac{19}{4}$ (ii) $\dfrac{29}{8}$ (iii) $\dfrac{38}{3}$

Using Letters and Numbers

Letters as variables or as unknown quantities

Suppose you are going to buy 3 apples and 5 oranges. If you know the price of both fruits, you can work out what the total cost will be. Suppose the apples cost 10 cents each and the oranges cost 12 cents each, then the total cost will be:

$$3 \times 10 + 5 \times 12$$

Using the correct order of working for arithmetic we can finish this:

$$\text{Total cost} = 3 \times 10 + 5 \times 12$$
$$= 30 + 60 = 90 \text{ cents}$$

But suppose we do not know the cost of the apples?

We can still do some of the work like this:

$$\text{Total cost} = 3 \times cost\ of\ an\ apple + 5 \times 12$$
$$= 3 \times cost\ of\ an\ apple + 60$$

This would take too much time to keep writing out.

If we use *apple* to mean the *cost of one apple*, the sum becomes

$$\text{Total cost} = 3 \times apple + 60$$

Even better we could shorten *apple* to a:

$$\text{Total cost} = 3 \times a + 5 \times 12$$
$$= 3 \times a + 60$$

We can make this look neater by using one of the rules of algebra, that $3 \times a$ can be shortened to $3a$.

Our final statement is:

$$\text{Total cost} = 3a + 60$$

Later, when we know the cost of the apples we can finish the sum.

The cost of one apple is a *variable* quantity. It could be 10 cents today and 12 cents tomorrow. The total cost is also variable, but depends on a.

Algebraic shorthand

To get started with algebra we must start to learn a few rules. We will often use x and y as our unknown quantities, but remember that we can use any letter. When two or more letters are different, we know that they are being used for *different* things.

We start with some shorthand.

Had you noticed that multiplication is a shorter form of addition?

There are two ways to work out $7 + 7 + 7 + 7 + 7 + 7 + 7 + 7 + 7$.

You can either go through and add sevens as you go along, or you can see that there are 9 sevens, and quickly get the answer:

$$9 \times 7 = 63.$$

You can do this with any number, not just 7, so we could call the number x.

$9 \times x$ can be shortened to $9x$ without any confusion, but 9×7 cannot be shortened to 97 which is completely different.

Remember:

- $x + x + x + x + x + x + x + x + x = 9 \times x$ and $y + y + y = 3 \times y$.
- $9 \times x$ can be shortened to $9x$, so $9 \times x = 9x$ and $3 \times y = 3y$.

Remember also:

- $x \times x = x^2$ and $y \times y \times y = y^3$.
- $1x$ can be written as just x, so $1x = x$ and $1y = y$.

These also agree with our work with numbers:

$$2 \times 2 \times 2 = 2^3 \text{ and } 1 \times 2 = 2$$

How algebra is similar to arithmetic (and how it is different)

When we are using letters and numbers, **simplify** is an instruction to write an answer in its simplest form. The answer will usually still contain letters.

Solve usually means to find a numerical answer to an equation.

When we are using numbers only, **calculate** is an instruction to find the solution to a numerical problem. The answer will be a number, or numbers.

We can *simplify* $y \times y \times y$ to y^3, but $2 \times 2 \times 2 = 2^3$ (= 8) can be *calculated* to give the numerical answer 8.

Much of your algebra will involve simplifying or, later, writing things in another form.

Work through this example, paying particular attention to the numerical questions which show the similarities between algebra and arithmetic, and the difference between simplifying and calculating.

Example 1

Simplify the following:

(a) $x + x + x + x + x$

(c) $y \times y$

(e) $4x + 2x$

(g) $3x + x$

(i) $3y - y$

(k) $x + x + y + y + y$

(m) $9x - 8x$

Calculate the following:

(b) $2 + 2 + 2 + 2 + 2$

(d) 8×8

(f) $4 \times 7 + 2 \times 7$

(h) $3 \times 5 + 5$

(j) $3 \times 12 - 12$

(l) $115 + 115 + 108 + 108 + 108$

(n) $9 \times 157 - 8 \times 157$

Answer 1

(a) $x + x + x + x + x = 5x$

(c) $y \times y = y^2$

(e) $4x + 2x = 6x$

(g) $3x + x = 4x$

(i) $3y - y = 2y$

(k) $x + x + y + y + y = 2x + 3y$

(b) $2 + 2 + 2 + 2 + 2 = 5 \times 2 = 10$

(d) $8 \times 8 = 8^2 = 64$

(f) $4 \times 7 + 2 \times 7 = 6 \times 7 = 42$

(h) $3 \times 5 + 5 = 4 \times 5 = 20$

(j) $3 \times 12 - 12 = 2 \times 12 = 24$

(l) $115 + 115 + 108 + 108 + 108$

$\qquad = 2 \times 115 + 3 \times 108$

$\qquad = 230 + 324$

$\qquad = 554$

(m) $9x - 8x = 1x = x$

(n) $9 \times 157 - 8 \times 157 = 1 \times 157 = 157$

If you feel tempted to go further with Example 1 (k) and attempt some sort of addition of the x's and y's try it with the numbers as well and see if it works.

For example, *if* you think that $2x + 3y$ could be shortened to $5xy$ (which of course you should not!), use numbers to check it.

$$2x + 3y: \quad 2 \times 115 + 3 \times 108$$
$$= 230 + 324$$
$$= 554$$

but $\qquad\qquad 5xy: \quad 5 \times 115 \times 108$
$$= 62100$$

which is clearly not the same as 554!

We can only arrive at a final answer when we know what numbers will replace x and y. Until then the question has to ask you to *simplify*, rather than *solve*.

Remember:

- **Simplify** means to write in the simplest form.
- **Solve** usually means to find a numerical answer to an equation.
- And **calculate** means to find a numerical answer.

Exercise 3.1

Copy and complete the following table by using algebraic shorthand to simplify, and the rules of arithmetic (BoDMAS) to calculate.

		Simplify	Answer		Calculate	Answer
(a)	(i)	$x + x + x$		(ii)	$30 + 30 + 30$	
(b)	(i)	$5y - 4y$		(ii)	$5 \times 154 - 4 \times 154$	
(c)	(i)	$z \times z \times z$		(ii)	$3 \times 3 \times 3$	
(d)	(i)	$x + x + x - y$		(ii)	$6 + 6 + 6 - 10$	
(e)	(i)	$x + x + y + y$		(ii)	$7 + 7 + 4 + 4$	
(f)	(i)	$y - y$		(ii)	$2 - 2$	
(g)	(i)	$x \times x + y \times y$		(ii)	$3 \times 3 + 4 \times 4$	
(h)	(i)	$5x + 3x - 2y$		(ii)	$5 \times 50 + 3 \times 50 - 2 \times 4$	

The Language of Algebra

Expressions, equations and terms

There are some other words that have a special meaning in algebra, and you must understand them as well. First of all, try to understand the difference between an *expression* and an *equation*. Have a look at this piece of algebra:

$$3x + 5y - 10z + 6$$

This is an **algebraic expression**. It is not an equation since it stands alone without an equals sign. It is made up of **terms** which are to be added or subtracted. The terms are $3x$, $5y$, $10z$ and 6.

$3x$ is a 'term in x', $5y$ is a 'term in y', $10z$ is a 'term in z', and 6 is a constant or a number term.

6 is a constant term because it is always 6, but $3x$ is not constant because it depends on what x stands for.

Each number in front of a term is the **coefficient** of that term.

Now look at the following expression:

$$2x + 7y - 3y + 4x$$

This is an expression that can be simplified. It has **like terms**. It has two terms in x and two terms in y. We can write:

$$2x + 7y - 3y + 4x$$
$$= 2x + 4x + 7y - 3y$$
$$= 6x + 4y$$

This is called **collecting like terms**.

Each of the two equals signs show that the next line is equivalent to the one before, but has been written in another way. They do *not* convert the expression into an equation.

But if we are given a bit more information, for example, that our expression is actually equal to something else, we have an equation.

For example, $6x + 4y = 34$ is an **equation**.

An expression is like a phrase in English, and an equation is more like a sentence. For example, 'hot and stormy weather' is a phrase in English. It means more when it becomes a sentence such as: 'Today we are having hot and stormy weather' and we have the extra bit of information that it is today that we are talking about.

An equation may be **solved** by finding replacements for the variables which make it a true statement. For example, we can solve $10z - 3 = 17$.

This is an equation which becomes true when z is replaced by 2.

$$10 \times 2 - 3 = 17$$

So the solution to the equation is $z = 2$, and in this case it is the only solution. The equation $6x + 4y = 34$, becomes true when we replace the x by 3 and the y by 4 because

$$6 \times 3 + 4 \times 4 = 18 + 16 = 34$$

So $x = 3$ and $y = 4$ is a solution to this equation.

In this case this is not the only possible solution.

For example, $x = 2.5$ and $y = 4.75$ is also a solution. Check it for yourself!

NOTE: NEVER try to turn an expression into an equation, for example, by making it equal to zero, unless the question asks you to. This is a common mistake made by students.

Remember that you may be able to simplify an expression, but not solve it. You may be able to simplify *and* solve an equation.

So far we have mainly used letters to represent unknown or variable quantities. But remember the example of buying apples and oranges?

We wrote: Total cost = $3 \times apple + 60$

Here *apple* represents the variable cost of one apple.

Variables can be represented by words, letters or symbols.

For example,

$$3 \times what = 21$$
$$3x = 21$$
$$3 \times ? = 21$$
$$3 \times \square = 21$$

In each case the unknown can be replaced by 7 to make the equation true. For simplicity it is usual to use letters.

Example 2

(a) $3x + 4y + y = 3x + 5y$ $7x + 10y = 37$ $3a - 4b$

From the above select:

   (i) a term in x,   (ii) a pair of like terms,

 (iii) an equation,   (iv) an expression which is then simplified,

  (v) another expression,   (vi) a constant term,

(vii) the coefficient of the term in b.

(b) Can you find replacements for x and y that would make $7x + 10y = 37$ true?

(c) Write word equations for the following situations:

  (i) A boarding kennels has room for cats and dogs. It has 14 kennels altogether. Write an equation to show the variable number of cats and dogs that can be accommodated at one time. Give three possible solutions to your equation.

 (ii) I give a shopkeeper 10 rupees. He gives me 4 mangoes, and 4 rupees change. Write a word equation to show this and so find the price of one mango.

(d) (i) Use the letters given to write an equation to represent the following statement.
'I buy 2 bags of crisps and 3 chocolate bars. I spend 10 rupees altogether.'
Use c = the cost of a bag of crisps and b = the cost of a chocolate bar.

   (ii) Find one pair of possible replacements for c and b which would make your equation true.

Answer 2

(a) (i) $3x$ or $7x$ are both terms in x.   (ii) $4y$ and y are like terms.

   (iii) $7x + 10y = 37$ is an equation.

   (iv) $3x + 4y + y$ is an expression which is simplified to $3x + 5y$.

   NOTE: Remember that $3x + 4y + y = 3x + 5y$ is not an equation.

   (v) $3a - 4b$ is another expression.   (vi) 37 is a constant term.

   (vii) 4 is the coefficient of the term in b.

(b) $7x + 10y = 37$

   By trying a few numbers we find that $x = 1$ and $y = 3$ would make this equation true.
   $7 \times 1 + 10 \times 3 = 7 + 30 = 37$

   If you use rational numbers there are an infinite number of solutions.
   For example, $x = 2$ and $y = 2.3$. Can you find some more?

(c) (i) If *cats* represents the number of cats and *dogs* represents the number of dogs then
   $cats + dogs = 14$
   Possible solutions are:
   $cats = 13$ and $dogs = 1$
   $cats = 10$ and $dogs = 4$ NOTE: We can only use counting numbers here! Why?
   $cats = 5$ and $dogs = 9$

   (ii) If *mangoes* = the cost of one mango,
   $10 = 4 \; mangoes + 4$
   so $4 \; mangoes = 6$
   so $mangoes = 1.5$
   Hence, mangoes cost 1.5 rupees each.

(d) (i) $2c + 3b = 10$

   (ii) $c = 2$ and $b = 2$ is one possible pair of values that would make this true.

   NOTE: Here c and b happen to be the same, but in all the other possible solutions they would be different.

Formulae and substitution

Formulae are equations that are used fairly frequently to calculate quantities and are arranged so that the required quantity is the **subject** of the formula. This makes them convenient to use. For example, in our original small problem of the cost of apples and oranges, we ended up with the formula:

$$\text{Total cost} = 3a + 60$$

This can be called a formula because it is arranged so that the quantity we want to find (total cost) is on its own on the left and so is the subject of the statement.

When we know the cost of an apple we will be able to substitute this in to replace a, and calculate the total cost. Suppose b is the cost of one orange, and T is the total cost, then our formula could become more useful:

$$T = 3a + 5b$$

T, a and b are the unknowns or variables because the prices may vary from day to day. When we know the cost of an apple and an orange on the day we can **substitute** these numbers for a and b and work out the total cost on the day.

Substitution is replacing unknowns or variables by numbers, usually in formulae, so that answers may be calculated. Calculating the answer when variables in an expression are substituted by numbers is often called **evaluating** the expression.

Example 3
(a) I think of a number (n), multiply it by 4, add 6, then take away the number I first thought of.
   Write a formula for the answer (A) in terms of n.
(b) Use the formula to find A when:
   (i) $n = 3$ (ii) $n = 100$ (iii) $n = 11$

Answer 3
(a) $A = n \times 4 + 6 - n$
   $A = 4n + 6 - n$
   $A = 3n + 6$
(b) (i) when $n = 3$ (ii) when $n = 100$ (iii) when $n = 11$
     $A = 3 \times 3 + 6 = 9 + 6$ $A = 3 \times 100 + 6$ $A = 3 \times 11 + 6$
     $A = 15$ $A = 306$ $A = 39$

Exercise 3.2

1. Maria is m years old. Her father, Bakari, is n years *older* than Maria.
   Write an expression for the sum of their ages.
2. A piece of wood is 6.5 metres long. Brian saws off and uses m metres.
   Write an expression for the length of wood which is left.
3. Amir starts his journey to school by walking for 10 minutes, and then takes a bus.
   The time (t minutes) the bus takes to get to the school depends on the traffic.
   (a) Write a formula for the total journey time (T minutes) in terms of t.
   (b) Find T when $t = 15$.
4. Substitute $y = 3$ and $z = 5$ into each of the equations below to find x.
   (a) $x = 2y + 3z$ (b) $x = yz + 2$ (c) $x = 4yz - 3z + 2y$
5.

   a cm a cm

   3 cm

The diagram shows a triangle with two sides of length a cm, and one side of length 3 cm.

(a) Write a formula for the total length (L cm) round the outside of the triangle.

(b) Use the formula to find L when $a = 10$.

(c) Why can a not be   (i) 1.5   (ii) 1 ?

6. A recipe requires 5 eggs, 0.5 kilograms of butter, 0.5 kilograms of tomatoes. Eggs cost e rupees per 10, butter costs b rupees per half kilogram and tomatoes cost t rupees per kilogram.

(a) Write a formula for the total cost (C rupees) of the recipe.

(b) Calculate C when e = Rs. 22, b = Rs. 58 and t = Rs. 12.

7. Evaluate the following expressions when $x = 2$ and $y = 3$.

(a) xy (b) $y - x$ (c) $y^2 - x^2$ (d) $3x + 9y$

Addition and Subtraction of Algebraic Quantities

As we have already seen, addition and subtraction of algebraic quantities is also called *collecting like terms*.

We could illustrate the process by thinking of a zoo which keeps antelopes and bears. Antelopes and bears are not variables or unknown quantities, so we will not replace them with letters.

The zoo starts with 10 antelopes and 5 bears, and then they trade 4 of their antelopes for 2 bears with another zoo. Later on they give away 2 more antelopes. How many antelopes and bears do they now have? The situation could be written like this:

10 antelopes + 5 bears − 4 antelopes + 2 bears − 2 antelopes

which can be rearranged to:

10 antelopes − 4 antelopes − 2 antelopes + 5 bears + 2 bears

This gives us:

6 antelopes + 7 bears

You should notice that we can rearrange the sum as long as we keep the sign with the animal that follows it.

Remember:

- The sign belongs to the term that follows it.
- Terms can be written in any order in the expression as long as they keep their signs.
- Only like terms can be added or subtracted.

Example 4

Simplify:

(a) $3x - 2y + 2x + 5y$

(b) $6x^2 + 2x + 3x + 5$ NOTE: The term in x^2 is *not* the same as the term in x.

(c) $1 + 6xy - 5x - yx + 3$ NOTE: The term in xy *is* the same as the term in yx.

(d) $x^2 + x^3 + 2x^2 + 3x^3$ (Remember that the multiplication can be done in either order.)

Answer 4

(a) $3x - 2y + 2x + 5y = 3x + 2x + 5y - 2y$
$= 5x + 3y$

(b) $6x^2 + 2x + 3x + 5$
$= 6x^2 + 5x + 5$

(c) $1 + 6xy - 5x - yx + 3$
$= 4 + 5xy - 5x$

(d) $x^2 + x^3 + 2x^2 + 3x^3$
$= 3x^2 + 4x^3$

Exercise 3.3

Simplify:

1. $3x + 10x$
2. $5x + 7x - x$
3. $2x + 5x - 4y$
4. $2a + 5a - 3a + 6b$
5. $6x - 3x + 2y - y$
6. $3x + 10 - 7$
7. $4z + 2w + 2z + w - 2z$
8. $7c - c + 6d - 3 - 6d$
9. $6 + 3a - 3 - a$
10. $4x + y - 3x - y$
11. $8x^2 + 4y^2 - 7x^2 - 2y^2$
12. $6x^2 + 2x - 3x^2 + x$
13. $x^2 + xy + 4x^2 + 2xy$
14. $2x^2 + y^2 - xy + x^2$
15. $3x^2 + 2xy - xy - 4y^2$
16. $3x^2 - 5x^3 + 2x^2y + x^2y$
17. $5x^2y^2 - 3x^2y - x^2y^2 + x^2y$

Multiplication and Division of Algebraic Quantities

When you multiply or divide algebraic quantities it is best to be systematic.

Remember:

- $c \times a \times b$ can be shortened to abc.
- It is usual to write the letters in alphabetic order, but it is not wrong if you do not.
- Deal with the numbers first, then the letters. $4x \times 5y = 20x \times y = 20xy$
- In division just as $2 \div 2 = 1$, so $x \div x = 1$.
- In division also deal with numbers first and then letters.

$$\frac{9y}{3xy} = \frac{3y}{xy} = \frac{3}{x} \qquad (y \div y = 1)$$

Example 5:

Simplify:

(a) $3a \times 4b \times 6c \times a$
(b) $5x \times 6x \times 4y$
(c) $6d \times 5c \div 3d$

(d) $x^2y \div x^2y$
(e) $3 \div 3x^2$

Answer 5

(a) $3a \times 4b \times 6c \times a = 72 \times a \times b \times c \times a$
$= 72a^2bc$

(b) $5x \times 6x \times 4y = 120 \times x \times x \times y$
$= 120x^2y$

(c) $6d \times 5c \div 3d$

$$= \frac{6d \times 5c}{3d} = \frac{30dc}{3d} = \frac{10dc}{d}$$

$$= 10c$$

(e) $3 \div 3x^2$

$$= \frac{3}{3x^2} = \frac{1}{x^2}$$

(d) $x^2y \div x^2y$

$$= \frac{x^2y}{x^2y}$$

$$= 1$$

NOTE: The x^2 is in the denominator, and must stay there. A common error is to move it to the numerator, which is definitely wrong.

Exercise 3.4

Simplify:

1. $5a \times 3b$
2. $6y \times 4z$
3. $3x \times 2x$
4. $10x \times 3x \times 2$
5. $x \times y \times 3z \times 2$
6. $5a \div 5$
7. $3d \times 4a \times d \times 5b$
8. $8 \div 2x$
9. $10cd \div 5c$
10. $4x \div 2$
11. $6 \div 12d$
12. $3ab \div 2ab$
13. $5ab \times 4cd$
14. $2d \times 6c \div dc$

Working with Directed Numbers

We shall meet directed numbers frequently in algebra, and it is essential that you can work with them with confidence. We first met directed numbers in Chapter 1, but now we need to work with them.

Have a look at the number line given below.

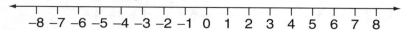

Remember:

- Numbers with a negative (−) sign attached to them are called negative numbers, so all the numbers to the left of the zero are negative numbers. All the numbers to the right of the zero are positive (+) numbers.
- A number or term at the beginning of an expression or equation that does not have a sign is taken to be positive.
- The sign belongs to the term that follows it, as we saw before.

Addition and subtraction of directed numbers

To **subtract** using your number line start at the first number and move to the *left*.
You know that $9 - 4 = 5$.
On your number line start at 9 and go back 4 steps to reach the answer 5.
How about $1 - 4$?
Start at 1 on the number line and go back 4 steps to reach −3 (negative 3), so

$$1 - 4 = -3$$

For $-3 - 4$ start at negative 3 on the number line and again go back 4 steps.

You will find that negative 3 take away 4 is negative 7.

$$-3-4=-7$$

The minus signs in front of the 3 and the 7 are both read as negative signs because they are directed number signs. The minus sign in front of the 4 means take away (subtract) four. So 'negative 3 take away 4 is negative 7'. In practice many people say 'minus' instead of 'negative', but we must be careful that this does not cause confusion.

To **add** you start at the first number and move to the *right*.

$$-6+15=9$$

Starting at − 6 and moving 15 steps to the right you arrive at positive 9, (or simply 9). There are two more ways that you will see negative numbers written in books or examination papers:

− 6 could be shown as ⁻6 or as (− 6),

and the positive numbers could be shown in four ways:

+ 9 or ⁺9 or (+ 9) or just 9.

In the examples and exercises that follow, you will find all these forms so that you get used to them.

Example 6

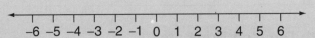

Using the given number line, calculate:

(a) − 4 + 3 + 2 − 6 + 5 − 1   (b) − 6 + 5 − 1 + 3 − 4 + 2

(c) 3 + 2 + 5 − 6 − 1 − 4   (d) Why do all these give the same answer?

Answer 6

(a) − 4 + 3 + 2 − 6 + 5 − 1   (b) − 6 + 5 − 1 + 3 − 4 + 2

    = − 1   = − 1

(c) 3 + 2 + 5 − 6 − 1 − 4   (d) You can perform addition and subtraction
    = − 1   in any order as long as you keep the sign
                                          with the term that follows it.

Exercise 3.5

Calculate:

1. 7 − 3 − 5 + 2 − 8   2. 10 − 2 − 1 − 3 − 4 + 7   3. ⁻2 − 3 − 4 − 1 + 10

4. 3 − 3 + 4 + 2 − 5 − 1   5. ⁻8 + 8   6. ⁺7 − 7

7. − 3 − 2 − 1   8. 6 − 4 + 5   9. − 8 + 9

10. 0 − 5 + 10

The next thing to understand is what it means when you see something like:

$$8-{}^-2,$$

which is read as '8 take away (or subtract or minus) negative 2'.

Taking away a negative number is the same as adding the same number.

There are many ways to try to explain this, but maybe we should just look at it as a double negative meaning a positive.

'You are *not* going to *miss* the train tonight' means that 'you *are* going to catch the train tonight'. This type of double negative is common, so it helps us to remember the rule.

If your bank balance is overdrawn, and you are in debt by, say $100, it would show as − $100. If, for some reason, someone took away the debt, you would end up with nothing in the bank, but no longer in debt. Taking away the negative amount is the same as adding it.

$$- 100 - {}^-100 = - 100 + 100 = 0$$

In the same way, adding a positive number is the same as addition, and adding a negative number is the same as subtraction.

Be sure that you remember these facts:

$$+ {}^+10 = + 10$$
$$- {}^-10 = + 10$$
$$+ {}^-10 = - 10$$
$$- {}^+10 = - 10$$

NOTE: Remember that two like signs (that is two signs which are the same) make a plus, and two unlike signs (different signs) make a minus.

$$+ + = +$$
$$- - = +$$
$$+ - = -$$
$$- + = -$$

It is best to simplify any two signs that are together (for example, −(−), −${}^+$ or + −) *before* you go on with the calculation. We could call them *double* signs, and will refer to them again later. So whenever you meet a double sign simplify it before carrying on.

Remember:

- Simplify double signs before adding or subtracting.
- Addition or subtraction can be done by taking steps along the number line in the appropriate direction.

Example 7

Calculate:

(a) $2 - 4$ (b) $- 19 + 7$

(c) ${}^-12 - 13$ (d) $17 - {}^-3$ (deal with the double sign first, a double negative is a positive)

(e) $21 + (-7)$ (double sign first)

(f) $14 - {}^+2$ (double sign first)

(g) $(-5) - (-2)$ (double sign first)

Answer 7

(a) $2 - 4$ (b) $- 19 + 7$ (c) ${}^-12 - 13$

   $= - 2$ $= -12$ $= - 25$

(d) $17 - {}^-3$
   $= 17 + 3$
   $= 20$

(e) $21 + (-7)$
   $= 21 - 7$
   $= 14$

(f) $14 - {}^+2$
   $= 14 - 2$
   $= 12$

(g) $(-5) - (-2)$
   $= -5 + 2$
   $= -3$

Multiplication and division of directed numbers

The same rules described for addition and subtraction with double signs apply here. Learn these!

+	×	+	=	+
−	×	−	=	+
+	×	−	=	−
−	×	+	=	−

+	÷	+	=	+
−	÷	−	=	+
+	÷	−	=	−
−	÷	+	=	−

The rules are not so difficult to learn if you remember that two like signs make a plus, and two unlike signs make a minus.

When multiplying or dividing directed numbers it pays to work in order as before. In fact you should always be systematic in algebra to avoid unnecessary mistakes. First deal with signs, next the numbers and lastly, if there are any, the letters (see the next section).

Example 8
Calculate:
(a) -2×-6,
(b) $-10 \div -5$,
(c) 20×-7,   NOTE: Remember the 20 is positive, so $+ \times - = -$
(d) $55 \div {}^-11$.

Answer 8
(a) -2×-6
   $= +2 \times 6$
   $= +12$

(b) $-10 \div -5$
   $= +10 \div 5$
   $= 2$

(c) 20×-7
   $= -20 \times 7$
   $= -140$

(d) $55 \div {}^-11$
   $= -55 \div 11$
   $= -5$

Remember:
- If addition or subtraction of directed numbers involves double signs, deal with them *first*, applying the rules. Then go to the number line to work out the answer.
- In multiplication and division of directed numbers apply the rules while multiplying and dividing.

To make this absolutely clear we will do one more mixed example.

Example 9

Calculate:

(a) $(-2) - (-3) + (-4)$ (b) $-3 + {}^-5$ (c) -1×-6 (d) $2 \div (-3)$

(e) $-2 \times (-2) \times (-2)$ NOTE: The first two negative 2s multiplied together make positive 4. Then what happens? That positive 4 has to be multiplied by the remaining negative 2 to make a negative number in the end.

Answer 9

(a) $(-2) - (-3) + (-4)$
   $= -2 + 3 - 4$
   $= -3$

(b) $-3 + {}^-5$
   $= -3 - 5$
   $= -8$

(c) ${}^-1 \times {}^-6$
   $= +1 \times 6$
   $= +6$

(d) $(+2) \div (-3)$
   $= -2 \div 3$
   $= -\dfrac{2}{3}$

(e) $-2 \times (-2) \times (-2)$
   $= +4 \times (-2)$
   $= -8$

Exercise 3.6

Calculate:

1. (a) $2 + 3$ (b) $-2 + 3$ (c) $2 - 3$ (d) $-2 - 3$
   (e) $-2 - {}^-3$ (f) $-2 + {}^+3$ (g) $-2 + (-3)$ (h) $(-2) + (-3)$

2. (a) 2×3 (b) -2×3 (c) $-2 \times (-3)$ (d) $2 \times {}^-3$ (e) $-2 \times +3$

3. (a) $6 \div 3$ (b) $-6 \div 3$ (c) $-6 \div (-3)$ (d) $6 \div {}^-3$

4. (a) $\dfrac{2}{4}$ (b) $\dfrac{-2}{4}$ (c) $\dfrac{-2}{-4}$ (d) $\dfrac{2}{-4}$

5. $-5 - -6 + -1 + 10$ 6. $4 - (-2) + (-3) - 5$

7. $10 - (+3) - (-3)$ 8. $5 \times (-2) \times (-1)$

9. $6 \times 3 \div (-2)$

10. (a) -1×-1 (b) $-1 + -1$ (c) $-1 \div -1$
    (d) $(-1) \times (-1) \times (-1)$ (e) $(-1) \times (-1) \times (-1) \times (-1)$

11. $-7 \times -4 \div +2$

Working with directed numbers in algebra

The same rules apply for working with directed numbers in algebra as we have just covered in the previous two sections where we used only numbers.

Example 10
Simplify, where possible.
(a) $2a + -3a - b - -6b$ (b) $-6x \times (-2y)$ (c) $-6x + (-2y)$

Answer 10
(a) $2a + -3a - b - -6b = 2a - 3a - b + 6b$
    $= -a + 5b$
(b) $-6x \times (-2y) = + 6x \times 2y$ (c) $-6x + (-2y)$
    $= + 12xy$ $= -6x - 2y$

Exercise 3.7

Simplify, where possible:
1. $-x \times 2y$ 2. $-x \times -2y$
3. $-x + (-2y)$ 4. $-x - {}^-2y$
5. $a^2 \div 2$ 6. $3a \div (-6b)$
7. $(-2x) \times (-4y)$ 8. $(-2x) \div (-4y)$
9. $xy \times (-3x)$ 10. $xy + (-3x)$
11. $xy \div (-3x)$ 12. $-3y \times {}^-2x \times {}^-x$
13. $-4z + (-2z) - (-3z^2)$ 14. $3a + 2b + (-a) - (-b)$

Working with Indices

We have already met the fact that $x + x + x$ can be shortened to $3x$, and $x \times x \times x$ can be shortened to x^3.

The 3 in x^3 is a **power** or **index**. The plural of index is indices.

Raising to a whole number power is a shorter form of multiplying something by itself many times. The number of times it is multiplied by itself is shown by the power or index.

If numbers are raised to a power the answer may be calculated, but if algebraic variables are raised to a power we can only simplify.

Two examples may help:
$$2^5 = 2 \times 2 \times 2 \times 2 \times 2 = 32$$
$$\text{but} \quad x \times x \times x \times x \times x = x^5$$

We can still work with powers of algebraic variables as we will see in the rest of this section.

Writing out $x^3 \times x^4$ in longhand:
$$x \times x \times x \times x \times x \times x \times x = x^7$$

In shorthand: $x^3 \quad \times \quad x^4 \quad = x^{3+4} = x^7$

When you **multiply** a number or a variable raised to a power by the *same* number or variable raised to another power you *add* the powers.

2^3	$\times$	2^2	$=$	2^{3+2}	$=$	2^5	$=$	32
or $2 \times 2 \times 2$	$\times$	2×2			$=$	2^5	$=$	32
or 2^3	$\times$	2^2	$=$	8	$\times$	4	$=$	32

A similar argument can be applied to division.

$$2^5 \div 2^2 = 32 \div 4 = 8$$
$$\text{or } 2^5 \div 2^2 = 2^{5-2} = 2^3 = 8$$

When you **divide** a number or variable raised to a power by the *same* number or variable raised to another power you *subtract* the second power from the first.

There is one more rule to learn for working with powers.

$$(2^4)^3 = (2^4) \times (2^4) \times (2^4) = 2^{4+4+4} = 2^{3\times4} = 2^{12}$$

So as you should see, raising to a *further* power means multiplying the powers.

Learn these three examples:

- $x^6 \times x^2 = x^{6+2} = x^8$
- $x^6 \div x^2 = x^{6-2} = x^4$
- $(x^6)^2 = x^{6\times2} = x^{12}$

NOTE: It is very important that you remember that the power *only* refers to the single variable before it *unless* it is outside a pair of brackets.

For example,

$$2xy^2 = 2 \times x \times y \times y = 2 \times x \times y^2$$
$$2(xy)^2 = 2 \times x^2 \times y^2$$
$$(2xy)^2 = 4 \times x^2 \times y^2$$

Example 11

(a) Calculate:

   (i) $3^2 \times 3^3$ (ii) $3^4 \div 3^3$ (iii) $(3^2)^2$ (iv) $3^2 + 3^4$

(b) Simplify, where possible.

   (i) $x^4 \times x^5$ (ii) $x^6 \div x^2$ (iii) $(x^7)^3$ (iv) $x^2 + x^3$ NOTE: These are unlike terms!

   (v) $x^3 + x^3$ NOTE: These are like terms.

Answer 11

(a)   (i) $3^2 \times 3^3$

        $= 9 \times 27 = 243$ (or 3^5)

   (iii) $(3^2)^2$

        $= 9^2 = 81$ (or 3^4)

   (ii) $3^4 \div 3^3$

        $= 81 \div 27 = 3$ (or 3^1)

   (iv) $3^2 + 3^4$

        $= 9 + 81 = 90$

(b)   (i) $x^4 \times x^5$

        $= x^{4+5} = x^9$

   (ii) $x^6 \div x^2$

        $= x^{6-2} = x^4$

   (iii) $(x^7)^3$

        $= x^{7 \times 3} = x^{21}$

   (iv) $x^2 + x^3$ No simplifying possible because *unlike* terms can not be added.

   (v) $x^3 + x^3$ Like terms *can* be added.

        $= 2x^3$

Exercise 3.8

1. Calculate:
   (a) $4^2 \times 4^3$ (b) $5^5 \div 5^3$ (c) $4^2 + 4^2$ (d) $2^3 - 2^2$ (e) $(2^3)^3$

2. Simplify, where possible.
   (a) $x^4 \times x^5$ (b) $x^4 + x^5$ (c) $x^3 + x^3$ (d) $x^3 \times x^3$
   (e) $x^6 \div x^2$ (f) $2x^5 \times 3x^6$ (g) $2x^5 + 3x^5$ (h) $(2x) \times (2x)$
   (i) $(2x)^2$ (j) $(2x^3) \times (2x^3)$ (k) $(2x^3)^2$ (l) $(x^5)^9$

3. Simplify, as far as possible.
   (a) $3x^4 \times x \times 2x^5$ (b) $2x^6 \times y^3 \times x^7$
   (c) $6x^5 \times 3x^4 \div 2x^3$ (d) $9x^3 \times 4x^2 \times 2y^3 \div 3y^2$

So far we have dealt with indices that come from the set of counting numbers. We also have to work with indices that are from the set of integers. These would include, for example, x^{-2}, x^0, x^{-1} and so on.

Look at this pattern:

$$x^3 = 1 \times x \times x \times x$$
$$x^2 = 1 \times x \times x$$
$$x^1 = 1 \times x$$
$$x^0 = 1$$
$$x^{-1} = \frac{1}{x}$$
$$x^{-2} = \frac{1}{x \times x} = \frac{1}{x^2}$$
$$x^{-3} = \frac{1}{x \times x \times x} = \frac{1}{x^3}$$

You should be able to see that $x^0 = 1$.

In fact, since x can stand for anything we could say that anything raised to the power zero is one.

$$1000000^0 = 1, \qquad 0.0000034^0 = 1, \qquad 123456789^0 = 1,$$

$$\left(\frac{576}{0.49}\right)^0 = 1, \qquad (x^{24} + y^{25})^0 = 1$$

Remember:

- Anything raised to the power zero $= 1$ (except zero itself)

This makes some of the questions you might meet rather easy!

You should also see that negative powers indicate that the term should be inverted to make the power positive.

Invert means turn upside down, so $x^{-1} = \dfrac{1}{x}$. (Remember that x^{-1} can be written as $\dfrac{x^{-1}}{1}$.)
For example,

$$x^{-2} = \frac{1}{x^2}, \qquad\qquad \frac{1}{x^{-3}} = x^3$$

and

$$\left(\frac{2}{3}\right)^{-1} = \frac{3}{2}, \qquad\qquad \left(\frac{2}{3}\right)^{-3} = \left(\frac{3}{2}\right)^3 = \frac{27}{8}$$

You may also meet the word **reciprocal** which is a fraction inverted. For example, the reciprocal of $\dfrac{2}{5}$ is $\dfrac{5}{2}$.

Remember:

- You can invert anything with a negative power to make the power positive.
- Negative powers can be added and subtracted in the same way as directed numbers.

Example 12

(a) Evaluate:

    (i) 2^{-1} (ii) 2^{-4} (iii) $\dfrac{1}{2^{-3}}$ (iv) $2^{-1} \times 2^4$

    (v) $2^{-1} \div 2^{-3}$ (vi) $\left(\dfrac{3}{5}\right)^{-1}$ (vii) $\left(1\dfrac{3}{5}\right)^{-2}$

(b) Simplify, giving the answers in a form with positive powers.

    (i) $x^{-3} \times x^7$ (ii) $x^{10} \times x^{-3} \times x^{-1}$ (iii) $x^{-8} \div x^{-6}$ (iv) $x^5 \div x^{-1}$
    (v) $(x^{1000})^0$ (vi) $(2x)^{-2}$ (vii) $2x^{-2}$

(c) In each case, find a replacement for n which will make the statement true.

    (i) $16 = 2^n$ (ii) $\dfrac{1}{4} = 2^n$ (iii) $\dfrac{3}{4} = \left(\dfrac{4}{3}\right)^n$

(d) Find a replacement for x in each of the following.

    (i) $27 = 3^x$ (ii) $x^3 = 64$ (iii) $\dfrac{1}{2^x} = 8$

Answer 12

(a)   (i) $2^{-1} = \dfrac{1}{2}$ (ii) $2^{-4} = \dfrac{1}{2^4} = \dfrac{1}{16}$ (iii) $\dfrac{1}{2^{-3}} = 2^3 = 8$

    (iv) $2^{-1} \times 2^4 = 2^{-1+4} = 2^3 = 8$ (v) $2^{-1} \div 2^{-3} = 2^{-1--3} = 2^{-1+3} = 2^2 = 4$

NOTE: Remember that to divide a number with a power by the same number to another power means you subtract the powers.

(Alternative method: $2^{-1} \div 2^{-3} = \dfrac{1}{2} \div \dfrac{1}{2^3} = \dfrac{1}{2} \div \dfrac{1}{8} = \dfrac{1}{2} \times \dfrac{8}{1} = 4$)

(vi) $\left(\dfrac{3}{5}\right)^{-1} = \dfrac{5}{3}$

(vii) $\left(1\dfrac{3}{5}\right)^{-2} = \left(\dfrac{8}{5}\right)^{-2} = \left(\dfrac{5}{8}\right)^{2} = \dfrac{25}{64}$

(b)   (i) $x^{-3} \times x^7 = x^{-3+7} = x^4$

(ii) $x^{10} \times x^{-3} \times x^{-1} = x^{10-3-1} = x^6$

(iii) $x^{-8} \div x^{-6} = x^{-8--6} = x^{-8+6} = x^{-2} = \dfrac{1}{x^2}$

(iv) $x^5 \div x^{-1} = x^{5--1} = x^{5+1} = x^6$

(v) $(x^{1000})^0 = 1$

(vi) $(2x)^{-2} = \left(\dfrac{1}{2x}\right)^2 = \dfrac{1}{4x^2}$

(vii) $2x^{-2} = \dfrac{2}{x^2}$

NOTE: Parts (c) and (d) can be answered by trying different numbers until the correct answer is found.

(c)   (i) $2^4 = 16$, so $n = 4$

(ii) $2^2 = 4$, and $2^{-2} = \dfrac{1}{4}$, so $n = -2$

(iii) The fraction is inverted (turned upside down), so $n = -1$

(d)   (i) $3^2 = 9$, $3^3 = 27$, so $x = 3$

(ii) Try $x = 2$
$2^2 = 4$, $2^3 = 8$ and $2^4 = 16$, so x cannot be 2
Try $x = 4$
$4^2 = 16$, $4^3 = 64$, so $x = 4$

(iii) $\dfrac{1}{2^3} = \dfrac{1}{8}$, and $\dfrac{1}{2^{-3}} = 8$, so $x = -3$

Exercise 3.9

Work through this exercise carefully, checking your answers as you go.

1. Evaluate:

   (a) 2^4
   (b) 2^{-4}
   (c) $2^{-1} \times 2^3$
   (d) $2^{-1} \div 2^3$

   (e) $5^{-2} \times 5^0$
   (f) $\left(\dfrac{5}{2}\right)^{-1}$
   (g) $\dfrac{5^{-1}}{2}$
   (h) $\left(\dfrac{5}{2}\right)^{-2}$

   (i) $\left(\dfrac{1}{2}\right)^{-3}$
   (j) $\left(2\dfrac{1}{2}\right)^{-2}$
   (k) $\left(3\dfrac{1}{2}\right)^{2}$
   (l) $\left(1\dfrac{3}{4}\right)^{-3}$

2. Simplify:

(a) $\left(\dfrac{x^2}{y}\right)^{-1}$ (b) $\left(\dfrac{x^2}{y}\right)^{-4}$ (c) $\left(\dfrac{x^2}{y}\right)^{-1} \div \left(\dfrac{x^2}{y}\right)^{-1}$

(d) $\left(\dfrac{x^2}{y}\right)^{0}$ (e) $\dfrac{x^4 y^2 z^3}{x^2 y^{-1} z^{-1}}$ **NOTE: Take one letter at a time:** $x^4 \div x^2 = x^2$ **and so on.**

(f) $(x^5 y^6) \div (xy)^{-2}$

3. In each question, find a replacement for n which will make the statement true.

(a) $2600 = 2.6 \times 10^n$ **NOTE: Remember standard form?**

(b) $\dfrac{1}{10000} = 10^n$ (c) $\dfrac{2}{3} = \left(\dfrac{3}{2}\right)^n$ (d) $\dfrac{4}{9} = \left(\dfrac{2}{3}\right)^n$

(e) $\dfrac{8}{27} = \left(\dfrac{2}{3}\right)^n$ (f) $\dfrac{27}{8} = \left(\dfrac{2}{3}\right)^n$ (g) $3\dfrac{3}{8} = \left(\dfrac{2}{3}\right)^n$

(h) $2^n = 2$ (i) $2^n = 1$ (j) $0.0015 = 1.5 \times 10^n$

(k) $x^2 = \dfrac{1}{x^n}$ (l) $2x = (2x)^n$ (m) $4x^2 = (2x)^n$

(n) $x^{-1} \times x^{-2} = \dfrac{1}{x^n}$ (o) $x^{-1} \div x^{-2} = x^n$ (p) $x^3 \times x = x^n$

(q) $x^k \times x = x^n$ (r) $x^k \div x^2 = x^n$

NOTE: In (q) and (r) k is any constant, and remember x is the same as x^1

4. Simplify, where possible, writing your answers in a form with positive powers.

(a) $3x^2 + 2x^3$ (b) $3x^{-2} \times 2x^{-3}$ (c) $\dfrac{1}{3x^{-2}}$ **NOTE: Only the x has a power**

(d) $\dfrac{1}{(3x)^{-2}}$ (e) $\dfrac{1}{(3x)^{-2}} \times \left(\dfrac{x^2}{2}\right)^{-1}$ (f) $(2x^2)^3 \times 3(x^{-1})^6$

Brackets and Common Factors

Dealing with brackets

In arithmetic we are told to work out brackets first (BoDMAS), but this is not so easy in algebra.

Look at the following.

$$2 \times (7 - 3)$$
$$= 2 \times 4$$
$$= 8$$

But what about $2 \times (x - y)$?

It is quite helpful to think of the pair of brackets as a bag, or bundle, containing x and $-y$. The 2 tells us that we have two of these bags, so if we emptied them on to the table we would have two xs and two $-y$s.

This shows that $2 \times (x - y) = 2x - 2y$.

Using algebraic shorthand we can say

$$2(x - y) = 2x - 2y$$

This is called **multiplying out the brackets**, (or commonly, getting rid of the brackets). Does this work with numbers as well?

Going back to our first example,

$$2 \times (7 - 3) = 2 \times 7 - 2 \times 3$$
$$= 14 - 6 = 8$$

This is the same answer as before, so it does work with numbers. However, with numbers it is usually quicker to work out the inside of the brackets first. (Remember BoDMAS.)

Remember:

- Multiplying out the brackets means multiplying every term inside the pair of brackets by the number (or letter) that is outside the brackets.

Once the brackets have been multiplied out we can continue with collecting like terms and so on.

Example 13
Multiply out the brackets.
(a) $2(c - 1)$ (b) $2(3a - 4b + c)$ (c) $5x(6x + 7y)$

Answer 13
(a) $2(c - 1) = 2c - 2$
(b) $2(3a - 4b + c) = 6a - 8b + 2c$
(c) $5x(6x + 7y) = 30x^2 + 35xy$

Exercise 3.10

Multiply out the brackets.

1. $2(a + b)$ 2. $6(3 + x)$ 3. $3(x - y)$
4. $5(6 - b)$ 5. $4(3x - 2)$ 6. $7(1 - 3c)$
7. $5(6x + 5y)$ 8. $8(x - y + 4z)$ 9. $5(x^2 + 4)$
10. $7(2x^2 - 3y^2)$ 11. $4(3xy + 5z)$ 12. $x(2 - 3y)$
13. $a(a + 2)$ 14. $x(x - y)$ 15. $2c(c + d)$
16. $3m(2m - n)$ 17. $4xy(2x - 9y)$ 18. $7x^2(3 - 2y + 4z)$

All the rules for multiplication with signs also apply to multiplying out brackets, and this is one of the most common areas where mistakes occur, so make doubly sure that you work through the next example and exercise and understand it clearly.

Example 14

Multiply out the brackets and simplify where possible.

(a) $-4(2z - 1)$ NOTE: Remember you are multiplying by negative 4! There is a MINUS sign so take care!

(b) $-(a - b + c)$ NOTE: You are taking away everything inside the pair of brackets. Taking away $-b$ is the same as adding b.

(c) $5(x - y) - 2(x + y)$

Answer 14

(a) $-4(2z - 1) = -8z + 4$ (b) $-(a - b + c) = -a + b - c$

(c) $5(x - y) - 2(x + y) = 5x - 5y - 2x - 2y = 3x - 7y$

You should see from this example that you must be careful with minus signs, particularly in front of brackets.

Exercise 3.11

Multiply out the brackets and simplify where possible.

1. $2(3 + 4x)$
2. $-2(3 + 4x)$
3. $x(3x + 4y)$
4. $-x(3x + 4y)$
5. $-x(3x - 4y)$
6. $-2x(7x - 6)$
7. $-(x + y)$
8. $-(2 - z)$
9. $6p(q + 3r - s)$
10. $-6p(q + 3r - s)$
11. $x^2y(y - 5)$
12. $-3x^2(2y - 3)$
13. $4a(-2 - 3a)$
14. $-4a(-2 - 3a)$
15. $(x + y) - (x - y)$
16. $3(a + 2b) + 2(a + 3b)$
17. $-5(2x + 3y) - 4(3x + 2y)$
18. $-3(-x - y) + 4(x - y)$
19. $-2(x - 3y) - (x - y)$
20. $x(x + y + z) - (x^2 + y + xz)$

Common factors

As usual, we have to be able to do the opposite to multiplying out the brackets, and this is called **taking out common factors** or **factorising**.

Remember that *common* means belonging to all, and *factors* are things that are multiplied together. We have already done some work with common factors in arithmetic when we found the HCF (highest common factor) of two or more numbers.

The following examples are designed to help you understand the process of factorising in algebra by comparing the process with the arithmetic work that you are familiar with.

Find the HCF of 20 and 35.

factors of 20 = {1, 2, 4, 5̲, 10, 20}
factors of 35 = {1, 5̲, 7, 35}
HCF = **5**
so 20 = **5** × 4 and 35 = **5** × 7

Find the HCF of $3ab$ and $6b^2$.

$$\text{factors of } 3ab = \{1, 3, a, b, 3a, \underline{3b}, 3ab\}$$
$$\text{factors of } 6b^2 = \{1, 2, 3, 6, b, b^2, 2b, \underline{3b}, 6b, 2b^2, 3b^2, 6b^2\}$$
$$\text{HCF} = \mathbf{3b}$$
$$\text{so } 3ab = \mathbf{3b} \times a \text{ and } 6b^2 = \mathbf{3b} \times 2b$$

In practice it is not as complicated as it looks.

If we are asked to factorise $3ab + 6b^2$, we would probably see quite easily that 3 would go into both terms, giving $3 \times (ab + 2b^2)$.

In other words,

$$3ab + 6b^2 = 3(ab + 2b^2)$$

However, this is only *partially* factorised as there is still a factor of b in both terms inside the pair of brackets, so we factorise that out as well.

$$3ab + 6b^2 = 3b(a + 2b)$$

This is now *fully* or *completely* factorised.

Remember:

- Always check that your answer is correct by multiplying out again. You should get back to the original expression.

Example 15

Factorise completely:

(a) $xy - 2y^2x$ (b) $abc + 4a^2b$ (c) $x^2 - 3x^2y$ (d) $10mn + 5m$

Answer 15

(a) $xy - 2y^2x$
$\quad = xy(1 - 2y)$

NOTE: Do not forget the 1 at the beginning of the pair of brackets. If you leave it out you will not be able to multiply out the brackets again to get back to the original expression.

(b) $abc + 4a^2b$
$\quad = ab(c + 4a)$

NOTE: Remember brackets come in pairs, so always draw the second (closing) bracket!

(c) $x^2 - 3x^2y$
$\quad = x^2(1 - 3y)$

(d) $10mn + 5m$
$\quad = 5m(2n + 1)$

Exercise 3.12

Factorise completely:

1. $8x + 4y$ 2. $15a - 25b$ 3. $4x - 20$ 4. $xy + 2x$

5. $x^2 - 2x$ 6. $x^2 - x$ 7. $3xy + 9x$ 8. $3x^3 - 9x^2y$

9. $3a^2 - 6ab$ 10. $xyz + 4yz$ 11. $10y + 100y^2$ 12. $5fg + 6fgh$

13. $3bx - 6xy$ 14. $3b^2x - 6bx$ 15. $4b^2 - 2b$ 16. $4b^2 - b$

17. $x^2y^2 - xy$ 18. $7c^2d^2 - 21cd^2$

Exercise 3.13

Mixed Exercise

1. Write down and simplify the product of these three factors: $2x$, $3x$ and $-4z$.
2. Write down and simplify the sum of these three terms: $2x$, $-5x$ and $3y$.
3. Using $x = 6$, $y = -1$ and $z = 2$, evaluate these expressions.
   - (a) $x^2 + y$
   - (b) $-x - y - z$
   - (c) $x^2 + y^2 - z^2$
   - (d) y^3
   - (e) $y(x + 4)$
   - (f) $x - 3z$
   - (g) $z^2(xy + yz - xz)$
   - (h) xyz
4. Seema writes down an expression which has a term in x^2 which has a coefficient of 2, a term in y which has a coefficient of -1, and a constant which is -5. Write down Seema's expression.
5. Simplify the following:
   - (a) x^0
   - (b) $x \div x$
   - (c) $\dfrac{x}{x}$
   - (d) $x \times 0$
   - (e) $x \times 1$
   - (f) $x - x$
   - (g) $x + x + x + x$
   - (h) $x^{-1} \times x$
   - (i) $(x^0)^2$
   - (j) $x \times x \times x \times x$
   - (k) $x^{-2} \times x^2$
   - (l) $(-x)^2$
   - (m) $(-x)^3$
   - (n) $-(x \times x)$
6. Find pairs of replacements for x and y which would make the following true. Write your answers as $x = \ldots$ and $y = \ldots$.
   Take your values from the set of integers. (Part (e) has only one possible answer.)
   - (a) $x + y = 4$
   - (b) $xy = -6$
   - (c) $\dfrac{x}{y} = 2$
   - (d) $x - y = -2$
   - (e) $x^y = 9$
   - (f) $2x + y = 5$
   - (g) $x^2 + y^2 = 25$
   - (h) $\sqrt{x} + \sqrt{y} = 5$
   - (i) $\sqrt{x + y} = 5$
7. Evaluate the following:
   - (a) $\left(\dfrac{3}{4}\right)^{-2}$
   - (b) $\left(\dfrac{19}{9}\right)^{-1}$
   - (c) $\left(\dfrac{1}{2}\right)^0$
   - (d) $2^{-1} \times \dfrac{1}{2}$
   - (e) $\left(2\dfrac{1}{5}\right)^2$
8. Simplify:
   - (a) $-x + 4y - -7x$
   - (b) $(-x)^3 \times 3x^2$
   - (c) $-2x \times {}^-3y$
   - (d) $\dfrac{-x}{-x^2}$
   - (e) $\dfrac{1}{x^{-2}} \times x^2$
   - (f) $\left(\dfrac{1}{x^{-2}}\right)^3 \times \dfrac{1}{x}$
9. Find replacements for n:
   - (a) $10^n = 10000$
   - (b) $3^n = \dfrac{1}{9}$
   - (c) $3^{2n} = 9$
   - (d) $2^{2n} = 16$
   - (e) $6 = \dfrac{2}{3^n}$
10. Multiply out the brackets and simplify where possible:
    - (a) $a(ab - bc)$
    - (b) $2(3x + 4y) - x(1 - 2y)$
    - (c) $2a(3b + 4c - 5a) + 10a^2$
    - (d) $6x^2 - 3x(2 + 3x)$

11. Factorise completely:
    (a) $ab^2 - a^2b$ (b) $2x^2 - 6xy + 4x$
    (c) $2xyz^2 + 4x^2y^2z$ (d) $2abc - 4a^2b^2c^2$

12. Simplify:
    (a) $x^{-1} \times x^3 \times x^0$ (b) $a^2 \times b^3 \times a^{-1} \div b^2$ (c) $x^2 \times x^3 + x^{-2} \times x^7$
    (d) $\dfrac{a^3 b^4 c^6}{a^2 bc}$ (e) $\left(a^3\right)^2$ (f) $\left(x^{-2}\right)^{-1}$

13. Fleur can carry no more than 10 kilograms home from the market. She has bought b bags of sugar already and would also like to buy some flour. Each bag of sugar weighs 0.5 kilograms.
    (a) Write a formula to express the amount of flour (f kilograms) she can buy in terms of b.
    (b) If $b = 6$, use your formula to calculate f.
    (c) If $f = 4$, how many bags of sugar has Fleur bought?

14. (a) Take 49 from 51   (b) Take 49 from 40
    (c) Take $(2x + 3)$ from $(x + 5)$ (d) Take $x - y$ from $x + y$
    (e) Divide 2 by 6   (f) Divide $2x$ by x^2

Examination Questions

15. Factorise completely $4xy - 6xz$.   (0580/01 May/June 2004 q 8)

16. $y = a + bc$
    Find the value of y when $a = -3$, $b = 2$ and $c = 8$.

                                        (0580/01 May/June 2004 q 16a)

17. When $x = 5$ find the value of:
    (a) $4x^2$ (b) $(4x)^2$ (0580/01 Oct/Nov 2004 q 4)

18. Simplify the following expressions.
    (a) $a^2 \times a^5$ (b) $b^4 \div b^3$ (0580/01 Oct/Nov 2004 q 10)

19. (a) Multiply out the brackets $5x(2x - 3y)$.
    (b) Factorise completely $6x^2 + 12x$.   (0580/01 Oct/Nov 2003 q 9)

20. Write down the value of n in each of the following statements.
    (a) $1500 = 1.5 \times 10^n$ (b) $0.00015 = 1.5 \times 10^n$
    (c) $5^n = 1$ (d) $\dfrac{1}{36} = 6^n$ (0580/01 Oct/Nov 2003 q 17)

21. Write down the value of $\left(1\dfrac{1}{2}\right)^{-2}$ as a fraction.

                                        (0580/01 May/June 2003 q 7)

22. (a) $y = 4uv - 3v$
    Find the value of y when $u = -3$ and $v = 2$.
    (b) Factorise $4uv - 3v$.   (0580/01 May/June 2003 q 8)

23. (a) (i) Rajeesh thought of a number.
    He multiplied the number by 2.
    He then added 10.
    The answer was 42.
    What was the number Rajeesh first thought of?
    (ii) Simon thought of a number x.
    He multiplied this number by 3 and then added 8.
    Write down an expression in x for his answer.
    (b) Simplify $-8a + 7b - a - 2b$.
    (c) Factorise fully $6a - 9a^2$.   (0580/03 Oct/Nov 2004 q 7a, b and c)

24. Work out 4^{-3} as a fraction.

                                                        (0580/01 May/June 2005 q 5)

25. When $x = -3$ find the value of $x^3 + 2x^2$.

                                                        (0580/01 May/June 2005 q 8)

26. (a) Expand the brackets and simplify the expression:
    $7x + 5 - 3(x - 4)$.
    (b) Factorise $5x^2 - 7x$.   (0580/01 May/June 2005 q 17)

27. An integer n is such that $60 \leqslant n \leqslant 70$.
    Write down the value of n which is:
    (a) a prime number,   (b) a multiple of 9,
    (c) a square number.   (0580/01 Oct/Nov 2005 q 10)

28. Simplify the following expressions:
    (a) $9r - 4s - 6r + s$ (b) $q^4 \div q^3$ (c) $p^6 \times p^{-2}$
                                                        (0580/01 Oct/Nov 2005 q 16)

29. Simplify:
    (a) $p^2 \times p^3$ (b) $q^3 \div q^{-4}$ (c) $(r^2)^3$
                                                        (0580/01 May/June 2006 q 10)

30. Factorise completely:
    (a) $7ac + 14a$ (b) $12ax^2 + 18xa^3$ (0580/02 Oct/Nov 2005 q 14)

31. (a) $4^p \times 4^5 = 4^{15}$. Find the value of p.
    (b) $2^7 \div 2^q = 2^4$. Find the value of q.
    (c) $5^r = \dfrac{1}{25}$. Find the value of r.
                                                        (0580/01 May/June 2007 q 13)

Chapter 4

Working with Numbers

This chapter will give you the skills needed when using Mathematics in everyday life. You will learn how to make calculations involving measurements, speed, money and time. You will learn more about using a calculator as well as working without a calculator.

Essential Skills

1. Write these numbers in standard form:
   (a) 12345   (b) 0.00034

2. Write these standard form numbers as normal numbers:
   (a) 3.45×10^6 (b) 5.123×10^{-3}

3. Look at the number 3764012.
   (a) Write down the digit in:
       (i) the tens place,   (ii) the thousands place.
   (b) What is the place value of:
       (i) the 7,   (ii) the 3?

4. (a) Calculate 27% of 510.
   (b) Find 42 as a percentage of 700.

5. Find (without using a calculator):
   (a) (i) 0.0645×1000 (ii) 83×10000
   (b) (i) $0.0059 \div 100$ (ii) $67932 \div 1000$
   (c) (i) 0.00259×10^4 (ii) $7015 \div 10^6$

Answers	
1. (a) 1.2345×10^4	(b) 3.4×10^{-4}
2. (a) 3450000	(b) 0.005123
3. (a) (i) 1	(ii) 4
(b) (i) hundred thousand	(ii) million

4. (a) 137.7   (b) 6%
5. (a) (i) 64.5   (ii) 830000
   (b) (i) 0.000059   (ii) 67.932
   (c) (i) 25.9   (ii) 0.007015

More Symbols

For this chapter you will need to know the following mathematical symbols:

- ± means 'plus or minus'.
- $\simeq$ and $\approx$ mean 'approximately equal to'.

Units of Measurement

You are already familiar with some units of measurement, such as centimetres, kilograms, hours, kilometres per hour and litres.

The systems of units that most of us use today are very much easier than the older systems. Most are based on powers of ten, which makes conversions much simpler. The old systems used in Britain were all different.

For example, for length:

$$12 \text{ inches} = 1 \text{ foot}$$
$$3 \text{ feet} = 1 \text{ yard}$$
$$1760 \text{ yards} = 1 \text{ mile.}$$

Money, mass and capacity were just as difficult and all these conversions had to be learned for examinations.

Some of the conversions that you need to know today are:

- Length   10 millimetres (mm) = 1 centimetre (cm)
                    100 centimetres = 1 metre (m)
                    1000 metres = 1 kilometre (km)
- Mass   1000 milligrams (mg) = 1 gram (g)
                    1000 grams = 1 kilogram (kg)
                    1000 kilograms = 1 tonne
- Capacity (volume) 1000 millilitres (ml) = 1 litre (*l*)

The one to watch out for is time!

Time is still measured by an old system and is not based on powers of ten. You have to be *very careful* about changing it to a decimal system.

- Time   60 seconds (s) = 1 minute (min)
                    60 minutes = 1 hour (h)
                    24 hours = 1 day
                    7 days = 1 week and so on.

Added to these are compound units, such as kilometres per hour (km/h). Also area (such as cm^2) and volume (such as cm^3) units derived from the units of length.

Simple areas and volumes

In this section we shall look at the simplest examples of areas and volumes by considering squares and rectangles, and cubes and cuboids. Wherever possible the size of a shape or object is described using standard **dimensions**, like length, width and height, measured in directions at right angles to each other.

The **area** of a shape is the amount of surface it covers in *two* dimensions. This means that to calculate an area we need *two* length measurements multiplied together. The diagram shows this for a square and a rectangle.

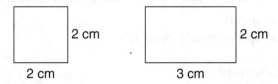

The **volume** of a solid is the amount of space it takes up in *three* dimensions. This means that to calculate a volume we need *three* length measurements multiplied together. The next diagram shows this for a cube and a cuboid.

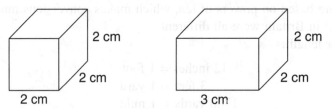

We will look at areas and volumes in more detail in Chapter 8.

NOTE: **The words width and breadth are both used to mean the same thing, so you might see the formula for the area of a rectangle as $A = l \times b$, or $A = l \times w$.**

The units of measurement for area and volume depend on which units of length have been used. If the lengths are centimetres the area will be in square centimetres, which is usually written as cm^2, and the volume will be in cubic centimetres or cm^3. Hence, the units of length must be the same in each calculation of area or volume.

Conversions for length, area and volume units

- Length $1 \text{ cm} = 10 \text{ mm}$
  $1 \text{ m} = 100 \text{ cm}$
  $1 \text{ km} = 1000 \text{ m}$
- Area $1 \text{ cm}^2 = 10 \text{ mm} \times 10 \text{ mm} = 100 \text{ mm}^2$
  $1 \text{ m}^2 = 100 \text{ cm} \times 100 \text{ cm} = 10000 \text{ cm}^2$
  $1 \text{ km}^2 = 1000 \text{ m} \times 1000 \text{ m} = 1000000 \text{ m}^2$
- Volume $1 \text{ cm}^3 = 10 \text{ mm} \times 10 \text{ mm} \times 10 \text{ mm} = 1000 \text{ mm}^3$
  $1 \text{ m}^3 = 100 \text{ cm} \times 100 \text{ cm} \times 100 \text{ cm} = 1000000 \text{ cm}^3$
  $1 \text{ km}^3 = 1000 \text{ m} \times 1000 \text{ m} \times 1000 \text{ m} = 1000000000 \text{ m}^3$

You will notice how the area and volume units are much larger than the corresponding units of length. The next diagram illustrates this.

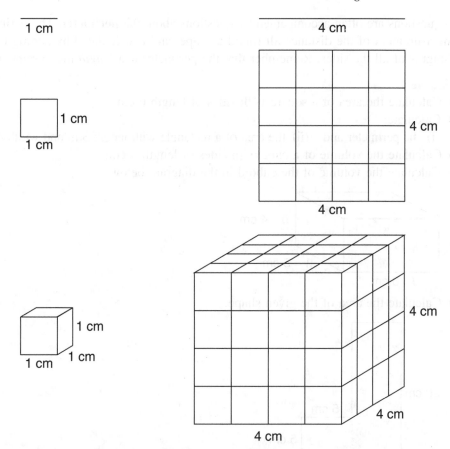

The diagram shows a line segment of length 1 cm, and compares it with another of length 4 cm. (A line segment is *part* of a line. A line could, in theory, go on forever). We know the second line segment is 4 times the length of the first.

The diagram then shows a square with a side of 1 cm (area = 1 cm^2) and compares it with a square of side 4 cm. You should be able to see that this square actually covers $4 \times 4 = 16$ centimetre squares. The area of this square is therefore 16 cm^2.

Subsequently, the diagram shows a cube of side 1 cm and compares it with a cube of side 4 cm.

PROJECT

Try to decide how many one centimetre cubes would fit in the 4 centimetre cube.
Either get some 1 centimetre cubes and stack them up to see how many are needed, or try to visualise it in the following way.
How many 1 centimetre cubes would fit along one edge of the 4 centimetre cube to make one row? How many rows of these cubes would be needed to make one layer? How many layers would be needed to make the whole 4 centimetre cube?
So what is the volume of the cube?
Now check using arithmetic. 4 cm $\times$ 4 cm $\times$ 4 cm = 4^3 cm^3 = 64 cm^3.

Area questions are often accompanied by questions about the **perimeter.** The perimeter is the measurement of the distance all round a shape, and is calculated by adding together the lengths of all the sides. Remember that the perimeter is a *length* measurement.

Example 1
(a) Calculate the area of a square with sides of length 6 cm.
(b) Calculate:
    (i) the perimeter and   (ii) the area of a rectangle with length 3 m and width 4 m.
(c) Calculate the volume of a cube with sides of length 3 cm.
(d) Calculate the volume of the cuboid in the diagram below.

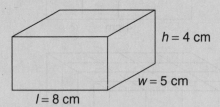

(e) Calculate the area of the given shape:

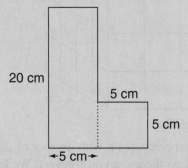

(f) Calculate the volume of the given solid:

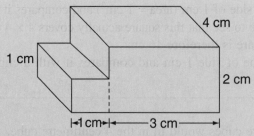

Answer 1
(a) Area of square = 6 cm × 6 cm = 36 cm^2
(b) (i) Perimeter of rectangle = 3 m + 4 m + 3 m + 4 m = 14 m
    (ii) Area of rectangle = 3 m × 4 m = 12 m^2
(c) Volume of cube = 3 cm × 3 cm × 3 cm = 27 cm^3
(d) Volume of cuboid = 8 cm × 5 cm × 4 cm = 160 cm^3
(e) Area = 5 cm × 5 cm + 5 cm × 20 cm = 125 cm^2
(f) Volume = 4 cm × 2 cm × 3 cm + 4 cm × 1 cm × 1 cm = 24 cm^3 + 4 cm^3
        = 28 cm^3

The units (centimetres) are included in these calculations to help you understand. Normally you would not put them in until you write the answer.

Exercise 4.1

1. Calculate the areas of the following:

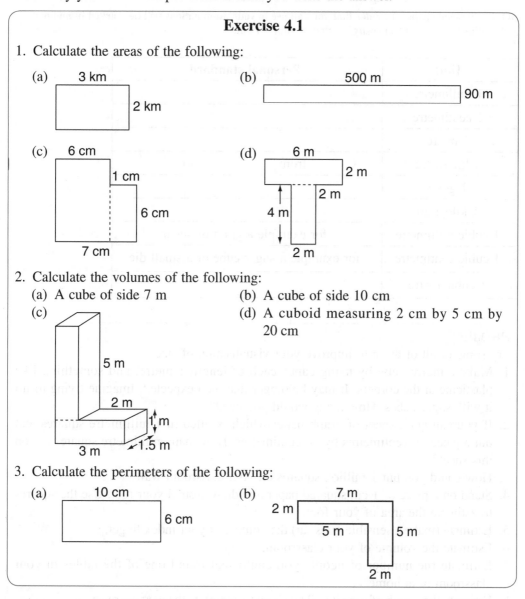

(a)   3 km
2 km

(b)   500 m
90 m

(c)   6 cm
1 cm
6 cm
7 cm

(d)   6 m
2 m
2 m
4 m
2 m

2. Calculate the volumes of the following:
   (a) A cube of side 7 m
   (b) A cube of side 10 cm
   (c)
   (d) A cuboid measuring 2 cm by 5 cm by 20 cm

   5 m
   2 m
   1 m
   3 m
   1.5 m

3. Calculate the perimeters of the following:
   (a)   10 cm
   6 cm

   (b)   7 m
   2 m
   5 m
   5 m
   2 m

Estimation

Before we look at the conversion of units of measurement it would be helpful for you to find your personal approximations for the sizes of the units. For example, a centimetre might be the width of your middle finger, a metre might be the length of your stride, and a kilometre the length of part of a well known journey. These 'personal standards' should be ones that are easy for you to remember. Copy and complete the table below, filling in your own personal standards.

Some examples have already been suggested, but use your own if you think of better ones. Add other measurements if you can think of any.

NOTE: Remember that the ruler that you can use in your examinations will be marked in millimetres and centimetres, so you can easily see that 10 mm = 1 cm.

Unit	Personal standard
1 millimetre	
1 centimetre	
1 metre	
1 kilometre	from to
1 gram	
1 kilogram	
1 cubic millimetre	for example a grain of sugar
1 cubic centimetre	for example a sugar cube or a small die
1 cubic metre	

PROJECT

Try some or all of these to improve your visualisation of size.
1. Make a metre cube by using canes each of length 1 metre, and something like plasticine at the corners. It may be bigger than you expected. Imagine trying to fill it with sugar cubes. How many would you need?
2. If you can get a sheet of graph paper which is ruled in 1 millimetre squares, cut out a piece 10 centimetres by 10 centimetres. How many millimetre squares are on this sheet?
3. How could you put 1 million squares on your classroom wall?
4. Stand on a piece of 1 cm squared paper and draw around your foot. Use the squares to estimate the area of your foot.
5. Estimate (make a sensible guess at) the volume of your index finger.
6. Estimate the volume of your classroom.
7. Estimate the number of people you could seat round one of the tables in your classroom or at home.
8. Estimate the length of time it will take you to complete the next exercise.
9. Estimate the weight of this book. Check its weight on some scales to see how close you were.

Estimation is an important skill. You will not always be able to measure things accurately. For example, if you are catering for a large number of people you will have to estimate the amount of food and drink you need to supply.
You probably find that lengths are the easiest to estimate, followed by areas, and finally volumes.

Approximation

It is a fact that measurements can never be completely accurate. You know that if your friend draws a line and says that it is exactly 15 cm long and you measure it yourself you could probably make it just a *little* more or less than 15 cm. The truth is that however accurate our measuring instruments we can never be certain of any measurement.

So what does it mean when we say that a line is 15 cm long?

We probably mean 15 cm *to the nearest centimetre.*

Look at the next diagram.

In each case decide whether the line is 14 cm, 15 cm or 16 cm *to the nearest centimetre.* You should find (a) is nearest to 14 cm, (b) is nearest to 15 cm, (c) is nearest to 15 cm and (d) to 16 cm. But what about (e) and (f)?

There is a fairly general agreement that if the measurement is half way between the two numbers then we take it to the larger number. So (e) will be 15 cm and (f) will be 16 cm to the nearest centimetre. This is the rule that will be used for your examinations.

This process is called **rounding to the nearest centimetre.** Incidentally it is also rounding **to the nearest whole number**.

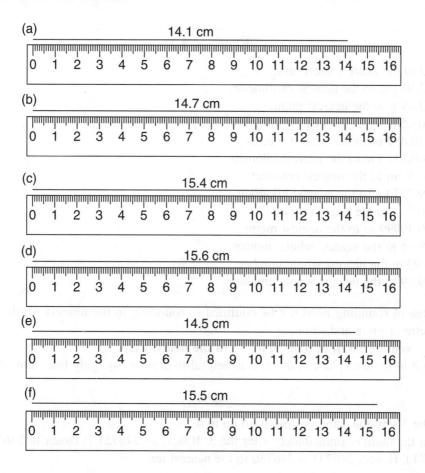

Example 2

Round to the nearest centimetre:

(a) 14.358 cm   (b) 14.4999 cm   (c) 15.099 cm

(d) 15.501 cm   (e) 16.5 cm

Answer 2

In each case ask yourself if the measurement is nearest to the whole number below it or the whole number above it.

(a) 14.358 is nearest to 14, so 14.358 cm = 14 cm to the nearest centimetre

(b) 14.4999 is close to 14.5, but it *is* still less than 14.5 so it is closest to 14.
      14.4999 cm = 14 cm to the nearest centimetre

(c) 15.099 is closer to 15 than 16, so 15.099 cm = 15 cm to the nearest centimetre

(d) 15.501 is closer to 16 than to 15, so 15.501 cm = 16 cm to the nearest centimetre

(e) 16.5 is exactly half way between 16 and 17, so using the rule we round to 17.
      16.5 cm = 17 cm to the nearest centimetre, by convention

NOTE: To avoid confusion, it may be better not to say 'round up' or 'round down' to yourself. For example, 14.3 never goes down to 13.

Exercise 4.2

Round to the nearest stated unit.

1. 32.94 cm to the nearest centimetre.
2. 63.49 g to the nearest gram.
3. 705.501 kg to the nearest kilogram.
4. 610.889 m to the nearest metre.
5. 500.471 km to the nearest kilometre.
6. 90.8 cm to the nearest centimetre.
7. 89.793 kg to the nearest kilogram.
8. 60.5 m to the nearest metre.
9. 60.49999 m to the nearest metre.
10. 799.5 to the nearest whole number.
11. 9.99 to the nearest whole number.
12. 99.99 to the nearest whole number.

This idea of rounding need not be confined to rounding to the nearest whole number, centimetre or gram and so on.

Suppose you are asked to round 246731 **to the nearest ten.**

The digit in the tens place is the 3. It is helpful to draw a 'squiggly line' after the 3 like this:

$$246731$$

Does the 3 stay the same or does it go up to 4?

Look at the number immediately after the 3. It is 1, so 246731 is closer to 246730 than to 246740. Hence, 246731 = 246730 to the nearest ten.

NOTE: **Remember that you consider only the digit immediately after the squiggly line. Ignore all the rest.**

You could be asked to round 5467 to the nearest thousand.
By the same method the squiggly line goes after the 5 because this is in the thousands place.

$$5\rangle467$$

Is this closer to 5000 or 6000?
The number after the 5 is 4, so the answer is: 5467 = 5000 to the nearest thousand.

NOTE: **Do remember to include the zeroes. Leaving them out is one of the commonest mistakes in rounding. In this example you do not want to say that 5467 is approximately equal to 5!**

Decimal places

You may also be asked to round to a given number of decimal places. Decimal places are counted to the right starting at the decimal point, so in the number 893.45 the 5 is in the second decimal place, and the 4 is in the first decimal place.

Example 3
Round these numbers to the stated accuracy.
(a) 20056 to the nearest ten.
(b) 20056 to the nearest hundred.
(c) 20056 to the nearest thousand.
(d) 3.4109 to 3 decimal places (dp).
(e) 3.4109 to 2 dp.
(f) 20.404 to 2 dp.
(g) 351.499 to the nearest whole number.

Answer 3
(a) $2005\rangle6$ = 20060 to the nearest ten

(b) $200\rangle56$ = 20100 to the nearest hundred

(c) $20\rangle056$ = 20000 to the nearest thousand

(d) $3.410\rangle9$ = 3.411 to 3 dp

(e) $3.41\rangle09$ = 3.41 to 2 dp

(f) $20.40\rangle4$ = 20.40 to 2 dp   NOTE: **The zero is here to show that the number has been rounded to 2 decimal places. Although 20.4 has the same numerical value it implies that you have rounded the number to only 1 decimal place.**

(g) $351.\rangle499$ = 351 to the nearest whole number

NOTE: **Never round progressively. This means *do not* round the 4 to 5 and then the 1 to 2. This is because 0.499 is less than 0.5, so the number 351.499 is nearer to 351 than 352.**

Exercise 4.3

Round to the stated accuracy.

1. 239 to the nearest ten.
2. 520.65 to the nearest ten.
3. 7381.3 to the nearest hundred.
4. 649 to the nearest hundred.
5. 3985.6 to the nearest ten.
6. 7959 to the nearest hundred.
7. 1234 to the nearest thousand.
8. 56.1358 to 1 decimal place.
9. 56.1358 to 2 decimal places.
10. 56.1358 to 3 decimal places.
11. 3.098 to 1 decimal place.
12. 3.098 to 2 decimal places.

Significant figures

There is one more method of approximation you need to know, and that involves the idea of *significant figures*.

To count significant figures you start at the beginning of the number and count to the right, past the decimal point if necessary. Looking at the number 25937.065, 2 is the first significant figure, 9 is the third significant figure, 6 is the seventh significant figure and so on.

But there is one thing you have to be careful of. Zeroes at the beginning of a number do not count as significant figures, so in the number 0.005096, the first significant figure is 5, the second is the 0 that comes after the 5, the third is 9, and the fourth is 6. The three initial zeroes are place holders only, to tell you where the decimal point belongs. An example will help you see the significance of this.

Suppose a distance is measured as 5096 metres to 4 significant figures, and then this length is given in kilometres. 5096 metres becomes 0.005096 kilometres. This length is *not* more accurate than the first, so must still be to 4 significant figures. The zeroes only tell you where the decimal point belongs, and do not count as significant figures.

Example 4

Round these numbers to the stated number of significant figures.

(a) 10.04 to 3 significant figures.   NOTE: **Remember here that the two zeroes are significant figures.**

(b) 0.0079 to 1 significant figure.   (c) 15637 to 2 significant figures.
(d) 19.998 to 3 significant figures.   (e) 0.01009 to 3 significant figures.

Answer 4

(a) 10.0   (b) 0.008   (c) 16000
(d) 20.0   (e) 0.0101

Exercise 4.4

Round to the stated number of significant figures (s.f.).

1. 215.67   3 s.f.
2. 215.67   2 s.f.
3. 350.49   3 s.f.
4. 350.49   1 s.f.

5. 6009.156	3 s.f.	6. 6009.156	2 s.f.
7. 80.964	3 s.f.	8. 0.19852	3 s.f.
9. 0.19852	2 s.f.	10. 1.00098	3 s.f.
11. 0.0003946	3 s.f.	12. 0.0003946	1 s.f.
13. 10.149	3 s.f.	14. 657280	3 s.f.
15. 657280	1 s.f.			

The general instructions on the front of your examination papers will say something like the following:

'If the degree of accuracy is not specified in the question, and if the answer is not exact, give the answer to three significant figures. Give answers in degrees to one decimal place.'

Make sure that you understand this, and follow these instructions in all your work so that it becomes normal for you to give your answers in this way. You will probably lose a mark if you merely 'truncate' the number without correct rounding. So for example, 3.4567… should be given as 3.46 (rounded) *not* 3.45 (truncated).

Remember that if the answer works out exactly you may not need to round it.

Example 5

Using your calculator:

(a) Calculate 34.1×47.3 giving your answer correct to 4 significant figures.

(b) Calculate $76.3 \div 14.2$.   (c) Calculate $31.52 \div 2$.

(d) Calculate one third of $302°$ (302 degrees).

Answer 5

(a) $34.1 \times 47.3 = 1612.93 = 1613$ correct to 4 significant figures

(b) $76.3 \div 14.2 = 5.373239437 = 5.37$ to 3 significant figures

(c) $31.52 \div 2 = 15.76$ exactly

(d) $\dfrac{1}{3} \times 302° = 100.6666666 \ldots = 100.7°$

NOTE: **If you are asked to give your answer to 'an appropriate degree of accuracy' do not give it to a higher degree of accuracy than that of the data given in the question. Answers to questions about money, for example in dollars, should not be given to more than two decimal places.**

Exercise 4.5

1. Calculate 354.1×67, giving your answer to 4 significant figures.
2. Calculate $278 \div 34$.
3. Calculate one third of $16.7°$.
4. Calculate $337.38 \div 6$.

Limits of accuracy

In mathematics, we must be able to think back as well as forward. So we need to be able to decide exactly what is meant when a measurement is given to us as, say, 14 cm to the nearest centimetre.

You should be able to see that it could have been, for example, 13.6 cm, or 14.4 cm. In fact it must lie somewhere between 13.5 cm and 14.5 cm. Using our rule about the half way point we know that 13.5 cm would round to 14 cm, but 14.5 cm would round to 15 cm. So we can say that the length is greater than *or equal to* 13.5 cm, but less than 14.5 cm.

If the length of the line is l cm we write:

$$13.5 \leqslant l < 14.5$$

Notice that, since the length is given as l cm we do not put the centimetres in the answer. It is only the l we are referring to. This is as close as you can get to the original number, and 13.5 and 14.5 are known as the **limits of accuracy**, where 13.5 is the **lower limit** or bound and 14.5 is the **higher limit** or bound.

Example 6

State the limits of accuracy for the following approximations:
(a) 735 to the nearest whole number,
(b) 23.56 to 2 dp (decimal places),
(c) length $x = 130$ cm to 2 significant figures,
(d) length $y = 130$ cm to 3 significant figures.

Answer 6
(a) $734.5 \leqslant 735 < 735.5$ (b) $23.555 \leqslant 23.56 < 23.565$
(c) $125 \text{ cm} \leqslant x < 135 \text{ cm}$ (d) $129.5 \text{ cm} \leqslant y < 130.5 \text{ cm}$

NOTE: It might help you to work out the limits of accuracy if you think of adding or subtracting half the smallest number of the given accuracy. This is much easier to explain by examples rather than words! So let us look back at these examples.
(a) Half the smallest whole number (1) is 0.5, so 735 ± 0.5 becomes 734.5 or 735.5.
(b) Half the smallest second decimal place (0.01) is 0.005, so 23.56 ± 0.005 becomes 23.555 or 23.565.
(c) 130 to 2 significant figures would mean 130 ± 5 ($5 = \frac{1}{2} \times 10$) which becomes 125 and 135.
(d) 130 to 3 significant figures would mean 130 ± 0.5 which becomes 129.5 and 130.5.

Example 7

A metal bar is measured and found to be 10.5 centimetres in length (l), correct to the nearest millimetre.
(a) what is the least possible measurement for the metal bar?
(b) state the upper bound of the measurement.
(c) copy and complete the statement below:

$$\text{-------} \leqslant l \text{ cm} < \text{--------}$$

Answer 7
(a) the least possible measurement is 10.45 centimetres.
(b) the upper bound of the measurement is 10.55 centimetres.

NOTE: The upper bound is always given as 10.55 even though we know that it actually has to be less than 10.55, because 10.55 would round up to 10.6.

(c) $10.45 \leqslant l$ cm < 10.55

This is the most accurate way of stating the bounds because it clearly shows that the measurement can be greater than *or equal to* 10.45, but it actually has to be *less than* 10.55.

Exercise 4.6

1. A wooden rod is 157 cm to the nearest cm.
   What is its least possible length?
2. A sheet of paper has a width of w cm and a height of h cm, both to the nearest centimetre.
   If $w = 10$ and $h = 19$, copy and complete the following inequalities.
   (a) ......$\leqslant w <$ ....
   (b) ......$\leqslant h <$ ....
3. State the limits of accuracy for the following approximations.
   In each case copy and complete the inequalities.
   (a) ......$\leqslant 15$ cm $<$ ....   to the nearest centimetre.
   (b) ......$\leqslant 23.6$ cm $<$ ....   to the nearest millimetre.
   (c) ......$\leqslant 3060 <$ ....   to the nearest ten.
   (d) ......$\leqslant 99.7 <$ ....   to 3 significant figures.
   (e) ......$\leqslant 678.9 <$ ....   to 1 decimal place.
   (f) ......$\leqslant 60000 <$ ....   to 1 significant figure.
   (g) ......$\leqslant 300 <$ ....   to the nearest hundred.
   (h) ......$\leqslant 99.9 <$ ....   to 3 significant figures.
4. A coin weighs 9 grams correct to the nearest gram.
   (a) What is its least possible weight?
   (b) State the upper bound of its weight.

Changing Units

You need to know how to change one unit of measurement to another. You may need to look back to the beginning of this chapter to remind yourself of the conversions of units of measurement.

For example, you are asked to change 0.75 km to centimetres. The first thing to ask yourself is 'would you need more or fewer centimetres?'

Remember that a centimetre is only about the width of your finger and a kilometre is an easy walking distance or whatever your personal standard distance is from the table you completed earlier. You obviously need many more centimetres, so the change will involve multiplying by a power of ten.

We know that   1 kilometre = 1000 metres
                                            1 metre = 100 centimetres
          so   1 kilometre = 1000 × 100 centimetres
                                        = 10^5 or 100000 centimetres

We need to change 0.75 kilometres to centimetres,

so $\qquad$ 0.75 km = 0.75 × 100000 cm

$\qquad\qquad\qquad\qquad = 75000$ cm

Alternatively make the change progressively.

$\qquad\qquad$ 0.75 km $\ = 0.75 \times 1000$ m

$\qquad\qquad\qquad\qquad = 750$ m

$\qquad\qquad\qquad\qquad = 750 \times 100$ cm

$\qquad\qquad\qquad\qquad = 75000$ cm

Example 8

(a) Change to the units stated.
  (i) 0.5 m to mm   (ii) 1565 g to kg
  (iii) 61 m^3 to cm^3 (iv) 1.39 mm^2 to cm^2

(b) Calculate the following, stating the units in the answer.
  (i) The area of a rectangle 3 metres by 4 centimetres.
  (ii) The volume of a cuboid with the following dimensions:
      length = 2.1 metres, breadth = 1.5 metres and height = 0.6 centimetres

NOTE: Before each calculation check that the units are consistent and change if necessary.

Answer 8

(a) (i) You need more millimetres than metres, so
      0.5 m = 50 cm = 500 mm

  (ii) You need fewer kilograms than grams, so
      1565 g = 1.565 kg

  (iii) 61 m^3 = 61 × (100 × 100 × 100) cm^3 = 61000000 or 6.1 × 10^7 cm^3

  (iv) 1.39 mm^2 $= 1.39 \div (10 \times 10)$ cm^2
          $= 1.39 \div 100$ cm^2
          $= 0.0139$ cm^2

(b) (i) 3 metres = 3 × 100 cm = 300 cm
      Area of rectangle = 300 cm × 4 cm = 1200 cm^2
      or   4 cm = 4 ÷ 100 m = 0.04 m
      Area of rectangle = 3 m × 0.04 m = 0.12 m^2

  (ii) 0.6 cm = 0.6 ÷ 100 m = 0.006 m
      Volume of cuboid = 2.1 m × 1.5 m × 0.006 m
              $= 0.0189$ m^3
      or   2.1 m = 2.1 × 100 cm = 210 cm
      and 1.5 m = 1.5 × 100 cm = 150 cm
      Volume of cuboid = 210 cm × 150 cm × 0.6 cm
              $= 18900$ cm^3

Exercise 4.7

1. Change to the stated units:
   (a) 3.5 m to cm
   (b) 581 mm to cm
   (c) 4096 cm to km
   (d) 0.57 km to mm
   (e) 0.812 kg to g
   (f) 3 cm^2 to mm^2
   (g) 50681 m^2 to km^2
   (h) 0.0067 m^3 to cm^3
   (i) 210 ml to l

2. Calculate the areas of the following shapes. State the units in your answers.

   (a) 30 cm / 6.5 cm

   (b) 81 mm / 12 mm / 2.7 cm / 67 mm

3. Calculate the volumes of the following solids. State the units in your answers.

   (a) 1 m / 98 cm / 3100 mm

   (b) 15 mm / 3 cm / 1.2 cm / 15 mm / 43 mm

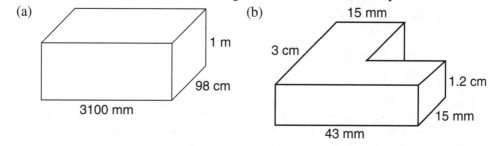

Working without a Calculator

Estimation is very useful in practical situations if you have to work without a calculator. For example, Rama arrives at Washington airport with Rs. 523.5 in her pocket. She is told that Re. 1 is worth \$0.02242. How can she quickly work out *approximately* how many dollars she should expect when she changes her rupees to dollars?

The best way is to round each number to 1 significant figure so that the calculation becomes easy. Re. 1 is worth about \$0.02, so Rs. 500 are worth about Rs. 500 × \$0.02 = \$10.

$$\text{Rs. } 523.5 \simeq \$10$$

If you are working without a calculator in an examination you may be asked to *estimate* the answer by rounding to 1 significant figure, or you may be asked to *calculate* the answer by using your skills of addition, subtraction, multiplication and division.

Example 9
(a) Estimate 7951 × 0.578.
(b) Calculate 35 × 16.
(c) Calculate $(3.4 \times 10^6) + (5.9 \times 10^5)$.
(d) Calculate $(2.1 \times 10^3) \times (9 \times 10^2)$.

Answer 9

(a) $7951 \times 0.578 \simeq 8000 \times 0.6 = 800 \times 6 = 4800$

(b)
$$
\begin{array}{r}
35 \\
16\times \\
\hline
210 \\
350 \\
\hline
560 \\
\hline
\end{array}
$$

NOTE: Show your working clearly in any non-calculator paper.

(c) The next answer has been written out in detail so that you can follow each step of the method.

$(3.4 \times 10^6) + (5.9 \times 10^5)$ *

$= (3.4 \times 10 \times 10^5) + (5.9 \times 10^5)$

$= (34 \times 10^5) + (5.9 \times 10^5)$ *

$= (34 + 5.9) \times 10^5$

$= 39.9 \times 10^5$ *

$= 3.99 \times 10 \times 10^5$

$= 3.99 \times 10^6$ *

NOTE: In practice you do not need to write down all these steps once you are sure that you know what you are doing, but you do need to make sure that you write down enough to make logical sense. Remember that with addition and subtraction the numbers must first be written with the same powers of 10. Essential steps are starred (*).

Alternative method:

$(3.4 \times 10^6) + (5.9 \times 10^5)$

$= 3400000 + 590000$

$$
\begin{array}{r}
3400000 \\
590000+ \\
\hline
3990000 \\
\hline
\end{array}
$$

$3990000 = 3.99 \times 10^6$

(d) NOTE: With multiplication and division the given powers of 10 may be used directly according to the normal rules of indices.

$(2.1 \times 10^3) \times (9 \times 10^2)$ *

$= 2.1 \times 9 \times 10^3 \times 10^2$

$= 18.9 \times 10^{3+2}$

$= 18.9 \times 10^5$ *

$= 1.89 \times 10 \times 10^5$

$= 1.89 \times 10^6$ *

Exercise 4.8

Do not use a calculator in this exercise.

1. Hank is going to explore parts of India, but he only has a limited time.
   He finds the distances between several cities by looking on the Internet.
   Bangalore to Delhi is 2039 km.
   Delhi to Mumbai is 1405 km.
   Mumbai to Kolkata is 1916 km.
   Kolkata to Bangalore is 1824 km.
   (a) Estimate the total length of this round trip.
   (b) Assuming Hank travels approximately 200 km per day, how many days travelling should he allow for this journey?

2. Estimate the answers to these calculations:
   (a) $238582 + 496 \times 1087$
   (b) $(3.987 + 1.05)^2 \div (6.93 - 1.87)$
   (c) $\dfrac{4182}{210} + \dfrac{3944}{52.28}$

3. Calculate, giving your answers in standard form:
   (a) $(4.1 \times 10^5) \times (2 \times 10^3)$
   (b) $(4.1 \times 10^5) \div (2 \times 10^3)$
   (c) $(7.12 \times 10^6) + (3.56 \times 10^7)$
   (d) $(9.012 \times 10^5) - (1.1 \times 10^2)$

Working with a Calculator

Carla attempts this calculation using her calculator:

$$\frac{3213 + 6156}{29 \times 52}$$

She tries three times and gets three different answers:

(a) 14251.34483   (b) 16799.58621   (c) 3217.082228

Chander says none of these is right!
He has estimated the answer by rounding each number to 1 significant figure, and doing the calculation mentally.

$$\frac{3213 + 6156}{29 \times 52} \simeq \frac{3000 + 6000}{30 \times 50} = \frac{9000}{1500} = 6$$

Carla tries once more.
This time she puts brackets round the numerator and the denominator and gets the correct answer.

(d) $\dfrac{(3213 + 6156)}{(29 \times 52)} = 6.212864721 = 6.21$ to 3 s.f.

She should have remembered that the line in a fraction acts like a bracket, tying the whole of the numerator together and the whole of the denominator together. Her calculator, of course, could not see this line and so in (a), (b) and (c) did not work out the numerator and denominator separately before doing the division.

Can you work out how Carla entered the sum into her calculator in each of her first three attempts? (She did in some of them insert brackets, but not in the best places.)
The moral of the story is that when you are asked to calculate this type of sum the first thing to do is to insert the brackets. You would be wise also to do an estimated calculation to check your answer.

Using your calculator for standard form

Calculators are changing all the time, and you need to get used to your own calculator before your examination.
For example, to find the square root of 16 do you press the square root key followed by 16 or do you have to type 16 first, then press the square root key?
How do you enter standard form in your calculator?

Try typing 1.5 $\boxed{\text{EXP}}$ 3

(some of the newer calculators have a button marked $\boxed{\times 10^x}$, if yours has one use that instead of the $\boxed{\text{EXP}}$ button).

Do you get 1.5×10^3? Or do you get something like 1.5^3 or 1.5^{03}?
Find out how your calculator shows standard form, so that you recognise it.
Type in 1234567891234, until you run out of space on the screen. Press the equals sign and have another look at the screen. Your calculator should give the number in standard form. The calculator may not show the $\times 10^{12}$ on the screen, but may just show a small number (^{12}). However your calculator displays the number, when you give the answer in standard form you *must write* $\times 10^{12}$.

Example 10
Estimate the answers and then use your calculator to calculate the following:

(a) $\sqrt{16.23 + 8.546}$ (b) $\dfrac{496.3 + 35.2 \times 34}{79 \times 23}$ (c) $5.12 \times 10^2 + 6.34 \times 10^3$

Answer 10

(a) $\sqrt{16.23 + 8.546} \simeq \sqrt{16 + 9} = \sqrt{25} = 5$

$\sqrt{16.23 + 8.546} = \sqrt{(16.23 + 8.546)} = \sqrt{24.776} = 4.977549 \ldots$
$= 4.98$ to 3 s.f.

> NOTE: The line in the square root sign also acts like a bracket. Once the brackets are in, your calculator should be able to calculate this without any intermediate steps. Start by pressing the square root key followed by the opening bracket, then the numbers followed by the closing bracket. But check this with your own calculator, because not all calculators follow the same logic. Let your calculator teach you its rules!

(b) $\dfrac{496.3 + 35.2 \times 34}{79 \times 23} \simeq \dfrac{500 + 40 \times 30}{80 \times 20} = \dfrac{1700}{1600} \simeq 1$

$$\frac{496.3+35.2\times34}{79\times23} = \frac{(496.3+35.2\times34)}{(79\times23)}$$

$$= 0.931810 \dots$$

$$= 0.932 \text{ to 3 s.f.}$$

NOTE: Check this calculation with your own calculator to make sure that your calculator uses BoDMAS and does not need an extra pair of brackets round the 35.2 × 34.

(c) $5.12 \times 10^2 + 6.34 \times 10^3 \simeq 500 + 6000 = 5600 \simeq 6.0 \times 10^3$

$5.12 \times 10^2 + 6.34 \times 10^3 = 6852 = 6.852 \times 10^3$

Exercise 4.9

Use your calculator to work out the following, first estimating the answer.

1. $34.93 + 356.1 \times 0.029$

2. $6.598 \times 3.111 - 24.701 + 17.3 \times 28$

3. $\dfrac{34.9+3.005}{13.1} + 28.35$

4. $\sqrt{\dfrac{16.01+19.49}{15.28-5.82}}$

5. $1.239 \times 10^4 - 5.87 \times 10^3$

6. $(3.967 \times 10^5) \times (7.65 \times 10^3)$

PROJECT

Use your calculator to investigate what happens when you multiply or divide a number by numbers that are larger or smaller than (or equal to) 1.

Number		1000	10	1	0.1	0.001	0.0001	0.000001
6	×	6000						
6	÷	0.006						

For example, you could copy and complete this table.

Can you summarise the results, or find a rule to remind yourself whether the answers get larger or smaller in each case?

If you use the table shown, how many zeroes are in each answer?

What happens as you divide by smaller and smaller numbers? Why is this?

What happens when you ask your calculator to divide by zero?

Ratio

Calculating a ratio is a method for comparing the sizes of two or more quantities. For example, Hamish wishes to make basmati rice for a party. To be on the safe side, he decides to use 2 cups of water to 1 cup of rice as he normally does.

This is the ratio, water : rice = 2 : 1.

It can be used for any volumes. For example, if it was a very big party he might have to use 2 buckets of water to each 1 bucket of rice! For a more moderate party he could use 6 cups of water to 3 cups of rice. The ratio will still remain the same.

Map scales and scale models

You may already be familiar with the scale of a map or model.

You may see a map with a scale of 1 : 20000. This means that every 1 unit on the map represents 20000 of these same units on the ground. So for example, 1 cm on the map represents 20000 cm (200 m) on the ground.

A model may be 'half scale', or 1 : 2. It means every 1 cm on the model represents 2 cm on the real object.

The 'Toy Train' Darjeeling Himalayan Railway has a gauge of 610 mm which means that the distance between the two parallel tracks is 610 mm. A standard, full-size train has a gauge of 1435 mm.

The ratio of: Toy Train gauge : Standard gauge
                  = 610 : 1435

We can simplify this by dividing both numbers by the smallest number, so

                  Toy Train gauge : Standard gauge
                  = 1 : (1435 ÷ 610)
                  = 1 : 2.35

If the Toy Train was a scale model all its length measurements would be in the same ratio as those of a standard train.

An architect is designing a large building for the Olympic Games. In order to show his clients what the building will look like in three dimensions he builds a scale model. The blocks that will be used in the actual building are 450 mm long. The model blocks are 20 mm.

The scale of the model is:

                  model : building
                  = 20 : 450
                  = 1 : 22.5

This ratio can be used to calculate all the other measurements in either the model or the real building.

Now that we have seen how ratios are used in everyday life we will see how to simplify and calculate with them.

Simplifying ratios

		Example
• If necessary rewrite measurements in the same units.		200 m : 4 km
		200 m : 4000 m
• Drop the units.		200 : 4000
• Divide both (or all) parts of the ratio by any common factor.		1 : 20

Example 11

Simplify the following ratios:

(a) $36 \text{ m}^2 : 120000 \text{ cm}^2$ (b) $500 \text{ g} : 1.5 \text{ kg}$ (c) $24 : 36 : 48$

Answer 11

(a) $36 \text{ m}^2 : 120000 \text{ cm}^2$

$= 36 \times (100 \times 100) \text{ cm}^2 : 120000 \text{ cm}^2$

$= 360000 : 120000$ (divide by 10000)

$= 36 : 12$ (divide by 12)

$= 3 : 1$

(b) $500 \text{ g} : 1.5 \text{ kg}$

$= 0.5 \text{ kg} : 1.5 \text{ kg}$

$= 0.5 : 1.5$ (divide by 0.5)

$= 1 : 3$

(c) $24 : 36 : 48$ (divide by 12)

$= 2 : 3 : 4$

Simplifying ratios of fractions

	Example
• Change any mixed numbers to improper fractions.	$1\dfrac{1}{2} : \dfrac{3}{4}$
	$= \dfrac{3}{2} : \dfrac{3}{4}$
• Write both fractions as equivalent fractions with the same denominators.	$= \dfrac{6}{4} : \dfrac{3}{4}$
• Multiply both by this common denominator to get rid of the denominator.	$= 6 : 3$
• Divide both fractions by any common factors.	$= 2 : 1$

Writing ratios as 1: *n* or *n* : 1

You may sometimes be asked to give your ratios in a specific way, particularly to have one part equal to 1. An example of this was when we changed the exchange rate from one based on American Dollars to one based on Indian Rupees earlier in this chapter.

• Simplify the ratio as usual.

• Divide both parts by the side you need to express as unity (one).

Example 12

(a) Simplify the following ratios:

(i) $\dfrac{3}{5} : \dfrac{1}{2}$ (ii) $1\dfrac{2}{3} : \dfrac{5}{6}$

(b) Write the ratio $4 : 5$ in the form:

(i) $1 : n$ (ii) $n : 1$

Answer 12

(a) (i) $\dfrac{3}{5}:\dfrac{1}{2} = \dfrac{6}{10}:\dfrac{5}{10} = 6:5$

(ii) $1\dfrac{2}{3}:\dfrac{5}{6} = \dfrac{5}{3}:\dfrac{5}{6} = \dfrac{10}{6}:\dfrac{5}{6} = 10:5 = 2:1$

(b) (i) $4:5$ in form $1:n$ (ii) $4:5$ in form $n:1$

       $= 4 \div 4 : 5 \div 4$ $(\div 4)$ $= 4 \div 5 : 5 \div 5$ $(\div 5)$

       $= 1 : 1.25$ $= 0.8 : 1$

Exercise 4.10

1. Simplify the following ratios:

(a) $3:48$ (b) $50:75:125$ (c) $45:360$

(d) 15 litres : 3 litres   (e) 14000 ml : 2.8 litres

(f) 2.5 cm : 5 km   (g) $\dfrac{3}{7}:\dfrac{1}{2}$ (h) $2\dfrac{3}{4}:1\dfrac{2}{5}$

2. A school has 25 teachers and 750 students. Write the teacher : student ratio in its simplest form.

3. A builder mixes up a mortar by mixing 3 shovels full of cement with 12 shovels full of sand. Write the cement : sand ratio in its simplest form.

4. Write the following ratios in the form $n:1$

(a) 81 litres : 90000 ml   (b) 10 km : 5 cm

5. Write the following ratios in the form $1:n$.

(a) $4:1$ (b) 6 kg : 72000 g

6. The area of India is approximately 3300000 km^2.

The area of the whole world is approximately 510000000 km^2.

Write the ratio area of India : area of whole world in its simplest form, giving your answer to 2 significant figures.

7. The rotation period of the planet Pluto is 6.4 Earth days. (This means that Pluto takes 6.4 of our days to rotate once.) The rotation period of the planet Venus is 243 Earth days. Write the ratio of the rotation of Pluto to that of Venus in the form:

(a) $1:n$ (b) $n:1$

Using ratios

An easy way to deal with ratios is to use columns. For example, divide **$450** between Jo and Sandy in the ratio **2 : 3**. This means dividing the $450 into **5 equal parts,** and then giving Jo 2 parts and Sandy 3 parts.

We need a column for Jo, another for Sandy, and another for the total number of parts.

	Jo	Sandy	Total
Parts	**2**	**3**	5

$450 \div 5 = 90$, so one part is $90. Use 90 as a multiplier ($5 \times 90 = 450$).

	Jo	Sandy	Total
amounts	2×90 $= 180$	3×90 $= 270$	**450**

Jo gets $180 and Sandy gets $270.
(It is also worth checking that $180 + 270$ adds up to 450.)

Example 13

(a) A map has a scale of 1 : 250000. Sanjeev measures the length of his bicycle ride on the map as 5 centimetres.
How far will he cycle? Give your answer in kilometres.

(b) The ratio of boys to girls in a class is 4 : 3. If there are 20 boys how many **students** are there in the class?
(In all the working shown on ratios the figures given or implied in the question are shown in **bold**, to make the working clearer.)

Answer 13

NOTE: Start by setting out your columns and rows, then enter the numbers from the question. Put a question mark where the answer will appear. There are then 2 ways to do the working. Either notice that in the Scale row 1 has to be multiplied by 250000, and do the same in the Length row. Or notice that in the Map column the 1 has to be multiplied by 5, and do the same in the Ground column. The second method has been applied here, and for consistency in most of the other worked examples. You can decide later which method you prefer, but a routine is very helpful in these questions. The 5 is called the *multiplier*.

(a)

	Map		Ground
Scale	**1**	:	**250000**
		$\times 5$	$\times 5$
Lengths	**5** cm		**?**

$250000 \times 5 = 1250000$ cm
1250000 cm $= (1250000 \div 100)$ m
$= 12500$ m
12500 m $= (12500 \div 1000)$ km
$= 12.5$ km

Sanjeev will cycle 12.5 kilometres.

(b) For this question we are asked to find the total number of students, so we need a Total column.

		Boys	Girls	Total students
Ratios	($20 \div 4 = 5$ so the multiplier is 5)	**4**	**3**	7
		$\times 5$		$\times 5$
Numbers		**20**		?

$7 \times 5 = 35$

There are 35 students in the class.

Exercise 4.11

1. The chemical formula for water is H_2O. This means that one water *molecule* is made of *atoms* of Hydrogen and Oxygen in the ratio
   $H : O = 2 : 1$.
   How many atoms of Hydrogen and how many atoms of Oxygen are present in 42 molecules of H_2O?

2. Divide the following quantities in the stated ratios.
   (a) Rs.605 in the ratio 4 : 1   (b) 831.6 g in the ratio 1 : 5
   (c) 4.5 m in the ratio 5 : 1   (d) $300 in the ratio 2 : 5 : 13
   (e) 216 in the ratio 1 : 4 : 10   (f) 0.98 in the ratio 3 : 1

3. The scale of a map is 1 : 150000. What distance on the map represents 15 km on the ground?
   Give your answer in suitable units.

4. A plan is drawn so that 5 cm on the plan represents 25 m on the ground. What is the scale of the plan?

5. A school has staff (teachers) and students in the ratio
   teachers : girls : boys = 1 : 10 : 9.
   How many boys are in the school if there are 700 staff and students?

   NOTE: You will need four columns, headed teachers, girls, boys and total.

6. A bag of sweets has yellow, green, red and purple sweets in the ratio
   yellow : green : red : purple = 2 : 3 : 5 : 1
   There are 15 red sweets.
   (a) How many sweets are yellow?
   (b) How many sweets are there all together?

7. Jane and Jill each receive some money in the ratio Jane : Jill = 3 : 5. Jane receives $15. How much does Jill receive?

8. Mortar is made in the ratio cement : sand = 1 : 4. How much of each are needed to make 5 tonnes?

9. Mina is using a map, but she does not know what is the scale of the map. She does know that the distance (in a straight line) between her home and her school is 5.5 km. She measures this distance on the map and finds it to be 2.2 cm.
   She measures the distance between her home and the railway station on the map. It is 4 cm. How far does she live from the railway station?

10. A classroom measures 2.5 metres high by 6 metres wide by 8 metres long.
    (a) Calculate the volume of the classroom.
        Nitrogen and Oxygen are the main constituents of air.
        Nitrogen and Oxygen are present in air in the ratio 4 : 1.
    (b) Calculate the volumes of (i) Nitrogen (ii) Oxygen
        present in the classroom. Give your answers in m^3.

Proportion

Direct and inverse proportion

A builder uses 400 bricks to make a wall 10 metres long. How many bricks would he need to build a wall 30 metres long?

The answer is clear: the wall would be three times as long as before so he would need three times as many bricks. He would need 1200 bricks. This is called **direct proportion** because as one quantity (the length of wall) increases the other (the number of bricks) also increases in the same ratio. Notice that only two quantities are changing, the length of the wall and the number of bricks.

The builder takes three days to build his wall. How long would it take if he employed two other builders and they all worked together at the same rate?

The two quantities that are changing are now number of builders and time taken. With three builders working on the wall it should take one third of the time, so it would take only one day. This is **inverse proportion** because as one quantity (the number of builders) increases the other quantity (the time taken) decreases in the same ratio.

The words 'in the same ratio' are important, but not easy to explain. A few examples should help.

   1. As children get older they get taller. This is *not* an example of proportion because they get older at an even rate, but they may grow taller over a period, and then stay the same height for a while, and then grow again. Hence, age and height are not in proportion.

   2. The further you walk the longer it takes. But sooner or later you will get tired and slow down so this distance and time are not in proportion.

   3. If you are driving at a *constant speed* the further you go the longer it takes in the same ratio, so this time the distance and time are in proportion. Notice that we are saying 'the further'... 'the longer', so this is *direct* proportion. In this example the quantities that are changing are the distance and time. The speed is staying the same.

   4. If you drive a *fixed* distance, say 10 km, at a constant speed, the faster this speed is the less time it takes to complete the journey. Here we are saying 'the faster' ... 'the less' so these two quantities are in *inverse* proportion.

Here the distance stays the same and the speed and time vary.

Example 14

(a) In 2005, in Britain, the costs for sending letters first class depended on weight. Are these quantities in proportion?

Weight	Cost
60 g	£0.30
100 g	£0.46
150 g	£0.64
200 g	£0.79

(b) (i) In one week a farmer uses 10 bales of hay for his 25 cows. How many bales would he need in a week for 65 cows if they ate them at the same rate?

(ii) How long would the 10 bales last for 2 cows at the same rate?

Answer 14

(a) This is an example of two quantities both increasing, but not in the same ratio, so they are not in proportion.

Compare just two of the increases:

Weight		Cost	
100		0.46	
	$\times 1.5$		$\times 1.5$
150		0.69	(not 0.64)
100		0.46	
	$\times 2$		$\times 2$
200		0.92	(not 0.79)

(b) (i) Quantities varying are bales of hay and number of cows. Time stays the same (1 week).

Bales of hay		Number of cows	
10		25	
	$\times 2.6$		$\times 2.6$
26		65	$(65 \div 25 = 2.6)$

Hence, the farmer needs 26 bales per week for 65 cows.

NOTE: When you are doing direct and inverse proportion questions you need to think about whether you expect the answer to be larger or smaller, and then divide or multiply by the multiplier to make this happen. Here we will definitely need more hay so we multiply by 2.6.

(ii) Quantities varying are time and number of cows. Number of bales (10) stays the same. Would you expect the time that the bales would last to be more or less for fewer cows?

With fewer cows to feed, the bales will last longer so you should check that your answer is longer than one IGECK.

Time		Number of cows	
1 week		25	
	$\times 12.5$		$\div 12.5$ $(25 \div 2 = 12.5)$
12.5		2	

The farmer's 10 bales will last 12.5 weeks with only 2 cows.

Exercise 4.12

1. State whether each of the following are examples of direct proportion, inverse proportion or neither.
   (a) Cost of one stamp. Cost of a number of the same stamps.
   (b) Age of a child. Size of the child's shoes.

(c) Number of cows in a field. Length of time the grass will last (assuming that they always eat at the same rate).

(d) The length of a video tape and the number of half-hour programmes that can be recorded on it.

2. A tin of paint will cover 5.2 m². How many whole tins will have to be purchased for a wall 20 m by 2.5 m?

3. Curtain material costs $25 for 5 metres. How much will 34 m cost?

4. It takes 25 minutes to do a particular journey at a steady speed of 55 km/h. How long would it take at 45 km/h? Give your answer to the nearest minute.

5. It takes 25 minutes to travel 16 kilometres. How long would it take to cover 40 kilometres at the same speed? Give your answer to the nearest minute.

6. 3 painters take 4 days to paint the outside of a house.

(a) How long would it take 6 painters to paint the same house at the same rate?

(b) How many painters would be needed to paint three similar houses in 4 days?

Time

As we mentioned above, time is not decimalised, so care is needed to work out problems involving time.

For example, to change time in hours, minutes and seconds to decimal part of an hour or minute *without* using a calculator we need to divide by 60 as in the example below.

Example 15

Change 9 hours 10 minutes to hours.

Answer 15

10 minutes = $10 \div 60$ hours = $\dfrac{10}{60} = \dfrac{1}{6} = 0.166666 \ldots = 0.167$ hours to 3 s.f.

9 hours and 10 minutes = 9.167 hours to 3 s.f.

This calculation can be done in the same way with a calculator.

Some calculators will also allow you to enter time in hours, minutes and seconds using a button marked $\boxed{\cdot\,,\,,,}$

The calculation will then change the hours, minutes and seconds to decimal parts of an hour.

Try these:

1. To enter 9 hours, 10 minutes and 15 seconds:

Enter	Calculator display
9 $\boxed{\cdot\,,\,,,}$ 10 $\boxed{\cdot\,,\,,,}$ 15 $\boxed{\cdot\,,\,,,}$	9°10°16°
Press $\boxed{\cdot\,,\,,,}$ once more	9.170833333

2. To enter 9 hours 10 minutes:

9 $\boxed{\cdot\,,\,,,}$ 10 $\boxed{\cdot\,,\,,,}$ 9°10°

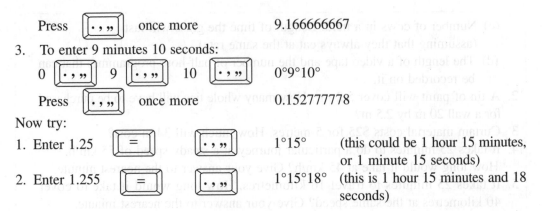

Press [·,,,]   once more   9.166666667

3. To enter 9 minutes 10 seconds:

0 [·,,,] 9 [·,,,] 10 [·,,,]   0°9°10°

Press [·,,,]   once more   0.152777778

Now try:

1. Enter 1.25   [=]   [·,,,]   1°15°   (this could be 1 hour 15 minutes, or 1 minute 15 seconds)

2. Enter 1.255 [=]   [·,,,]   1°15°18° (this is 1 hour 15 minutes and 18 seconds)

If your calculator does not work exactly like this you might have to experiment with the shift button as well until you know how to use it.

You might need to change decimal parts of an hour to hours, minutes and seconds, without a calculator.

Example 16

Change 3.75 hours to hours and minutes.

Answer 16

0.75 hours = 0.75 × 60 minutes = 45 minutes

3.75 hours = 3 hours and 45 minutes

12-hour and 24-hour clocks

You should be able to convert between the 12-hour and 24-hour clocks. To change from the 12-hour to 24-hour clock, follow these steps.

- If the time is a.m. add a zero at the beginning if necessary to write as four figures, and drop the 'a.m.'
- If the time is p.m. add 1200 and drop the 'p.m.'

To change from the 24-hour to the 12-hour clock, follow these steps.

- If the time is less than 1200, add 'a.m.' and remove any zero at the beginning.
- If the time is later than 1200, subtract 1200 and add 'p.m.'

Example 17

(a) Change to the 24-hour clock.
    (i) 8.15 a.m.   (ii) 2.45 p.m.

(b) Change to the 12-hour clock.
    (i) 0800   (ii) 1514

Answer 17

(a) (i) 8.15 a.m.   (ii) 2.45 p.m.
     = 0815   = 1445

(b) (i) 0800   (ii) 1514
     = 8 a.m.   = 3.14 p.m.

You should be able to work with timetables and international time.

> **Example 18**
> (a) A train leaves Edinburgh at 0845 and arrives in London at 1515.
>   How long does the journey take?
> (b) A television program starts at 6.55 p.m. and runs for 1 hour and 10 minutes.
>   At what time does it finish?
> (c) How much time has elapsed between 1845 on Monday and 0830 on Tuesday?
>
> **Answer 18**
> (a) Counting-on is often the best way to deal with time.
>
> | | |
> |---|---|
> | From 0845 to 0900 is | 15 minutes |
> | From 0900 to 1500 is | 6 hours |
> | From 1500 to 1515 is | 15 minutes |
>
> Hence, total time taken is 6 hours and 30 minutes, or $6\frac{1}{2}$ hours.
>
> (b) 6.55 p.m. plus 10 minutes is 7.05 p.m.
>   7.05 p.m. plus 1 hour is 8.05 p.m.
>   Hence, the program finishes at 8.05 p.m.
> (c) Again, counting-on is probably best.
>
> | | |
> |---|---|
> | 1845 to 1900 is | 15 minutes |
> | 1900 to midnight is | 5 hours |
> | midnight to 0800 is | 8 hours |
> | 0800 to 0830 is | 30 minutes |
> | Total | 13 hours and 45 minutes |

Rate

Rate is a measure of how one quantity changes as another changes.

Speed

Speed is probably the easiest rate to understand. Speed measures how far you go in a given time. The faster the speed the further you go in this given time, so speed is calculated by dividing distance gone by time taken.

If you travel 32 km in 1 hour your speed is 32 km/h (32 kilometres per hour). A speed of 90 km/h means that you travel 90 kilometres in one hour. Light travels at a speed of nearly 3×10^8 m/s (metres per second). This means that light travels 300000 kilometres in one second. A snail may travel at a speed of 0.5m/h (or much less!). This is 0.5 metres in 1 hour.

Exchange rates

You will also come across the word 'rate' when you see Exchange Rates, which represent how the currency of one country relates in value to that of another country.

Below is a table of exchange rates taken from a particular time on a particular day.

Currency	
American Dollars	1
Indian Rupees	44.61322
British Pounds	0.50537
Singapore Dollars	1.54075
South African Rands	7.18385
Euros	0.75050

This table is based on the American Dollar so it shows how many units of each currency you would get for 1 Dollar. The table can also be based on the Rupee.

According to the table, the ratio of Dollars to Rupees is 1: 44.61322.

To change this ratio to make the ratio Rupees (Rs.) to Dollars we need to divide both parts of the ratio by 44.61322, as shown below:

Dollars ($)   Rupees (Rs.)

1   **44.61322**

$\div 44.61322$ $\div 44.61322$

?   **1**

$1 \div 44.61322 = 0.02241$

So Re. 1 is equivalent to $0.02241.

Copy and complete the next table showing the exchange rates based on the Indian Rupee. Notice that all the rates given to 5 decimal places. The table shows how many units of each currency you would get for Re. 1.

The Exchange Rates are constantly changing according to what is happening in the world financial markets, and the rates given above might be completely different by the time you come to read this book. Try to find out some current exchange rates by looking in a newspaper, or on the Internet.

The other thing to remember about exchanging currencies is that the bank or bureau that changes the currency for you will charge a commission to cover the costs of their work, so you will never get the total amount you calculate.

Currency	
Indian Rupees	1
American Dollars	0.02241
British Pounds	0.01123
Singapore Dollars	
South African Rands	
Euros	

Gradient

Slope, or gradient is another example of rate. The gradient of a hill is measured by finding how much the hill rises (vertical measurement) for each unit in a horizontal direction (horizontal measurement). So you might say that a gradient is 50 metres rise for every kilometre horizontally. The gradient is 50 m/km.

However, in practice, it is usual to use the same units of measurement both horizontally and vertically which makes it unnecessary to state the units.

So 50 metres per kilometre would be 50 metres per 1000 metres or 1 metre per 20 metres, a gradient of 1 in 20.

$$\text{gradient} = \frac{50 \; metres}{1 \; kilometre}$$

$$= \frac{50 \; metres}{1000 \; metres} = \frac{50}{1000} = \frac{1}{20}$$

Gradients are also given as percentages, so 1 in 20 would become $\frac{1}{20} \times 100 = 5\%$.

Average rates

You may see a sign at the top or bottom of a steep hill warning drivers to take care. As the diagram shows, a gradient of a hill given as 20% does not necessarily mean that the gradient is the same all the way up the hill. The gradient would be given as an average gradient, calculated by dividing the total height risen vertically by the total distance covered horizontally.

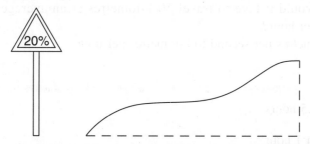

The same applies to speed on a journey. You may be asked to calculate the average speed on a journey, which always means total distance travelled divided by total time taken. Parts of the journey might have been faster, and parts slower than the average speed.

Compound Units

Compound units are units of measurement that are derived from other units. Examples of compound units are km/h or g/cm^3. The units themselves tell you how the measurements are calculated. For example, speed measured in kilometres *per* hour means that you find how many kilometres are travelled in one hour, so you have to divide distance gone (in kilometres) by time taken (in hours).

Grams *per* cubic centimetre gives the mass of a body per unit volume, and is calculated by dividing the mass of the body (in grams) by the volume of the body (in cubic centimetres). This is known as the **density** of the body and depends on the material it is made from. Lead has a high density, and quite a small volume has a large mass. Feathers have a low density, and quite a large volume has a low mass.

A table of the densities of some metals is shown below.

metal	kg/m^3
Lead	11389
Brass	8400
Iron	7850
Copper	8930
Silver	10490

Rupees *per* metre of material is the cost of each metre of the material. It would be calculated by dividing the total cost in rupees by the total length in metres.

Metres per rupee is the number of metres you could buy for each rupee and is calculated by dividing the length in metres by the total cost in rupees.

Example 19

(a) Find the average speed in kilometres per hour for a car which travels 100 kilometres in 1.5 hours.

(b) How far could you go at an average speed of 65 kilometres per hour for 2.5 hours?

(c) How long would it take to travel 50 kilometres at an average speed of 80 kilometres per hour?

(d) Change 15 metres per second to kilometres per hour.

Answer 19

(a) NOTE: **Remember average speed in km/h is the distance in kilometres travelled in one hour**.

100 km in 1.5 hours

$= \dfrac{100}{1.5}$ km in 1 hour

$= 66.7$ km in 1 hour

Average speed $= 66.7$ kilometres per hour.

(b) 65 km in 1 hour

$= 65 \times 2.5$ km in 2.5 hours

$= 162.5$ km in 2.5 hours

Distance gone $= 162.5$ kilometres

(c) 80 km in 1 hour

$= 1$ km in $\dfrac{1}{80}$ hour

$$= 50 \text{ km in } 50 \times \frac{1}{80} \text{hours} = 0.625 \text{ hours}$$

$$= 0.625 \text{ hours} = 0.625 \times 60 \text{ minutes} = 37.5 \text{ minutes}$$

Time taken = 37.5 minutes

(d) Distance gone   Time taken

15 metres in   **1** second

$15 \times 60 \times 60$ metres in $1 \times 60 \times 60$ seconds

54000 metres in   1 hour

54 kilometres in   1 hour

So 15 metres per second = 54 kilometres per hour

Exercise 4.13

1. Wood costs $4 per cubic metre. Find the cost of a piece of wood 5 m by 20 cm by 50 cm.
2. The world turns through 360° every 24 hours.
   (a) How long does it take to turn through 80°? Give your answer in hours and minutes.
   (b) What angle has the world turned through in 6.5 hours?
3. Water has a mass of 1000 kg per cubic metre. Calculate the mass of water in a swimming pool measuring 30 m × 20 m × 2 m.
4. A TV film starts at 11.35 p.m. on Wednesday and finishes at 1.20 a.m. on Thursday. How long is the film?
5. A train leaves Delhi at 7.15 a.m. and arrives in Agra at 9.45 a.m. The distance from Delhi to Agra is 206 km. Calculate the speed of the train.

   NOTE: **Remember that speed is distance gone in one hour.**

6. A train leaves Kolkata at 2015 and arrives in Patna 10 hours later. At what time does it arrive in Patna?
7. Change 35 grams per cubic centimetre into kilograms per cubic metre.
8. 4 cubic metres of a compound has a mass of 5 tonnes. Calculate its density in grams per cubic centimetre. (There are 1000 kg in 1 tonne.)

Personal and Household Finance

Best buys

It can often happen that manufacturers will offer two different sizes of a commodity at two different prices, and you may wish to work out which one is the best buy. There are two different ways of doing this, either:

- choose the one with the *lowest* cost per unit volume (or mass), or
- choose the one with the *highest* volume (or mass) per unit cost.

An example should make this clear.

Example 20
A shop is offering two sizes of a commodity:
A is a pack with a mass of 100 g costing Rs. 75, and **B** is a pack of mass 75 g costing Rs. 50. Which is the better buy? You must show all your working.

Answer 20

A
100 g
Rs. 75

B
75 g
Rs. 50

First method

A: 100 g costs Rs. 75

$\qquad$ 1 g costs Rs. $\dfrac{75}{100}$

$\qquad$ 1 g costs Rs. 0.75

So **B** is the better buy (1 g costs less)

B: 75 g costs Rs. 50

$\qquad$ 1 g costs Rs. $\dfrac{50}{75}$

$\qquad$ 1 g costs Rs. 0.67

Alternative method

A: Rs. 75 buys 100 g

$\qquad$ Re. 1 buys $\dfrac{100}{75} g$

$\qquad$ Re. 1 buys 1.33 g

So **B** is the better buy (Re. 1 buys more)

B: Rs. 50 buys 75 g

$\qquad$ Re. 1 buys $\dfrac{75}{50} g$

$\qquad$ Re. 1 buys 1.5 g

Profit and Loss

Profit or loss may be expressed as either an amount of money, or as a percentage *of the original cost price*. Always calculate the actual profit or loss first and then, if required, express as a percentage of the cost price, as shown in the example below.

Example 21
Pierre bought 50 pens for €135, and sold 48 of the pens for €4 each. He kept the remaining two pens to use in his shop. Calculate his percentage profit.

Answer 21

Cost price of pens		= €135.00
Selling price of pens (48 at €4 each)		= €192.00
Actual profit	(Selling Price − Cost Price)	= €57.00
Percentage profit	(€57.00 ÷ €135 × 100%)	= 42.2%

Simple and Compound Interest

If you put money into a savings account at a bank you will receive **interest**, which is calculated as a percentage of the **amount** you have in the account. The percentage varies according to the bank and the type of account and is called the **interest rate**.

There are two ways to calculate this interest, called **Simple Interest** and **Compound Interest**.

Simple Interest

When simple interest is calculated, the interest earned at the end of each year is not taken into account in the following year. For example, it may be removed from the account for other purposes.

Compound Interest

When compound interest is calculated, the interest earned each year is left in the account and the interest for the following year is calculated using this new (larger) sum.

In each of these cases you may be asked to calculate either the total interest earned, or the total sum in the account at the end of the period. You should read the question particularly carefully to make sure that you give the required answer.

An example should make this clear.

Example 22

Calculate the *total interest* earned on an investment of $150 for 3 years at a rate of 4% per annum,

(a) using simple interest (b) using compound interest.

Answer 22

     amount = $150　　　　　　　rate = 4%　　　　　　time = 3 years

(a) Simple interest

    At the end of the first year:　　　Interest = 4% of $150　　= $6.00

    At the end of the second year:　　Interest = 4% of $150　　= $6.00

    At the end of the third year:　　Interest = 4% of $150　　= $6.00

    Total interest earned ($6 + $6 + $6, or 3 × $6)　　　　　= $18.00

(b) Compound interest

    At the end of the first year:　　　Interest = 4% of $150　　= $6.00

    New amount ($150 + $6.00)　　= $156.00

    At the end of the second year:　　Interest = 4% of $156　　= $6.24

    New amount ($156 + $6.24)　　= $162.24

    At the end of the third year　　　Interest = 4% of $162.24 = $6.4896

    New amount ($162.24 + $6.49)　　= $168.73

    Total interest earned ($168.73 − $150)　　= $18.73

NOTE: Hence, the interest earned using compound interest is more than the interest earned using simple interest. The difference is not very much for these small amounts, but it could be considerable if very large sums were invested.

Depreciation

The value of many consumer goods decreases as time goes by. For example, a car with a cost when new of, say, $12000 might have lost 30% of this value by the end of the first year. If it were to carry on depreciating at the same rate the calculation would be the same as that for compound interest, but with the value falling each year instead of increasing. In this case the rate of depreciation is 30% in the first year. It is likely to be less, say 20% in the succeeding years. The second-hand car trade has tables showing how the values of makes of cars fall each year. The depreciation rate is a factor to take into account when deciding which car to buy.

Electricity bills

Electricity and other Utility bills (for example, gas, water, and telephone) are usually based on the number of units used as shown on a meter. The units will be a measure used for the particular Utility, and we normally refer to them as 'units', without having to worry about what exactly they are. There are then various methods to calculate the total sum to pay, and you will need to read each question carefully and follow the instructions. For example, the cost of the 'first' units used may be different from the remaining units, as you will see in the example given. There could also be a 'standing charge' which you have to pay just for having the service connected, even if you have not used any units.

Example 23

I have just read my electricity meter, one month after the last reading.

Nov 12th	9	9	1	5	3
Dec 12th	9	9	6	7	6

Units cost £0.1276 for the 'first' 170 units, and £0.0931 for the remaining units. There is no standing charge, but VAT is added at 5%. Calculate my total bill for this month.

Answer 23

NOTE: Remember to take the old reading away from the new reading

Units used $99676 - 99153 = 523$ units

First units   170 at £0.1276 = £21.69

Remaining units $(523 - 170) = 353$ units

                              353 at £0.0931 = £32.86

Total before VAT $(£21.69 + £32.86) = £54.55$

VAT at 5% $(5 \times 54.55 \div 100) = £2.73$

Total after VAT $(£54.55 + £2.73) = £57.28$

Total Bill = £57.28

Examination Percentages

You will probably be very used to calculating your percentage marks in an examination.

Example 24
Calculate the percentage mark for 45 marks out of a possible 65.

Answer 24

$$\frac{45}{65} \times 100 = 69\%$$

Exercise 4.14

1. For each question   (i) state whether there is a profit or loss,
   Work out   (ii) the actual profit or loss,
                   (iii) the profit or loss as a percentage of the cost price.

	Cost Price	Selling Price
(a)	$50	$45
(b)	$50	$55
(c)	Rs.365	Rs.456.25
(d)	10 for Rs.786	Rs.85 each
(e)	150 for £75	£0.55 each

2. Fred puts $1000 into the bank. The bank pays 5% interest per annum. Calculate the amount Fred has in the bank after 2 years if
   (a) Fred takes out the interest earned at the end of the first year,
   (b) Fred leaves the interest in the bank.

3. On the Island of Equality everyone earns the same amount (€25) per hour and everyone pays annual income tax at the same rate (25%).
   There is a personal allowance of €2500 per annum which is tax free.
   Calculate the amount each of these people pays in tax per annum.
   (a) Raj works for 20 hours a week and has three weeks unpaid holiday a year.
   (b) Tamara works for 175 hours a month and has one month unpaid holiday per year.
   (c) Anet works for 796 hours per year, and has no extra holiday.

4. An approximation used to convert between kilometres and miles is 8 kilometres ≈ 5 miles.
   (a) Work out the approximate equivalent of 20 kilometres in miles.
   (b) Work out the approximate equivalent of 35 miles in kilometres.

5. Change the following to the units stated.
   (a) 5.5 km/h to m/s
   (b) 60 miles per hour to kilometres per hour (use the approximate conversion 5 miles ≈ 8 kilometres)

6. Zac is making a paperweight. He will choose between the following.
   (a) A cuboid 10 cm × 2 cm × 4 cm, made of iron.

   (b) A cube of side 4 cm, made of copper.
      Use the table of densities on page 105 to work out which will be the heavier paperweight.

7. Assuming the light from a flash of lightening arrives instantaneously, but that the sound travels at 300 m/s, work out how far away a thunder storm is if you count 10 seconds between seeing the flash and hearing the thunder. Give your answer in kilometres.

8. Supriti buys 10 toy cars for Rs.50 each. She sells 7 for Rs.75 each, and then reduces the remaining three and sells them in her sale at 20% discount on her previous selling price. Calculate her profit as a percentage of the cost price.

9. A large (1 litre) bottle of lemonade costs £1.80, and a small (250 ml) bottle costs £0.44. Which is the better buy?

10. Anita is going to buy a used car. She is making a choice between a Penti Hatchback and a Quadri Saloon.
The Penti uses 10 litres of fuel to travel 90 kilometres, and the Quadri uses 15 litres to travel 180 kilometres. Which of these two cars would be the most economical to run? You must show all your working.

11. Tomas got 36 marks out of a possible 60 in his Mathematics test. He got 52 marks out of 75 in his Science test. By calculating his percentage marks for each test find out which was his better test.

12. Niraj bought a new car in 2000 costing $35000.
The rate of depreciation was 46% in the first year and 19% in the second year. Calculate the value of the car at the end of 2002.

Exercise 4.15

Mixed Exercise

1. A mineral known as 'fool's gold' because it glitters like gold has a chemical formula FeS_2. Fe stands for Iron and S stands for Sulphur.
This means that 1 molecule of FeS_2 contains 1 atom of Iron and 2 atoms of Sulphur. How many atoms of (a) Iron and (b) Sulphur are present in 72 molecules of fools gold?

2. The air in the classroom contains 0.03% CO_2.
A classroom is 2.5 metres high, 8 metres long and 6 metres wide.
(a) Calculate the volume of the room.
(b) Work out the volume of CO_2 in the room.

3. A cake requires 275 grams of flour, 250 grams of sugar and 4 eggs.
Amrit wants to make enough cakes to use up all the eggs in his refrigerator.
He finds he has 10 eggs.
(a) How much flour and sugar will he need to make the cakes?
(b) How many whole cakes can he make?

4. A flight leaves London at 2135 on February 1st and arrives in Bangkok at 1540
   London time on February 2nd.
   (a) How long has the flight taken?
   The distance between London and Bangkok along the plane's route is 8038 miles.
   (b) Calculate the speed of the plane in miles per hour.
       1200 in London is 1800 in Bangkok.
   (c) What is the time in Bangkok when the plane arrives?

5. Change the following to the stated units.
   (a) 2 hours and 43 minutes to minutes   (b) 7.15 hours to hours and minutes
   (c) 3 hours 36 minutes to hours   (d) 27 minutes to hours

6. The mass of Earth is approximately 5.97×10^{24} kg.

   The mass of the Moon is approximately $\dfrac{1}{81}$ of the mass of the Earth.

   Calculate the mass of the moon giving your answer to 3 significant figures and
   in standard form.

7. The distance of the Moon from Earth is 384.4×10^3 km.
   (a) Write this in standard form.
   (b) Change to metres, giving your answer in standard form.

Examination Questions

8. A model of a car has a scale of 1:25.
   The model is 18 cm long. Calculate, in metres, the actual length of the car.
   (0580/01 May/June 2004 q 5)

9. Sergio's height is 142 cm, to the nearest centimetre.
   Copy and complete the statement about the limits of his height.
   ....cm $\leqslant$ height < ....cm
   (0580/01 May/June 2004 q 7)

10. Alix changed a traveller's cheque for 200 euros (€) into dollars ($)
    when she visited the USA. The exchange rate was 1 dollar = 1.05 euros.
    How many dollars did she receive?
    (0580/01 May/June 2004 q 9)

11. $218 \div 39$
    (a) (i) Write both numbers in the calculation above correct to one significant
             figure.
         (ii) Use your answer to **part (i)** to estimate the value of the calculation.
    (b) Use your calculator to find the value of the calculation correct to two
        significant figures.
    (0580/01 May/June 2004 q 13)

12. Shampoo is sold in two sizes, *A* and *B*.
    A contains 800 ml and costs $1.30. *B* contains 1.5 litres and costs $2.30.
    Which is the better value for money? **Show your working clearly.**
    (0580/01 May/June 2004 q 14)

13. Carlos buys a box of 50 oranges for $8.
    He sells all the oranges in the market for 25 cents each.
    (a) Calculate the profit he makes.
    (b) Calculate the percentage profit he makes on the cost price.
                                            (0580/01 May/June 2004 q 18)

14. The time in Dubai is 3 hours ahead of Birmingham.
    (a) If it is 2115 on Sunday in Birmingham, what time on Monday is it in
        Dubai?
    (b) An aircraft leaves Birmingham at 2115 on Sunday and arrives Dubai on
        Monday at 0745 **local time**.
        (i) How long did the journey take?
        (ii) The distance from Birmingham to Dubai is 5620 km.
             Calculate the average speed of the aircraft.
                                            (0580/01 May/June 2004 q 21)

15. Antonia is making a cake.
    She uses currants, raisins and sultanas in the ratio
    currants : raisins : sultanas = 4 : 3 : 5.
    The total mass of the three ingredients is 3.6 kilograms.
    Calculate the mass of sultanas.
                                            (0580/01 Oct/Nov 2004 q 5)

16. Ferdinand's electricity meter is read every three months.
    The reading on 1st April was 70683 units and on 1st July it was 71701 units.
    (a) How many units of electricity did he use in those three months?
    (b) Electricity costs 8.78 cents per unit.
        Calculate his bill for those three months.
        Give your answer in dollars, correct to the nearest cent.
                                            (0580/01 Oct/Nov 2004 q 17)

17. Write 0.4 kilograms in grams.
                                            (0580/01 Oct/Nov 2003 q 1)

18. The price of a book is $18. Sara is given a discount of 15%.
    Work out this discount.
                                            (0580/01 Oct/Nov 2003 q 2)

19. The length of a road is 1300 metres, correct to the nearest 100 metres.
    Copy and complete this statement.
                ....m ⩽ road length < ....m.
                                            (0580/01 Oct/Nov 2003 q 8)

20. (a) Work out:
        $2.7 \times 8.3 \div (12 - 2.7)$
        writing down:
        (i) your full calculator display,
        (ii) your answer to two decimal places.

(b) Work out $\left(6-\sqrt{11}\right)^2$.

(0580/01 Oct/Nov 2003 q 13)

21. Mahesh and Jayraj share $72 in the ratio 7 : 5.
     How much does Mahesh receive?

(0580/01 May/June 2003 q 4)

22. Areeg goes to a bank to changes $100 into riyals.
     The bank takes $2.40 and then changes the rest of the money at a rate of
     $1 = 3.75 riyals.
     How much does Areeg receive in riyals?

(0580/01 May/June 2003 q 6)

23. The diagram shows a pole of length *l* centimetres.

     ←- - - - - - - - - - - - - - - - - - - *l* cm - - - - - - - - - - - - - - - - - - - →

     (a) Hassan says that $l = 88.2$. Round this to the nearest whole number.
     (b) In fact the pole has a length 86 cm, to the nearest centimetre.
          Copy and complete the statement about *l*.
          ....$\leqslant l <$ ....

(0580/01 May/June 2003 q 12)

24. On a journey a bus takes 35 minutes to travel the first 10 kilometres.
     It then travels a further 20 kilometres in the next 40 minutes.
     (a) The bus started the journey at 1850.
          At what time did it complete the journey?
     (b) Calculate the average speed of the whole journey in
          (i) kilometres/minute,   (ii) kilometres per hour.

(0580/01 May/June 2003 q 13)

25. Work out $\sqrt{7.1^3 + 2.9^3}$, giving:

     (a) your full calculator display,   (b) your answer to 2 decimal places.

(0580/01 May/June 2003 q 1)

26. (a) A mobile phone company changes its rental charge from $80 per **year** to
          $7.50 per **month**. Work out the percentage increase.

     (b) George's phone card lasts for 300 minutes. He has used $\frac{3}{5}$ of this time.

          Work out how many minutes are left on his phone card.

(0580/03 Oct/Nov 2003 q 8a and b)

27. Anne took a test in chemistry.
     She scored 20 marks out of 50.
     Work out her percentage mark.

(0580/01 May/June 2005 q 3)

28. Write, in its simplest from, the ratio
    3.5 kilograms : 800 grams.

    (0580/01 May/June 2005 q 4)

29. Yasmeen is setting up a business.
    She borrows $5000 from a loan company.
    The loan company charges 6% per year simple interest.
    How much interest will Yasmeen pay after 3 years?

    (0580/01 May/June 2005 q 11)

30. $\dfrac{8.95 - 3.05 \times 1.97}{2.92}$

    (a) (i) Write the above expression with each number rounded to one
        significant figure.

      (ii) Use your answer to find an **estimate** for the value of the expression.

    (b) Use your calculator to work out the value of the **original** expression.
        Give your answer correct to 2 decimal places.

    (0580/01 May/June 2005 q 19)

31.
    | SALE |
    | All items |
    | 35% Reduction |

    Abdul bought a spade in this sale. Its **original** price was $16.

    (a) How much did Abdul save?

    (b) The next day, all items were sold at half the **original** price.
        How much **more** would Abdul have saved if he had waited until
        the next day to buy the spade?

    (0580/01 May/June 2005 q 21)

32. (a) Write 0.48 correct to 1 significant figure.

    (b) (i) Find an approximate answer for the sum

        $9.87 - 5.79 \times 0.48$

        by rounding each number to 1 significant figure. Show your working.

      (ii) Use your calculator to find the exact answer for the sum in **part (b) (i)**.
        Write down all the figures on your calculator.

    (0580/01 Oct/Nov 2005 q 15)

33. Lorenzo saves money for a motorbike.
    The marked price of the motorbike is $900.
    He pays a deposit of 35% of the marked price.
    (a) Calculate his deposit.
    (b) He then makes 12 monthly payments of $60 each.
        How much more than the $900 marked price does he pay altogether?

    (0580/01 Oct/Nov 2005 q 20)

34. Rodriguez puts $500 into a bank account. The bank pays 5% compound interest per year.

    (a) How much is the interest after one year?

    (b) Work out the **total amount** he has in his bank account after two years.

    (0580/01 May/June 2006 q 11)

35. A train sets off at 11 53 on a journey to Mumbai. The journey takes 2 hours 30 minutes.

    (a) Write down the time when the train arrives in Mumbai.

    (b) The distance to Mumbai is 235 kilometres.
    Calculate the average speed of the train.

    (0580/11 Oct/Nov 2009 q 9)

36. The scale on a map is 1:250000. A road is 4.6 cm long on the map. Calculate the actual length of the road in kilometres.

    (0580/01 Oct/Nov 2008 q 6)

Chapter 5

Working with Algebra

This chapter will provide you with more skills in the use of algebra. You should begin to feel more confident in the language of algebra and the use of letters to replace numbers. Remember to work through each worked example carefully yourself before going on to the exercise that follows it. Check your answers to each exercise as you proceed. If you get an answer wrong go back, read the question carefully, check that you have not made an error in copying data from the question, and then *start your answer to the question again*. It can be very difficult to check through existing work to find a mistake, so start afresh.

The most likely errors usually involve signs, particularly minus signs! Remember the rules for plus and minus signs. Use the number line you made in Chapter 1 when necessary.

Essential Skills

1. Simplify:
   (a) $-1-2+4$ (b) -2×5 (c) -1×0 (d) 3×-6
   (e) $(-5)^2$ (f) $-2(x-y)$ (g) $-(3+a)$

2. Simplify:
   (a) $3+0$ (b) 3×0 (c) $0 \div 2$ (d) $2+1$ (e) 2×1
   (f) $\dfrac{0}{1}$ (g) $\dfrac{3}{1}$ (h) $\dfrac{2}{-3}$ (i) $\dfrac{-2}{-3}$ (j) $3-3$
   (k) $\dfrac{3}{3}$ (l) $4 \div 4$ (m) $x \div x$ (n) $0 \times x \times y$
   (o) $0-a$ (p) $\sqrt{4}$ (q) 3^2 (r) $x-x$

3. Which of the following are expressions and which are equations?
   (a) $3+x-y$ (b) $6(x-5)=1$ (c) $3+x-y=0$ (d) $6(x-5)$

4. $2x + 3z - 5y - 3x + w$
   (a) Which is the term in y?
   (b) Which two terms are like terms?
   (c) List the variables.
   (d) What is the coefficient of the first term?
   (e) Which sign belongs to the term in z?

5. List pairs of these operators that are inverses of (undo) each other.
   square   **divide**   **add**   **subtract**   **multiply**   **square root**

Answers

1. (a) 1　　　　　(b) −10　　　　　(c) 0　　　　　(d) −18
   (e) 25　　　　　(f) −2x + 2y　　　　(g) −3 − a

2. (a) 3　　　　　(b) 0　　　　　(c) 0　　　　　(d) 3　　　　　(e) 2

   (f) 0　　　　　(g) 3　　　　　(h) $-\dfrac{2}{3}$　　　　(i) $\dfrac{2}{3}$　　　　(j) 0

   (k) 1　　　　　(l) 1　　　　　(m) 1　　　　　(n) 0　　　　　(o) −a
   (p) 2　　　　　(q) 9　　　　　(r) 0

3. Expressions:　(a) and (d)
   Equations:　　(b) and (c)

4. (a) −5y　　　　(b) 2x and −3x　　　　(c) x, y, z, w　　　（d) 2　　　（e) +

5. square, square root;　divide, multiply;　add, subtract

Solution of Equations

You will remember that we looked at the difference between expressions and equations in Chapter 3. We noted that expressions may be simplified but not solved, but that we can often find solutions to equations. A solution to an equation is a number that can replace the variable (for example, x) and make the equation a true statement.

A simple example of an equation would be $2x + 3 = 11$.

A little thought, or perhaps trial and error, will help you see that if the variable x is replaced by the number 4 the equation becomes $2 \times 4 + 3 = 11$, which is a true statement. This means that the solution to this equation is $x = 4$.

For simple equations like these it is often easy to just look at the equation and see what the solution must be (you are solving 'by inspection'), but you will rapidly find that the equations become too difficult to solve in this way, so we need to develop a systematic method to find the solution.

It is useful to think of equations as items in a set of balancing scales. Our equation above has two sides, a left side and a right side, and the equals sign in the middle tells us that the two sides are indeed equal to each other. We can use this equation to work out a systematic method for solving equations.

Remembering that x is at this moment an unknown quantity, let us imagine that the two sides represent quantities which have mass. If we put the two sides of the equation in the two sides of a set of scales, then the equation tells us that they must balance.

They would still stay in balance if we removed 3 from both pans, so:
$$2x + 3 - 3 = 11 - 3$$
$$2x = 8$$

We can now easily see that if $2x = 8$ then x must be one half of 8.

$$2x = 8$$
$$x = 8 \div 2$$
$$x = 4$$

We have arrived at the same solution by this method as we did by inspection.

The basis for our method is: **the equation will stay in balance if we do the same thing to *both* sides.** You can add, subtract, multiply or divide as long as you do the same to *both sides*.

It is good practice to keep your equals signs in a straight line down the page when you are solving equations, and always write only one statement on each line, as in the worked examples below.

Example 1

Clearly showing your working solve the following equations.

(a) $3x + 1 = 7$ (b) $5x - 11 = 4$ (c) $5 + 2x = 8$

(d) $3x + 6 = 2$ (e) $7 - 2x = 13$

Answer 1

(a) $3x + 1 = 7$

$\qquad 3x = 7 - 1$ (-1 from both sides)

$\qquad 3x = 6$

$\qquad x = 6 \div 3$ ($\div 3$)

$\qquad x = 2$

(b) $5x - 11 = 4$

$\qquad 5x = 4 + 11$ ($+ 11$ to both sides)

$\qquad 5x = 15$

$\qquad x = 15 \div 5$ ($\div 5$)

$\qquad x = 3$

(c) $5 + 2x = 8$ NOTE: If there is no sign in front of the first term (in this case 5) it is positive ($+ 5$), so to remove the 5 take (subtract) 5 from both sides.

$\qquad 2x = 8 - 5$ (-5)

$\qquad 2x = 3$

$\qquad x = 3 \div 2$ ($\div 2$)

$\qquad x = \dfrac{3}{2}$ (or 1.5)

(d) $3x + 6 = 2$

$\qquad 3x = 2 - 6$ (-6)

$\qquad 3x = -4$

$\qquad x = -4 \div 3$ ($\div 3$)

$\qquad x = -\dfrac{4}{3}$

(e) $7 - 2x = 13$

$\quad\quad -2x = 13 - 7 \quad\quad\quad (-7)$

$\quad\quad -2x = 6$

$\quad\quad\quad x = \dfrac{6}{-2} \quad\quad\quad\quad (\div -2)$

$\quad\quad\quad x = -3 \quad\quad\quad\quad \text{NOTE: } + 6 \div -2 = -3$

The steps in these equations are all leading to the line where the x has been isolated by itself on the left hand side of the equation, leaving the right hand side to be simplified if necessary. It is as if we are slowly unpicking the equation to get to the unknown or variable x.

Simple equations may also be solved by thinking of number machines. Look at the number machine below, which represents the equation from Example 1 (d): $3x + 6 = 2$.

The input is x and the output is 2.

input x ⟶ ⟩ $\times 3$ ⟩ ⟶ ⟩ $+ 6$ ⟩ ⟶ output 2

We now solve the equation by running the machine backwards, so that the input is 2.

output $-\dfrac{4}{3}$ ⟵ ⟨ $\div 3$ ⟨ ⟵ ⟨ -6 ⟨ ⟵ input 2

The output is $-\dfrac{4}{3}$, which is the solution to the equation.

This can be a useful technique for rearranging formulae which you will meet later in the chapter, so practise now by writing number machines for the other three equations in the example, and running them in reverse to get the solutions.

Exercise 5.1

Solve the following equations either by inspection if they are easy enough or by our systematic method *showing your working* as in the answers to Example 1.

1. $3x + 4 = 10$

2. $5x = 20$

3. $x + 3 = 11$

4. $6x - 15 = 3$

5. $4x + 3 = 1$

6. $7x - 15 = 13$

7. $11x - 10 = 12$

8. $7 + 4x = 8$

9. $-2x = -10$

    NOTE: Remember that $- \div - = +$

10. $-8 - 3x = 10$

11. $-x = 9$

    NOTE: $-x$ means $-1x$, so divide both sides by -1.

12. $-x = -7$

13. $4 - x = 7$

14. $-x - 10 = 12$

15. $16 + 2x = 31$

    NOTE: You can leave the answer as a fraction, or give it as a

16. $2 - 3x = 7$

    decimal if it is an exact decimal.

The next step is to see what happens when the variable appears on both sides of the equation. For example, $3x + 2 = 4 - x$.

The same method will apply, and it may help to gather your terms in x on the left hand side of the equation, and the number terms on the right hand side. The x term is subtracted on the right hand side so adding x to both sides will remove it from the right hand side.

$$3x + 2 = 4 - x$$
$$3x + 2 + x = 4 \qquad (+ x) \qquad \text{(this takes the } x \text{ to the left hand side)}$$
$$4x + 2 = 4 \qquad \qquad \text{(this is the line above simplified)}$$
$$4x = 2 \qquad \qquad (-2 \text{ from both sides)}$$
$$x = \frac{2}{4} \qquad \qquad (\div 4)$$

$$x = \frac{1}{2} \qquad \qquad \text{(this is the fraction simplified by}$$
$$\text{dividing top and bottom by 2)}$$

Example 2

Solve the following equations:

(a) $5x - 3 = 7 + x$ 　　(b) $2 - 4x = 9 + 9x$ 　　(c) $5x - 7 + 3x = 10x + 6 - 6x - 3$

Answer 2

(a) $5x - 3 = 7 + x$
$$5x - 3 - x = 7 \qquad (- x)$$
$$4x = 7 + 3 \qquad (+ 3)$$
$$4x = 10$$
$$x = 2.5 \qquad (\div 4)$$

(b) $2 - 4x = 9 + 9x$
$$2 - 4x - 9x = 9$$
$$2 - 13x = 9$$
$$-13x = 9 - 2$$
$$-13x = 7$$
$$x = -\frac{7}{13}$$

(c) $5x - 7 + 3x = 10x + 6 - 6x - 3$　　NOTE: It may be easier to collect like terms
$$8x - 7 = 4x + 3 \qquad \text{on each side of the equation first.}$$
$$8x - 4x = 3 + 7$$
$$4x = 10$$
$$x = 2.5$$

NOTE: When you are really sure that you know what you are doing you can leave out some of the steps while still showing your working. The first two answers above could be shown as follows:

(a) $5x - 3 = 7 + x$
$$5x - x = 7 + 3$$
$$4x = 10$$
$$x = 2.5$$

(b) $2 - 4x = 9 + 9x$
$$-4x - 9x = 9 - 2$$
$$-13x = 7$$
$$x = -\frac{7}{13}$$

Exercise 5.2

Solve the following equations.

1. $7x + 3 = 2x + 7$

2. $5x - 1 = 6x + 3$

3. $4 - 8x = 6 - x$

4. $-10 - x = 7 - 6x$

5. $15 + 2x = 17 - 6x$

6. $23 - 3x = -7x + 11$

7. $100 + 2x = 50 - 25x$

8. $-8x + 4 = -16x + 8$

9. $11 - x = 11 + x$ NOTE: Zero divided or multiplied by anything
is still zero.

10. $12a + 6 = 6a - 17$ NOTE: The variable does not have to be x.

11. $7x + 3 - 5x + 2 = 6x$

12. $-8a - 1 + 3a = 7a - 6 + 5a - 3$

13. $-11 - 12y + 10 + 10y = 6y + 7 - 9y$

14. $10 + 3b - 5 = 12 + b - 3b$

The next type of equation you may have to solve involves brackets, but otherwise is no more difficult. The first thing to do is to multiply out the brackets and then proceed as above.

Example 3

Solve the following equations.

(a) $7(x - 3) = 3(x + 7)$

(b) $2(x + 3) = -5(2x - 1)$

(c) $9(3x - 1) - 6(2x - 1) = 5(3 + 5x)$

Answer 3

(a) $7(x - 3) = 3(x + 7)$

$\quad 7x - 21 = 3x + 21$

$\quad 7x - 3x = 21 + 21$

$\quad\quad 4x = 42$

$\quad\quad x = \dfrac{42}{4} = \dfrac{21}{2} = 10\dfrac{1}{2}$ or 10.5

(b) $2(x + 3) = -5(2x - 1)$

$\quad 2x + 6 = -10x + 5$

NOTE: Beware of the minus sign!

$\quad 2x + 10x = 5 - 6$

$\quad\quad 12x = -1$

$\quad\quad x = -\dfrac{1}{12}$

(c) $9(3x - 1) - 6(2x - 1) = 5(3 + 5x)$

$\quad 27x - 9 - 12x + 6 = 15 + 25x$

$\quad 27x - 12x - 25x = 15 + 9 - 6$

$\quad\quad -10x = 18$

$\quad\quad x = -1.8$

Exercise 5.3

Solve the following equations.

1. $5(x + 7) = 4(1 - x)$

2. $3(x - 1) = 8$

3. $9(3x - 2) = 7x$

4. $4(5 - x) = 3(2 - 3x)$

5. $-2(7 - 2x) = 7(3 + 2x)$

6. $-3(-4x + 5) = 2(-x + 6)$

7. $10(3x + 2) - (x - 1) = -7(x + 5)$ NOTE: Remember that $-(x - 1)$ is the same as
$-1(x - 1)$

8. $-(3x - 2) = 6(4 + 2x)$

9. $3(x - 2) - 7(x + 1) = -2(2x + 1) - 3(x + 2)$

10. $a - 8(7 + a) = 16(2a - 1) - 5(a + 3)$

Equations are often used to solve problems. The unknown quantity or quantities are given a letter or letters and equations are written using the given information. The equations may then be solved algebraically to find the unknown quantities.

Example 4

Alex is 2 years older than Bernard and half Callista's age. Bernard is 42 years younger than Callista. Find the ages of Alex, Bernard and Callista.

Answer 4

Let Alex be x years old.

Then Bernard is $x - 2$ years old, and Callista is $2x$ years old.

Since Bernard is 42 years younger than Callista, Bernard is $2x - 42$ years old.

We now have two expressions for Bernard's age, and both must, of course, be equal.

$2x - 42 = x - 2$ $(-x + 42)$

$\quad 2x - x = -2 + 42$

$\qquad\quad x = 40$

So, Alex is 40, Bernard is 38 and Callista is 80 years old.

You will find more examples of writing and using equations throughout the book.

NOTE: For the rest of this chapter you need to know that the perimeter of a shape is the distance all the way round the outside of the shape. So the perimeter is the sum of all the sides.

Exercise 5.4

Solve the following.

1. $3x = -10$
2. $15 = 5x$
3. $4 = 2x - 3$
4. $11 - 9x = 7x + 23$
5. $7x = 18 - x$
6. $6(a - 5) = 7(a + 2)$
7. $-3(y + 2) = 4(y - 1)$
8. $6(b - 3) + 2(b + 1) = 3(b - 5)$
9. $5x - 3 = 2x$
10. $16x - 10 + 2x = 18x - 12 - 3x$
11. $4(c - 5) - 3(c - 1) = 2(c + 1)$
12. $-(x - 1) + 2(x + 1) = 5$
13. $2(3x - 5) = 5(x - 2)$
14. $-(x + 1) = 2(-x - 1)$
15. $21x - 3 = 4$

16. A triangle has two sides each of length $3x$ centimetres, and one of length $(2x + 5)$ centimetres. The perimeter (sum of all the sides) of the triangle is 33 centimetres.
    (a) Form an equation in x.
    (b) Solve your equation.
    (c) Hence write down the lengths of the sides of the triangle.

17. Tomas has a pencil case containing only red, blue and green pencils. There are twice as many red pencils as blue pencils and two more red pencils than green pencils. There are 23 pencils altogether in the pencil case.
    Let the number of red pencils be x.
    (a) Write down expressions for the number of blue pencils and the number of green pencils in terms of x.
    (b) Form an equation in x.

(c) Solve the equation.

(d) Hence write down the number of each colour of pencil.

18. An examination paper is to have 20 questions altogether. It has to cover algebra, shape and graphs.

    There will be 4 more questions on algebra than on graphs, and twice as many questions on shape as on graphs.

    (a) Let the number of questions on graphs be x.

    Write down expressions for the numbers of questions on algebra and shape in terms of x.

    (b) Form an equation in x.

    (c) Solve the equation to find x.

    (d) How many questions will there be on algebra?

19. The sum of three consecutive numbers is 114.

    Let the first of the numbers be x.

    Form an equation in x and solve it to find the three numbers.

20. The sum of three consecutive odd numbers is 135.

    Find the three numbers.

    NOTE: If x is the first odd number then $x + 2$ is the next.

Rearranging Formulae

We have already met some formulae. The formula below is used to find the speed of an object such as a car (v) when it was initially travelling at a certain speed (u) and then accelerated (a) for a certain time (t). We will assume that the units for speed, acceleration and time are correct so that we do not have to worry about them.

The formula is: $v = u + at$ which we could read as 'the final speed of the car is equal to its initial speed plus the acceleration multiplied by the time for which the car has been accelerating'.

The formula is arranged so that it is easy to find v. We say that v is the **subject** of the formula.

For example, if $u = 10$, $a = 2$ and $t = 4$,

$$\text{using } v = u + at$$
$$v = 10 + 2 \times 4$$
$$v = 10 + 8$$
$$v = 18$$

But suppose we wanted to find, say, u? The best way is to *rearrange* the formula so that u is the subject. This means getting u on its own.

It can be helpful to underline the variable that is to become the subject of the formula.

$$v = \underline{u} + at$$

Then the first step could be to write the equation the other way round.

$$\underline{u} + at = v$$

Then, just as in the solution of equations we do the same thing to both sides of the formula, in this case take away at

$$\underline{u} = v - at \qquad\qquad (-at)$$

and this is the answer.

Now we can find u, given $v = 20$, $a = 10$ and $t = 0.5$.

$$u = v - at$$
$$u = 20 - 10 \times 0.5$$
$$u = 20 - 5$$
$$u = 15$$

This was an easy rearrangement, but suppose we were asked to make a the subject of the formula. We will work through the method at the same time as solving a similar equation with numbers instead of letters.

$8 = 2 + 3a$	compare with	$v = u + \underline{at}$
$3a + 2 = 8$		$u + \underline{at} = v$ (turning round)
$3a = 8 - 2$ (-2)		$\underline{at} = v - u$ $(-u)$

At this stage we must divide both sides by t to leave the a on the left hand side. We must divide the whole of the right hand side by t so it is safest to put brackets round the right hand side. This is equivalent to working out the numbers $(8 - 2)$ in the example.

$$3a = 6 \qquad\qquad\qquad \underline{at} = (v - u)$$

$$a = \frac{6}{3} \qquad\qquad\qquad \underline{a} = \frac{(v - u)}{t}$$

The formula has been rearranged to make a the subject.
Now we can find a given that $v = 32$, $u = 14$ and $t = 4$.

$$a = \frac{v - u}{t}$$

$$a = \frac{32 - 14}{4}$$

$$a = \frac{18}{4}$$

$$a = 4.5$$

In these examples we have been using mainly whole numbers or simple fractions, but if we have to use numbers that require a calculator, then we should give the answer rounded, probably to 3 significant figures.

Example 5

Solve the equations and rearrange the corresponding formulae to find a in each case.

(a) (i) $2a + 3 = 17$ (ii) $2a + b = c$

(b) (i) $7 = \dfrac{a}{5}$ (ii) $x = \dfrac{a}{b}$

(c) (i) $14 = \dfrac{28}{a}$ (ii) $p = \dfrac{q}{a}$

(d) (i) $5(a - 1) = 3(a + 1)$ (ii) $b(a - c) = d(a + 1)$

NOTE: In this question the variable appears twice, and terms containing it must be collected first.

Answer 5

(a) (i) $2a + 3 = 17$

$2a = 17 - 3$

$2a = 14$

$a = 7$

(ii) $2a + b = c$

$2a = c - b$

$2a = (c - b)$

$a = \dfrac{(c - b)}{2}$

(b) (i) $7 = \dfrac{a}{5}$

$\dfrac{a}{5} = 7$

$a = 7 \times 5$

$a = 35$

(ii) $x = \dfrac{a}{b}$

$\dfrac{a}{b} = x$

$a = x \times b$

$a = bx$

(c) (i) $14 = \dfrac{28}{a}$

$14 \times a = 28$

$a = \dfrac{28}{14}$

$a = 2$

(ii) $p = \dfrac{q}{a}$ ($\times a$ both sides)

$p \times a = q$

$a = \dfrac{q}{p}$

(d) (i) $5(a - 1) = 3(a + 1)$

$5a - 5 = 3a + 3$

$5a - 3a = 3 + 5$

$2a = 8$

$a = \dfrac{8}{2}$

$a = 4$

(ii) $b(a - c) = d(a + 1)$

$ab - bc = ad + d$

$ab - ad = d + bc$ (factorise out a)

$a(b - d) = (d + bc)$

$a = \dfrac{(d + bc)}{(b - d)}$

The next exercise provides some practice in rearranging formulae by comparing with simple equations.

Exercise 5.5

Solve the equations and rearrange the corresponding formulae to make x the subject.

1. (a) $x + 2 = 5$
   (b) $x + b = d$

2. (a) $\dfrac{x}{3} = 6$
   (b) $\dfrac{x}{y} = z$

3. (a) $2x - 5 = 7$
   (b) $ax - b = c$

4. (a) $2x + 4x = 9$
   (b) $ax + bx = c$ **NOTE: Factorise!**

5. (a) $4x - 2 = x$
   (b) $ax - 2 = bx$

6. (a) $3 + x = 4x$
   (b) $3 + x = ax$

7. (a) $5 + 9x = x - 7$
   (b) $a + bx = xc$

8. (a) $\dfrac{1}{3} = \dfrac{2}{x}$
   (b) $\dfrac{1}{a} = \dfrac{b}{x}$

9. (a) $7 = \dfrac{15}{x}$ (b) $y = \dfrac{3}{x}$

10. (a) $\dfrac{1}{2}x = 10$ (b) $\dfrac{1}{a}x = b$

As was mentioned earlier, another way to look at rearranging simple formulae is to use number machines. The following examples show both methods.

Example 6

(a) $S = \dfrac{1}{2}n(a+l)$

(i) Find S when $n = 10$, $a = 5$ and $l = 15$.
(ii) Make a the subject of the formula.
(iii) Find a when $S = 100$, $n = 8$ and $l = 23$.

(b) $t = 2\pi\sqrt{\dfrac{l}{g}}$

(i) Find t when $l = 15$ and $g = 9.8$, giving your answer correct to 3 significant figures.
(ii) Make l the subject.
(iii) Find l when $t = 5$ and $g = 9.8$, giving your answer correct to 3 significant figures.

Answer 6

(a) $S = \dfrac{1}{2}n(a+l)$

(i) $S = \dfrac{1}{2} \times 10 \times (5 + 15)$

$S = 100$

(ii) $S = \dfrac{1}{2}n(a+l)$

$\dfrac{1}{2}n(a+l) = S$

$n(a+l) = 2S$

$a + l = \dfrac{2S}{n}$

$a = \dfrac{2S}{n} - l$

(iii) $a = \dfrac{2 \times 100}{8} - 23$

$a = 2$

An alternative answer to part (ii):
Rearranging the formula using a number machine. Always start by inputting the letter you want to make the new subject.

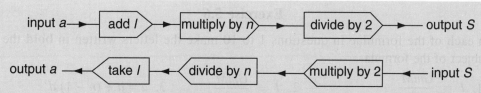

Result: $a = \dfrac{2S}{n} - l$

NOTE: The number machine method only works when the new subject appears only once.

NOTE: Rearranging this formula in a slightly different order can lead to an answer that at first appears to be different.

$$n(a + l) = 2S$$
$$na + nl = 2S$$
$$na = 2S - nl$$
$$a = \dfrac{2S - nl}{n}$$

In fact this is the same as before, but expressed differently.

(b) $t = 2\pi\sqrt{\dfrac{l}{g}}$

    (i) $t = 2 \times \pi \times \sqrt{\dfrac{15}{9.8}}$

        $t = 7.77$

    (ii) $t = 2\pi\sqrt{\dfrac{l}{g}}$

        $2\pi\sqrt{\dfrac{l}{g}} = t$

        $\sqrt{\dfrac{l}{g}} = \dfrac{t}{2\pi}$

        $\dfrac{l}{g} = \left(\dfrac{t}{2\pi}\right)^2$

        $l = \left(\dfrac{t}{2\pi}\right)^2 \times g$

NOTE: Remember that squaring is the inverse of (undoes) finding the square root.

    (iii) $l = \left(\dfrac{5}{2\pi}\right)^2 \times 9.8$

        $l = 6.21$

An alternative answer for part (ii):
Number machine method for rearrangement.

| input l → | divide by g | → | square root | → | multiply by 2π | → output t |

| output l ← | multiply by g | ← | square | ← | divide by 2π | ← input t |

Result: $l = \left(\dfrac{t}{2\pi}\right)^2 \times g$

Exercise 5.6

In each of the formulae in questions 1 to 10 make the letters written in **bold** the subject of the formula.

1. $F = \dfrac{GmM}{d^2}$

2. $F = \dfrac{GmM}{d^2}$

3. $u = a + (n - 1)\,\boldsymbol{d}$

4. $u = a + (\boldsymbol{n} - 1)\,d$

5. $c = \dfrac{a}{h}$

6. $c = \dfrac{a}{\boldsymbol{h}}$

7. $s = \dfrac{\boldsymbol{d}}{t}$

8. $s = \dfrac{d}{\boldsymbol{t}}$

9. $P = \dfrac{F}{\boldsymbol{A}}$

10. $F = \dfrac{9}{5}\,C + 32$

11. $S = ut + \dfrac{1}{2}at^2$

    Show that $a = \dfrac{2(s - ut)}{t^2}$

12. $\dfrac{a}{b} = \dfrac{c}{d}$ make: (a) a (b) c (c) b (d) d the subject.

13. $2y + 3x = 5$
    (a) Make y the subject of the formula.   (b) Find y when $x = 6$.

14. $A = \pi r^2$
    (a) Make r the subject of the formula.
    (b) Find r when $A = 20$, giving your answer to 3 significant figures.

15. $I = \dfrac{b}{c}$
    (a) Find I when $b = 5$ and $c = 3$.
    (b) Rearrange the formula and find c when
        $I = 1.5$ and $b = 20$.

16. $A = \dfrac{1}{2}(a + b) \times h$
    (a) Make h the subject of the formula.
    (b) Find h when $A = 15$, $a = 12$ and $b = 9$.

Sequences

Sequences are patterns of numbers that follow some rule so that, once the rule is known, any member of the sequence may be calculated.

Each member of the sequence is called a **term**, and has its own place in the sequence. For example, think of the sequence of square numbers. The rule would be 'square each member of the set of counting numbers' and the terms of the sequence would be

$$1, \quad 4, \quad 9, \quad 16, \quad 25, \quad 36, \, \ldots$$

As usual, the dots show that the sequence goes on and on.
In this sequence the *first* term is 1, the *second* term is 4, the *third* term is 9 and so on.

Can you see that the *ninth* term would be 81?

It can help to arrange the sequence vertically:

Term Number	Term	Calculation
1	1	1^2
2	4	2^2
3	9	3^2
4	16	4^2
5	25	5^2
6	36	6^2
7	49	7^2
8	64	8^2
9	81	9^2

We now think about a general term. We will call it term number n or the nth term.

Term Number	Term	Calculation
n	n^2	n^2

The formula for this sequence is: nth term $= n^2$. Using this formula you can find any term in the sequence.

Questions about sequences will give you the start of a sequence, probably ask you to find the next term or two, and then ask you to find the formula for the nth term.

The questions will often start by giving you a set of diagrams from which you can find the sequence of numbers.

These questions on sequences can be treated like puzzles. You think of a possible formula, test it, and adjust it until you get it right. The sequences themselves provide the clues you need.

Finding the clues

One way to find a formula for the nth term is to write down the differences between successive terms. (How much do you need to add to or subtract from a term to get the next term in the sequence?)

Take the sequence: 5, 8, 11, 14, 17, 20, 23, ...

Write it in a vertical table:

Term Number	Term	Difference
1	5	
		+3
2	8	
		+3
3	11	
		+3
4	14	
		+3
5	17	
		+3
6	20	
		+3
7	23	

For this sequence the difference between each term is a constant (+3).

This means that the formula for the nth term is based on the three times table, or $3n$.

This is the basis for the formula, but you now need to test one or two terms because you need to add another number to obtain the correct starting point. (The three times table would normally start with 3.)

Using the formula $3n$:

when $n = 1$ (the first term) the term would be $3 \times 1 = 3$, but it is actually 5.

when $n = 2$ (the second term) the term would be $3 \times 2 = 6$, but it is 8.

We have to add 2 each time to make the correct sequence.

The nth term is $3n + 2$.

Check by calculating the 7th term.

The 7th term = $3 \times 7 + 2 = 21 + 2 = 23$, which is correct, so we have found the correct formula.

Sometimes you might have to find a second set of differences before they become constant. This changes the form of the formula.

Look at the sequence: 4, 7, 12, 19, 28, 39, 52, ...

Term Number	Term	First Difference	Second Difference
1	4		
		+3	+2
2	7		+2
		+5	
3	12		+2
		+7	
4	19		+2
		+9	
5	28		+2
		+11	
6	39		+2
		+13	
7	52		

The differences have settled down and become constant in the second column.

This means that the formula is based on n^2.

Testing the first two terms:

$$1^2 = 1, \text{ but the term is 4}$$
$$2^2 = 4, \text{ but the term is 7}$$

We need to add 3 each time to generate the correct sequence, so the formula is:

$$n\text{th term} = n^2 + 3$$

Check the 7th term: $7^2 + 3 = 49 + 3 = 52$, so the formula is correct.

As you get used to doing these questions you will probably not have to write the terms in a vertical table, but it can help at first. The examples show the setting out horizontally to save space.

Example 7

(a) Write down the first three terms and the 100th term when nth term $= 2n^2 - 3$.

(b) Find the nth term for the following sequences:

    (i) $-4, 1, 6, 11, 16, 21, 26, \ldots$

    (ii) $6, 9, 14, 21, 30, \ldots$

    (iii) $10, 9, 8, 7, 6, \ldots$

(c) For each of the sequences in part (b) find the 30th term.

Answer 7

(a) $2 \times 1^2 - 3 = -1$

$2 \times 2^2 - 3 = 5$

$2 \times 3^2 - 3 = 15$

$2 \times 100^2 - 3 = 19997$

(b) (i)

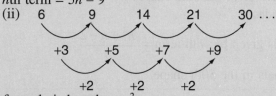

    formula is based on $5n$

    first term: $-4 = 5 \times 1 - 9$

    nth term $= 5n - 9$

    (ii)

	6		9		14		21		30 $\cdots$

    $+3 \quad\quad +5 \quad\quad +7 \quad\quad +9$

    $+2 \quad\quad +2 \quad\quad +2$

    formula is based on n^2

    first term: $6 = 1^2 + 5$

    nth term $= n^2 + 5$

    (iii) $\quad$ 10 $\quad\quad$ 9 $\quad\quad$ 8 $\quad\quad$ 7 $\quad\quad$ 6 $\cdots$

    $-1 \quad\quad -1 \quad\quad -1 \quad\quad -1$

    formula is based on $-1n$ (or just $-n$)

    first term: $10 = -1 + 11$

    nth term $= -n + 11$

(c)   (i) 30th term $= 5 \times 30 - 9 = 141$

    (ii) 30th term $= 30^2 + 5 = 905$

    (iii) 30th term $= -30 + 11 = -19$

Exercise 5.7

1. $4, 5, 6, 7, \ldots$

    For the sequence shown above find:

    (a) the 50th term,   (b) the nth term.

2. $7, 8, 9, 10, \ldots$

    Find: (a) the 50th term,   (b) the nth term.

3. −6, −3, 0, 3, 6, 9, …
   Find: (a) the next two terms,   (b) the *n*th term,   (c) the 103rd term.

4. 16, 25, 36, 49, …
   Find: (a) the next two terms,   (b) the *n*th term,   (c) the 55th term.

5. 3, 6, 11, 18, 27, 38, …
   Find: (a) the next two terms,   (b) the *n*th term,   (c) the 19th term.

6.

shape number	1	2	3	4	5	6
	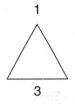					
number of sides	3	4	5	6	7	....
number of diagonals	0	2	5	9	....	....

   (a) Copy the diagram above and fill in the blank spaces.
   (b) How many **sides** would there be in the 50th shape?
   (c) Find the *n*th term for the sides.

   The *n*th term for the diagonals is given by: *n*th term = $\dfrac{n^2 + n - 2}{2}$

   (d) Find the number of diagonals in the 50th shape.

Simultaneous Equations

We have solved equations with one variable (usually *x*), and now need to study equations with two variables, for example, $y = 2x + 3$.

This equation is called a *linear equation* for reasons you will see in a later chapter. There is not just one value of *x* which will satisfy this equation because the value of *y* needs to be taken into account as well.

We can usually find pairs of values for *x* and *y* which satisfy the equation.

Suppose we make $x = 4$?
Then $y = 2x + 3$
becomes $y = 2 \times 4 + 3$
                       $y = 8 + 3$
                       $y = 11$

and we can say that $x = 4$ and $y = 11$ satisfy the equation, or make it true.
These values can be written as a pair of numbers in brackets: (4, 11).
Notice that the *x* value is always written first.

We can find an infinite number of solutions to the equation, all of them pairs of values for x and y.

Another example, suppose $x = 100$?

$$y = 2x + 3$$
$$y = 2 \times 100 + 3$$
$$y = 203$$

So (100, 203) is also a solution to the equation.

We will pick a few solutions to the equation, and you should check if they are correct.

$$y = 2x + 3$$

Some solutions: (4, 11), (1, 5), (−1, 1), (0, 3), (5, 13), (100, 203)

Looking at another equation:

$$y = 3x - 2$$

and some of its solutions, choosing the same values of x to try:

$$(4, 10), (1, 1), (-1, -5), (0, -2), (5, 13), (100, 298)$$

Comparing the two lists you should see that one solution, (5, 13), appears in both lists. You will see in a later chapter that for any pair of linear equations there will be only one solution which satisfies them both.

Solving a pair of linear equations to find the solution which belongs to them both is called solving them *simultaneously*. The two equations are called **simultaneous equations**. Sometimes two equations have no common solution, but all the simultaneous equations you will be given to solve will have a solution that is true for both of them.

We need to find a method for solving simultaneous equations because just trying solutions might take forever before we find the correct pair of values.

The method we will use is called *elimination* because we put the two equations together in a way which *eliminates* one of the variables, either x or y, to give an equation with only one variable, which we can solve.

For example, consider the pair of equations

$$y + 2x = 7$$

and

$$y + \ \ x = 4.$$

If we take the second equation away from the first we will eliminate the y terms.

$$y + 2x = 7$$
$$\underline{y + \ \ x = 4} \qquad \text{subtract}$$
$$0 + \ \ x = 3$$
$$x = 3$$

Now we know x we can substitute $x = 3$ into either of the original equations.

Substituting $x = 3$ in the first equation

$$y + 2 \times 3 = 7$$
$$y + 6 = 7$$
$$y = 1$$

We would get the same result if we substituted $x = 3$ into the second equation. Try it yourself to check this.

The solution for these two equations is (3, 1).

You may have to add or subtract the two equations as the example shows.

Example 8

In each case solve the given pairs of equations simultaneously.

(a) $y - x = 7$
    $2y + x = 5$

(b) $3y + 2x = 6$
    $-3y - x = 3$

Answer 8

(a) $y - x = 7$
    $\underline{2y + x = 5}$ add
    $3y + 0 = 12$
    $y = 12 \div 3$
    $y = 4$

NOTE: In this case adding the two equations will eliminate x because $-x + +x = 0$.

Substituting $y = 4$ into the first equation
    $4 - x = 7$ (-4)
    $-x = 3$ $(\times -1)$
    $x = -3$

Solution for these two equations is $(-3, 4)$.

(b) $3y + 2x = 6$
    $\underline{-3y - x = 3}$ add
    $0 + x = 9$

NOTE: Adding the two equations will eliminate the $3y$ terms.

Substituting $x = 9$ into the first equation
    $3y + 18 = 6$ (-18)
    $3y = -12$
    $y = -4$ $(\div 3)$

Solution is $(9, -4)$

Now do this exercise, but remember that you are trying to eliminate x or y. If for example, you add the equations when you should have subtracted then you will end up with another equation in both x and y, which will get you nowhere. If this happens try again!

Exercise 5.8

Solve these pairs of simultaneous equations.

1. $x + 5y = 6$
    $x + 3y = 2$

2. $3x + y = 8$
    $2x - y = 2$

3. $y + 4x = 1$
    $3y - 4x = 3$

4. $5y + 7x = 8$
    $5y + 4x = 2$

The two equations may not always have a term that can be eliminated straight away, and you may have to multiply one of them by a number first to get two terms that are same. For example, if we were asked to solve:

$$y + 3x = 11$$
$$2y + x = 2$$

we could multiply the first equation all through by 2 first so that both equations have the term $2y$ in them. We could then subtract to eliminate the $2y$ terms, leaving an equation in x only.

The following method makes your working very clear and easy to follow.

$$y + 3x = 11 \xrightarrow{\quad \times 2 \quad} 2y + 6x = 22$$
$$2y + x = 2 \xrightarrow{\hspace{3cm}} 2y + x = 2 \quad \text{subtract}$$
$$\overline{ 0 + 5x = 20}$$
$$x = 4$$

Substitute $x = 4$ into the first equation.

$$y + 3 \times 4 = 11$$
$$y = -1$$

Solution is $(4, -1)$.

It is easier to add the two equations than subtract because the chances of making a mistake with the signs are less. If the two terms you are trying to eliminate have the same sign there is something you can do to help prevent sign errors.

For example, look at these two equations:

$$y + 3x = 8$$
$$3y - 2x = 2$$

Set out the working as above, and instead of subtracting, multiply the whole of the second equation by -1 so that you can add. This is easy because all it means is that you change *every* sign in the second equation.

$$y + 3x = 8 \xrightarrow{\quad \times 3 \quad} 3y + 9x = 24 \xrightarrow{\hspace{2cm}} 3y + 9x = 24$$
$$3y - 2x = 2 \xrightarrow{\hspace{2cm}} 3y - 2x = 2 \xrightarrow{\quad \times -1 \quad} -3y + 2x = -2 \quad \text{add}$$
$$\overline{ 0 + 11x = 22}$$
$$x = 2$$

Substituting in the first equation gives $y = 2$, so the solution is $(2, 2)$.

Points to watch out for:

- Remember that the aim is to eliminate one of the terms from each equation.
- Look for terms in either x or y which have the same coefficients in each equation.
- If you need to multiply one equation by a number to equalise the coefficients then remember to multiply the *whole equation* (including the right hand side).
- If the signs of the terms that you want to eliminate are the *same*, it is easier to multiply one equation by -1 and *add* the equations.
- If you multiply one equation through by -1, just change *every* sign (including the right hand side) in that equation.
- If the signs of the two terms you need to eliminate are *different* you just add the two equations.

Example 9

Solve each pair of equations simultaneously.

(a) $7x - 3y = 11$
    $x + y = -3$

(b) $2y + x = 5$
    $2x + y = 7$

Answer 9

(a) $7x - 3y = 11$ $\xrightarrow{\hspace{2cm}}$ $7x - 3y = 11$

 $x + y = -3$ $\xrightarrow{\times 3}$ $3x + 3y = -9$ $\quad$ add

$$\underline{} \\ 10x = 2$$

$$x = \frac{1}{5}$$

Substitute $x = \dfrac{1}{5}$ in the second equation

$\dfrac{1}{5} + y = -3$

$\phantom{\dfrac{1}{5}}y = -3 - \dfrac{1}{5}$

$\phantom{\dfrac{1}{5}}y = \dfrac{-16}{5}$ $\qquad$ Answer: $\left(\dfrac{1}{5}, \dfrac{-16}{5} \right)$

(b) $2y + x = 5$

 $2x + y = 7$

Rearrange the first equation

$x + 2y = 5$ $\xrightarrow{\times 2}$ $2x + 4y = 10$ $\xrightarrow{\hspace{2cm}}$ $2x + 4y = 10$

$2x + y = 7$ $\xrightarrow{\hspace{1cm}}$ $2x + y = 7$ $\xrightarrow{\times -1}$ $\underline{-2x - y = -7}$ $\quad$ add

$$3y = 3 \\ y = 1$$

Substitute $y = 1$ in the first equation

$x + 2 \times 1 = 5$

$ x = 3$ $\qquad$ Answer: (3, 1)

Exercise 5.9

Solve each pair of equations simultaneously.

1. $-y + x = 15$ $\qquad$ 2. $y = 6x - 10$ $\qquad$ 3. $16x = 3y - 5$ $\qquad$ 4. $5x + 6y = 7$
 $2y + 3x = 5$ $\qquad\quad$ $3y = x + 4$ $\qquad\qquad$ $4x = y - 1$ $\qquad\qquad$ $x + 2y = 1$

5. $2x + y = 5$ $\qquad\;\,$ 6. $2p - q = 4$ $\qquad\;\;$ 7. $3x + y = -10$ $\qquad$ 8. $3x + 2y = 10$
 $x + 2y = 4$ $\qquad\quad$ $p - 3q = 2$ $\qquad\qquad$ $x + 2y = -5$ $\qquad\quad$ $5x - 4y = 2$

9. $5x - 2y = 17$ $\qquad$ 10. $7x - 2y + 5 = 0$
 $3x + 4y = 5$ $\qquad\quad$ $3x - 8y + 45 = 0$

Simultaneous equations may be used to solve problems with unknown quantities. A letter is assigned to each quantity that is unknown. The information available is used to write down equations using these letters. The equations are then solved simultaneously to find these unknown quantities.

NOTE: There must be at least as many independent equations as there are unknown quantities. In other words, if there are two unknowns, such as x and y, there must be at least two different bits of information linking x and y. If you work through the next example you will see what is meant by independent equations.

Example 10

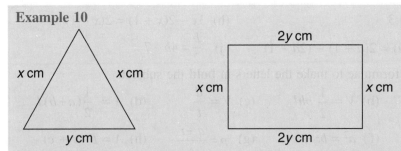

The diagrams show a triangle and a rectangle with sides of unknown lengths.
The perimeter of the triangle is 10 centimetres, and the perimeter of the rectangle is 28 centimetres.
  (a) Write down two independent equations in x and y using the above two separate pieces of information.
  (b) Solve these two equations simultaneously to find x and y.
  (c) Write down the lengths of the sides of the rectangle.

Answer 10

(a) $2x + y = 10$ (from the triangle)
      $2x + 4y = 28$ (from the rectangle)

NOTE: These two equations are derived from different pieces of information, and so are independent. You *could* write the equation for the rectangle as $x + 2y = 14$ (that is, two of the sides added together are equal to half the perimeter), which is a perfectly true statement. It would give the correct answer when used with the equation for the perimeter of the triangle. However, $x + 2y = 14$ and $2x + 4y = 28$ are not independent so you could not obtain any information about x and y by trying to solve these two simultaneously.

(b) $2x + y = 10$
      $\underline{2x + 4y = 28}$ subtract
      $-3y = -18$ $(\div\ ^-3)$
      $y = 6$
    substitute $y = 6$ into the first equation
      $2x + 6 = 10$ (-6)
      $2x = 10 - 6$
      $2x = 4$ $(\div 2)$
      $x = 2$
      $x = 2$ and $y = 6$

(c) The rectangle has sides 2 centimetres and 12 centimetres.

Exercise 5.10

Mixed Exercise

1. Solve these equations:
   (a) $3x = x + 2$ (b) $4x - 5 = 2x - 8$ (c) $6 = 5\,(2x + 3)$
   (d) $7x + 6 - 2x = 9x - 2$ (e) $3(x - 5) = 2(4 - 3x)$ (f) $5(x - 3) = -15x$

(g) $2x + 3 = x - 3$

(h) $3x - 2(x + 1) = 2(x - 1)$

(i) $4a + 3(3 - a) = 2(a + 1) - (2a - 1)$

(j) $\dfrac{b}{2} = 4b - 7$

2. Rearrange these formulae to make the letters in **bold** the subject.

(a) $A = \pi r^2$

(b) $V = \dfrac{1}{2}\,bhl$

(c) $V = \dfrac{d}{t}$

(d) $A = \dfrac{1}{2}(a+b)l$

(e) $D = \dfrac{M}{V}$

(f) $a^2 = b^2 + c^2$

(g) $p = \dfrac{q+r}{s}$

(h) $A = B\,(x + c)$

(i) $\dfrac{x}{a} - b = c$

(j) $3a^2 + x^2 = b^2$

(k) $V = ah$

(l) $a = \dfrac{b+c}{d}$

(m) $A = \dfrac{1}{2}(a+b)l$

3. (a) Copy the pattern below, and draw the next shape.

shape number	1	2	3	4

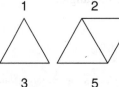 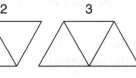 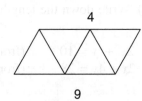

number of lines	3	5	7	9

(b) Find the *n*th term for the number of lines.

(c) Find the number of lines in the 99th shape.

4. Find the *n*th term for the following sequences.

(a)   2   6   10   14 ...

(b) −3   0   3   6 ...

(c)   4   2   0   −2 ...

5. Solve the following pairs of simultaneous equations.

(a) $y = 3x + 5$

$y = -x + 1$

(b) $2y + 3x = 6$

$y + x = 3$

(c) $5y + 3x = 19$

$y + x = 4$

6. Yasmin thinks of two numbers. She says that the sum of the two numbers is 57, and the difference between the two numbers is 15.

(a) Let the numbers be *x* and *y*.
Form two equations in *x* and *y*.

(b) Solve the two equations simultaneously to find the two numbers that Yasmin thought of.

7.

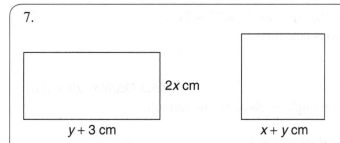

2x cm

y + 3 cm   x + y cm

The perimeter of the rectangle is 36 centimetres, and the perimeter of the square is 48 centimetres.

Use the information given to calculate the dimensions of the rectangle.

Examination Questions

8. (a) A pattern of numbers is shown below.

row

1 ------►   1

2 ------►   2   3   4

3 ------►   5   6   7   8   9

4 ------►   10   11   12   13   14   15   16

5 ------►   17   18   19   20   21   22   23   24   25

6 ------► 26 ---- ---- ---- ---- ---- ---- ---- ---- ----

   (i) Copy the diagram, and complete row 6.
 (ii) The last numbers in each row form a sequence.
      1, 4, 9, 16, 25, ...
     (a) What is the special name given to these numbers?
     (b) Write down the last number in the 10th row.
     (c) Write down an expression for the last number in the nth row.
(iii) The numbers in the middle column of the pattern form a sequence
      1, 3, 7, 13, 21, 31, ...
     (a) Write down the next number in this sequence.
     (b) The expression for the nth number in this sequence is
       $n^2 - n + 1$.
       Work out the 30th number.

                  (0580/03 Oct/Nov 2004 q 9)

9. (a) $2y = 75 - 7x$
   (i) Find y when $x = 7$.
  (ii) Find x when $y = 6$.

(b) Make x the subject of the equation $2y = 75 - 7x$.

(c) Solve these simultaneous equations.

$4x - y = 45$

$7x + 2y = 75$

(0580/03 Oct/Nov 2008 q 6)

10. (a) The perimeter, P, of a triangle is given by the formula

$P = 6x + 3$

(i) Find the value of P when $x = 4$.

(ii) Find the value of x when $P = 39$.

(iii) Rearrange the formula to find x in terms of P.

(b) The perimeter of another triangle is $(9x + 4)$ centimetres. Two sides of this triangle are of length $2x$ centimetres and $(3x + 1)$ centimetres.

(i) Find an expression, in terms of x, for the length of the third side.

(ii) The **perimeter** of this triangle is 49 cm. Find the length of each side.

(0580/03 Oct/Nov 2003 q 6)

11. Make s the subject of the formula $p = st - q$.

(0580/01 May/June 2005 q 12)

12. Solve the equation $5x - 7 = 8$.   (0580/01 Oct/Nov 2005 q 5)

13. The formula for the perimeter, P, of a rectangle with length a and width b is

$P = 2a + 2b$.

Make a the subject of this formula.   (0580/01 Oct/Nov 2005 q 8)

14. The diagram below shows a pattern made from a sequence of dots and lines.

(a) Copy the pattern and draw the next pattern in the sequence.

(b) Copy and complete the table below for the number of dots and lines.

(c) How many lines are there in the pattern with 99 dots?

(d) How many lines are there in the pattern with n dots?

(e) Copy and complete the following statement.

There are 85 lines in the pattern with .... dots.

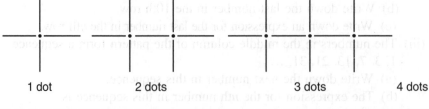

| 1 dot | 2 dots | 3 dots | 4 dots |

Dots	1	2	3	4	5	6
Lines	4	7	10			

(0580/03 Oct/Nov 2005 q 8)

15. (a) Solve the equations:

    (i) $3x - 4 = 14$,

    (ii) $\dfrac{y+1}{5} = 2$,

    (iii) $3(2z - 7) - 2(z - 3) = -9$.

  (b) Donna sent p postcards and q letters to her friends.

    (i) The total number of postcards and letters she sent was 12.

       Write down and equation in p and q.

    (ii) A stamp for a postcard costs 25 cents and a stamp for a letter costs 40 cents. She spent 375 cents on stamps altogether.

       Write down another equation in p and q.

    (iii) Solve these equations to find the values of p and q.

(0580/03 May/June 2008 q 4)

16. (a) Kinetic energy, E, is related to mass, m, and velocity, v, by the formula

$$E = \frac{1}{2}mv^2$$

    (i) Calculate E when $m = 5$ and $v = 12$.

    (ii) Calculate v when $m = 8$ and $E = 225$.

    (iii) Make m the subject of the formula.

  (b) Factorise completely $xy^2 - x^2y$.

  (c) Solve the equation $3(x - 5) + 2(14 - 3x) = 7$.

(0580/03 May/June 2007 q 3a, b and c)

17. In the pattern below each diagram shows a letter **E** formed by joining dots.

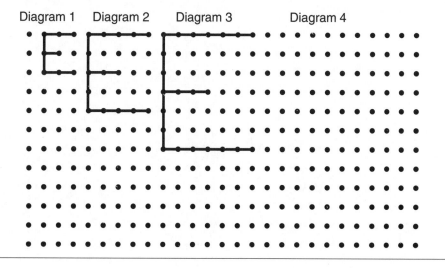

Diagram 1   Diagram 2   Diagram 3   Diagram 4

(a) Draw the next letter E in the pattern.

(b) Complete the table showing the number of dots in each letter E.

Diagram	1	2	3	4	5
Dots	8	15			

(c) How many dots make up the letter E in
   (i) Diagram 10,
   (ii) Diagram n?

(d) The letter E in Diagram n has 113 dots.
   Write down an equation in n and use it to find the value of n.

(0580/03 May/June 2007 q 9)

18. Solve the simultaneous equations:
   $3x - y = 18$,
   $2x + y = 7$.

(0580/01 May/June 2006 q 13)

6

Geometry and Shape

You will already be familiar with much of what is in the earlier parts of this chapter, but we will be looking at shape in more detail. There are a lot of words to learn, which you will need to know for your examination.

For this chapter you will need a protractor, a pair of compasses, a ruler, a pair of scissors and some tracing paper.

The easiest protractor to use is the circular type illustrated here, as simple as possible, with angles up to 360°. With a 360° protractor you can measure angles greater than 180° without having to do a calculation.

When you use the protractor make sure that the zero lies accurately on one of the arms of the angle, and that you count round the scale which starts at zero, round to the other arm, as shown in the illustrations.

The centre of the protractor must coincide with the point of the angle if you are measuring an angle, as shown in the first diagram. It must coincide with the end of the line on which you want to draw the angle as shown in the second diagram. If you have any doubts about your use of the protractor check the examples carefully and make sure that you are within the range given in the answers.

Measuring an angle

Drawing an angle

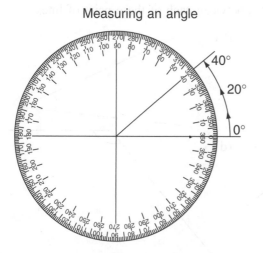

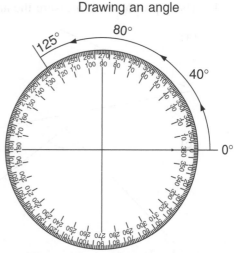

The angle measures 40°

The angle being drawn is 125°

The best pair of compasses have a small wheel in the centre which you rotate to open the compasses to the required length, and a piece of lead rather than a pencil to draw the curve. These will not slip while you are using them, which frequently happens with other types. The illustration shows an example of this type of compass, and a 180° protractor.

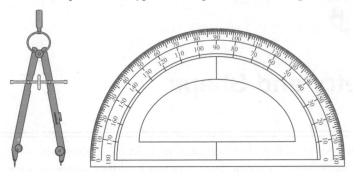

Make sure that your pencil is sharp to minimise experimental error!

There is quite a lot of practical work and accurate drawing in this chapter. This is designed to help you understand the shapes and their properties and thus memorise them. You will find out for yourself many of the facts that you need to know. However, bear in mind that these are not *proofs* of the facts. For this course you do not need to know the proofs, but you do need to be able to recall the facts.

NOTE: In examination questions on shape and geometry, you may be asked to *construct*, to *draw* and *measure*, or to *calculate* an answer. If the question says *calculate* you will get *no marks* for drawing and measuring.

Most of the diagrams in this chapter are **not** drawn to scale, so unless you are asked to measure an angle or side you will need to calculate it.

Essential Skills

1. Use a protractor to measure the angles between each of these pairs of lines.

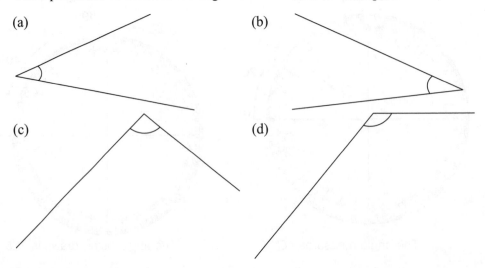

2. Use a ruler to measure these lines.

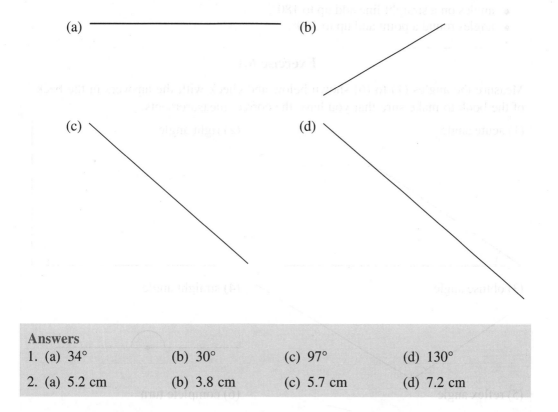

(a)

(b)

(c)

(d)

Answers

1. (a) 34° (b) 30° (c) 97° (d) 130°

2. (a) 5.2 cm (b) 3.8 cm (c) 5.7 cm (d) 7.2 cm

Lines and Angles

Angles are measured in degrees. One degree is a complete turn divided into 360 equal parts, so 360° is a complete turn. If you stand in a room, facing the door and turn round until you are facing the door again you have turned through 360°. This might sound a strange number to use when we are used to the metric system in which we might expect a complete turn to be divided into 100 or 1000 equal parts. However, 360 turns out to be a very good choice because the number 360 has so many different factors. This means that more fractions of complete turns may be written as a whole number of degrees. For example, one sixth of a complete turn is 60°, but one sixth of 100 is 16.6666666....

To start with, we must make sure that we are all talking about the same types of angles, which are illustrated below.

- An angle between 0° and 90° is called an **acute angle,**
- an angle of 90° is a **right angle,**
- an angle between 90° and 180° is an **obtuse angle,**
- an angle of 180° is a **straight angle,**
- an angle between 180° and 360° is a **reflex angle,**
- an angle of 360° is a **complete turn.**

Also note:

- angles on a straight line add up to 180°,
- angles round a point add up to 360°.

Exercise 6.1

Measure the angles (1) to (6) shown below and check with the answers in the back of the book to make sure that you have the correct measurements.

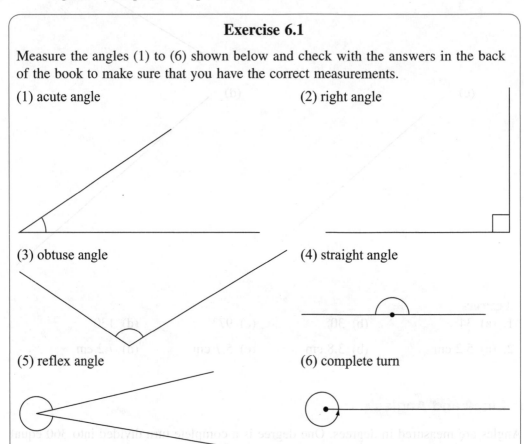

(1) acute angle

(2) right angle

(3) obtuse angle

(4) straight angle

(5) reflex angle

(6) complete turn

Angles and lines are usually given capital letters to name them in a diagram as shown below.

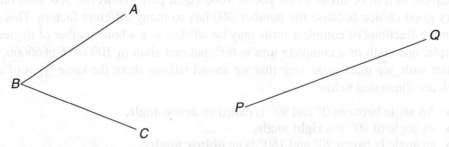

The angle shown is angle *ABC* (or *CBA*). Follow the letters round so that the letter at the point of the angle is in the centre of the three letters. (In this case the centre of *ABC* is *B*, and the sides of the angle are the lines *AB* and *BC*.)

Notation:

- Abbreviations for angle *ABC* are ∠*ABC*, or *AB̂C*.
- If there is **no** doubt about which angle we are referring to it may be called just ∠*B* or *B̂*.
- The line is more simply referred to as the line *PQ* (or *QP*).

The order of the letters does not matter except, as stated above, angles must have the letter at the point of the angle in the centre of the group of 3 letters.

Example 1

In the triangle shown below name the angles marked *x*, *y* and *z*.

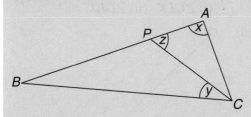

Answer 1

x is either ∠*CAB* or ∠*CAP*, or, because there is no doubt about which angle we are referring to it can be called ∠*A*.

y is ∠*PCB*. It cannot be called ∠*C* because there are 3 angles at *C* (∠*ACP*, ∠*PCB* and ∠*ACB*).

z is ∠*APC*, (to distinguish it from ∠*BPC*).

Exercise 6.2

1. Name the angles and sides shown by small letters in the diagrams below.

(a)

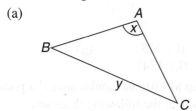

(b)

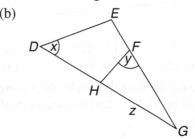

(c)

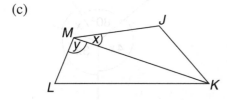

(d)

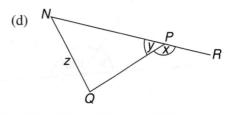

2. Measure the angles and sides named in the diagrams below.

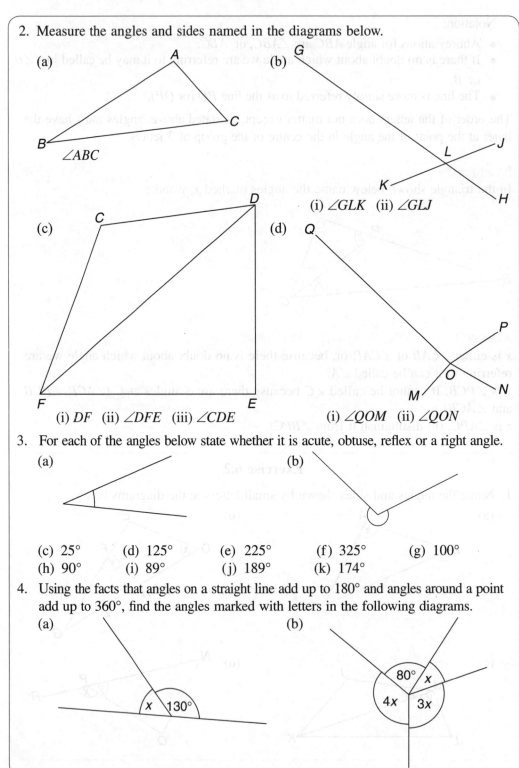

(a)

∠ABC

(b)

(i) ∠GLK (ii) ∠GLJ

(c)

(i) DF (ii) ∠DFE (iii) ∠CDE

(d)

(i) ∠QOM (ii) ∠QON

3. For each of the angles below state whether it is acute, obtuse, reflex or a right angle.

(a)

(b)

(c) 25°   (d) 125°   (e) 225°   (f) 325°   (g) 100°
(h) 90°   (i) 89°   (j) 189°   (k) 174°

4. Using the facts that angles on a straight line add up to 180° and angles around a point add up to 360°, find the angles marked with letters in the following diagrams.

(a)

x 130°

(b)

80° x
4x   3x

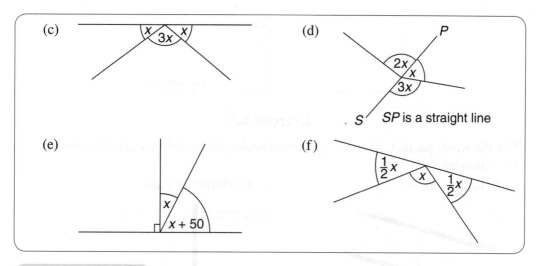

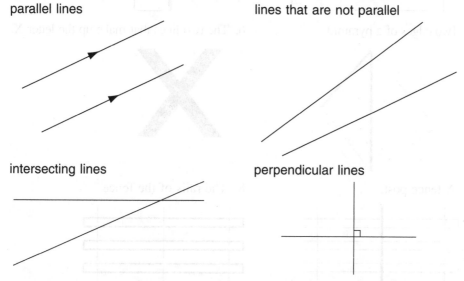

Pairs of Lines

A pair of lines drawn on paper can either be **parallel** (they never meet), or if they are not parallel they must **intersect** at some point. This point of intersection may not be on the paper, but if the lines were **extended** they would eventually intersect somewhere.

If the lines intersect at right angles they are said to be **perpendicular** to each other.

The diagrams below show: parallel lines and the arrows that we use to signify that the lines are parallel; lines that are not parallel but do not intersect on the page; intersecting lines; perpendicular lines with the little square which indicates a right angle.

parallel lines   lines that are not parallel

intersecting lines   perpendicular lines

Lines may also be **horizontal** (parallel to the surface of the earth, or to the horizon if it is not hilly), or **vertical** (at right angles, or perpendicular, to the ground).

It is usual to draw horizontal lines across the page and vertical lines up and down the page.

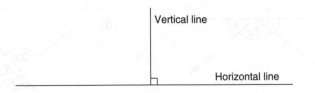

Vertical line

Horizontal line

Exercise 6.3

Use the words parallel, intersecting, perpendicular, horizontal or vertical to describe the following.

1. Railway lines.

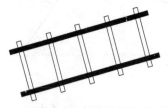

2. An electricity pole.

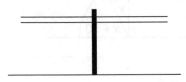

3. The side of a house.

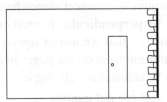

4. The top and side of a door.

5. Two edges of a pyramid.

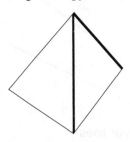

6. The two lines that make up the letter X.

7. A fence post.

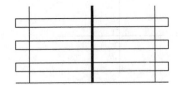

8. The rails of the fence.

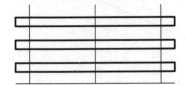

If one line divides another line into two equal parts it is called the **bisector** of the line. In the next diagram the line *AB* bisects the line *PQ* at the point *X*. The two dashes on each side of *X* show that *PX* and *XQ* are equal in length.

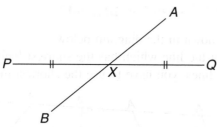

If one line divides another line into two equal parts *and* is at right angles it is called the **perpendicular bisector** of the line. This is shown below where *AB* is the perpendicular bisector of *RS*.

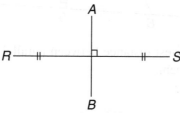

If the lines are both divided into equal parts they are the bisectors of each other as shown below (they could also be the perpendicular bisectors of each other as shown in the second diagram). The single dashes on one of the lines show that both parts are equal in length, but are not equal to the lines with double dashes.

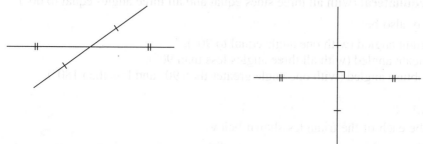

Investigation

The shortest distance from a point to a line

Measure the lengths of all the lines between the point *A* and the line *XY* in the diagrams below.

In which diagram is the line the shortest?

What could you say about the two lines (or the angles) in the diagram you have chosen?

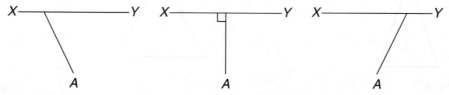

You should find that the shortest distance from a point to a line is the perpendicular from the point to the line.

The shortest distance between two parallel lines

Measure the lengths *AB* shown in the diagram below.
What could you say about the line which has the shortest length? If you are to measure the distance between two lines, you have to use the shortest distance.

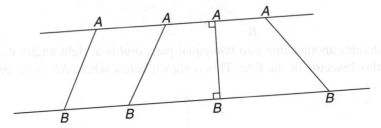

You should find that the shortest distance between parallel lines is along the line which is perpendicular to both of them.

Triangles

Triangles may either be:

- **scalene** (with no sides or angles equal to each other),
- **isosceles** (with two sides and two angles equal),
- **equilateral** (with all three sides equal and all three angles equal to 60°).

They may also be:

- right angled (with one angle equal to 90°),
- acute angled (with all three angles less than 90°),
- obtuse angled (with one angle greater than 90° and less than 180°).

Example 2
Describe each of the triangles shown below.

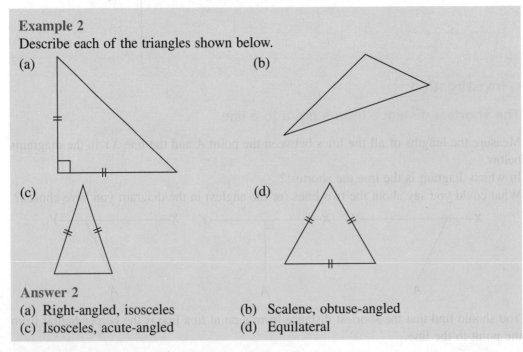

Answer 2
(a) Right-angled, isosceles   (b) Scalene, obtuse-angled
(c) Isosceles, acute-angled   (d) Equilateral

Practical Work

Sum of the angles of a triangle

1. For each of the triangles shown below measure the angles and find the sum of the angles.

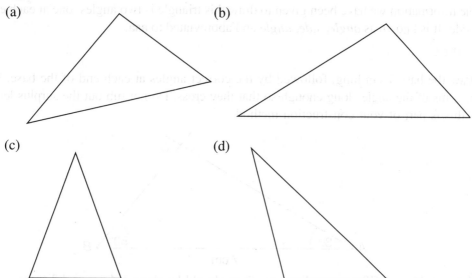

(a)

(b)

(c)

(d)

2. Draw any triangle, and cut it out.
    Tear off each angle as shown in the diagram below.
    Rearrange the torn-off angles to fit on a straight line.

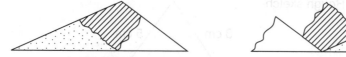

What is the angle sum of a triangle?

In each of these cases you should find that the angles add up to about 180°.
In fact, the angle sum of a triangle is 180°.

You will be asked to draw triangles accurately using either a ruler and protractor, or a ruler and compasses, or maybe all three. You will be asked to leave in your 'construction' lines to show how you have drawn the triangle. This shows your working and you will probably lose marks if you rub them out.

Methods for Constructing Triangles Accurately

Follow the constructions below to make sure that you are able to reproduce the triangles yourself.

1. In triangle *ABC*, *AB* = 7 cm, ∠*BAC* = 30°, ∠*CBA* = 50°.

NOTE: Before drawing any diagram accurately make a rough sketch with the given measurements marked on it. It does not matter which way up you draw the triangles, but in this case it is convenient to make the given side the base of the triangle.

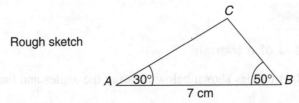

Rough sketch

The information we have been given to draw this triangle is two angles, one at each end of a side. It is known as *angle, side, angle* and abbreviated to **asa**.

Method

Draw the base 7 cm long, followed by the correct angles at each end of the base. Make the arms of the angles long enough so that they cross. Do not rub out the surplus lengths, as this is part of your construction method.

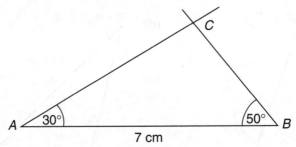

Measure *AC* and *CB* in your diagram. They should be about 5.4 cm and 3.6 cm.

2. In triangle *DEF*, *DE* = 5 cm, *EF* = 4 cm and *DF* = 3 cm.
    This information is known as *side, side, side* and is abbreviated to **sss**.

Rough sketch

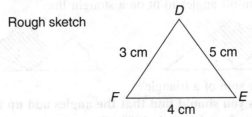

Method

Draw *FE* 4 cm long. Open the compasses to 3 cm and, with the point at *F* draw an arc above the line as shown. Repeat with the other side, with the point of the compasses at *E*, making sure that the arcs cross. Join the sides. Do not rub out your construction lines.

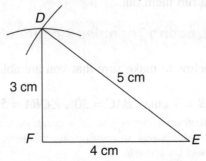

NOTE: This accurate drawing does not look the same as the rough sketch, but this is quite usual as the rough sketch is only drawn to see the relationship of the sides and/or angles, and no attempt is made to draw it in proportion.

Measure your angle *DFE*. It should be about 90°.

3. In triangle *GHJ*, *GH* = 8 cm, *HJ* = 9 cm and ∠*GHJ* = 70°.
    This information is *side, angle, side* and is abbreviated to **sas**. It is important to note that the angle given is *between* the two given sides. This is known as the **included angle**.

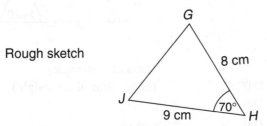

Rough sketch

This method is straightforward, and you should draw your own. Check that it is correct by measuring *GJ*. It should be about 9.8 cm.

Example 3
Try to draw each of the following triangles accurately. If it is impossible, explain why.

(a)

(b)

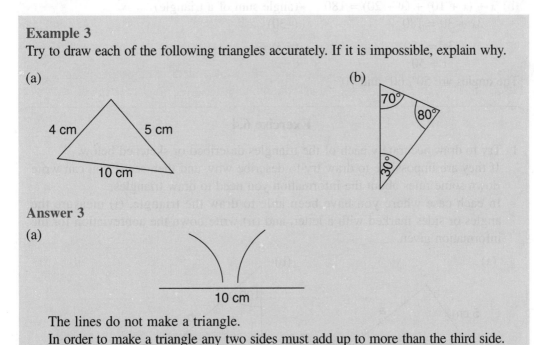

Answer 3
(a)

The lines do not make a triangle.
In order to make a triangle any two sides must add up to more than the third side.

(b) We cannot even start on this one because there are no length measurements given, so we do not know how large to make it.

Example 4

Calculate the angles marked with letters in these triangles.

(a) 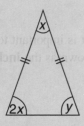   (b)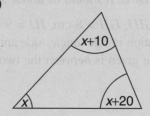

Answer 4

(a) $y = 2x$ (isosceles triangle)

    $x + 2x + y = 180°$ (angle sum of a triangle)

so $x + 2x + 2x = 180$

        $5x = 180$ ($÷5$)

        $x = 36$

The angles are 36°, 72° and 72°

(b) $x + (x + 10) + (x + 20) = 180$ (angle sum of a triangle)

      $3x + 30 = 180$ (-30)

        $3x = 150$ ($÷3$)

        $x = 50$

The angles are 50°, 60° and 70°

Exercise 6.4

1. Try to draw accurately each of the triangles described or sketched below.

   If they are impossible to draw try to describe why, and then see if you can write down some rules about the information you need to draw triangles.

   In each case where you have been able to draw the triangle, (i) measure the angles or sides marked with a letter, and (ii) write down the abbreviation for the information given.

   (a)   (b)

   (a) triangle with 5 cm side, 50° angle, angle a, and 30° angle

   (b) triangle with 100° angle, angle b, 80° angle, and 7 cm side

   NOTE: Calculate the third angle first.

(c)

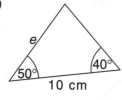

2 cm   7 cm

c

10 cm

(d)

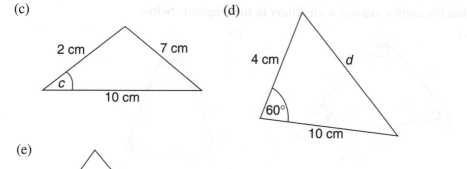

4 cm   d

60°

10 cm

(e)

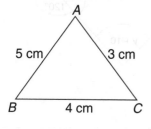

e

50°   40°

10 cm

Before you carry on with this exercise check your answers to question 1 with the answers at the back of the book.

2. Construct accurately each of these triangles. Measure the remaining sides and angles.

   (a) Use compasses and ruler only for construction.

   (b) Use compasses and ruler only for construction.

A

5 cm   3 cm

B   4 cm   C

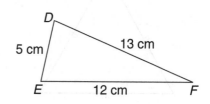

D

5 cm   13 cm

E   12 cm   F

(c)

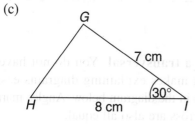

G

7 cm

30°

H   8 cm   I

(d)

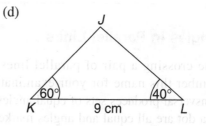

J

60°   40°

K   9 cm   L

(e)

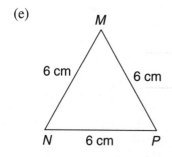

M

6 cm   6 cm

N   6 cm   P

3. Find the angles marked with letters in the diagrams below.

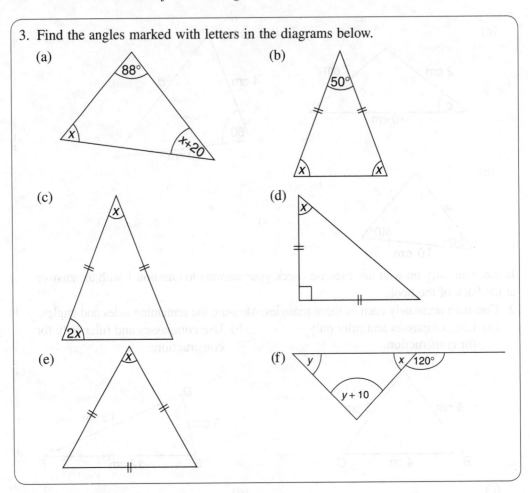

(a)   88°   x   x+20

(b)   50°   x   x

(c)   x   2x

(d)   x

(e)   x

(f)   y   x   120°   y + 10

Angles in Parallel Lines

A line crossing a pair of parallel lines is called a **transversal**. You do not have to remember this name for your examination, but it makes explaining diagrams easier. A transversal produces sets of equal angles as shown in the diagram below. Angles marked with a dot are all equal and angles marked with a cross are also all equal.

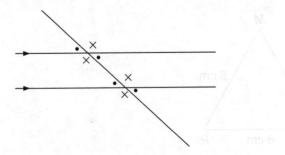

You *do* need to remember the names of pairs of angles in parallel lines.

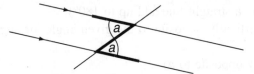

The angles marked with an *a* in the diagram above are called **alternate angles**. You can recognise them because they are in a Z shape, or because they are on *alternate* sides of the transversal.

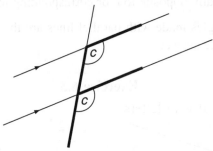

The angles marked with the letter *c* in the diagram above are **corresponding angles**. They can be recognised by being in an F shape, or that they are in corresponding positions between the transversal and the parallel lines.

Finally there are angles which are not associated with parallel lines. These are shown in the diagram below, marked with a *v* and are called **vertically opposite angles**. They are recognised as angles in an X shape, or because they are opposite at a vertex of two angles made between intersecting *straight* lines.

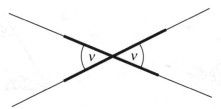

Of course, the two obtuse angles in the intersection are also vertically opposite and equal.

Example 5
Find all the angles marked with letters in the diagram below. Give reasons for your answers.

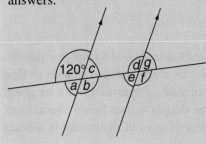

Answer 5

$a = 60°$ (angles on a straight line add up to 180°)

$b = 120°$ (either vertically opposite to the given angle, or angles on a straight line with a)

$c = 60°$ (vertically opposite to a)

$d = 120°$ (b and d are alternate angles, or d is corresponding with the given angle)

$e = 60°$ (c and e are alternate angles, or a and e are corresponding angles)

$f = 120°$ (either vertically opposite to d or corresponding to b)

$g = 60°$ (either vertically opposite to e or corresponding to c)

As you can see most angles made with parallel lines are the same! There is often more than one way to reach the answer.

Exercise 6.5

Find the angles marked with letters.

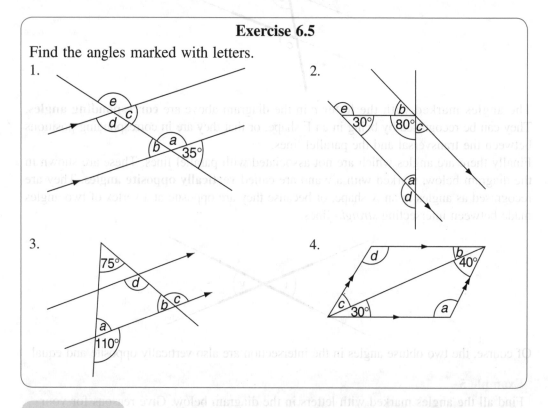

1.

2.

3.

4.

Symmetry

Line symmetry

A shape is said to have line symmetry if you can fold it over along a line so that one side fits exactly on top of the other. The line along which you fold the shape is called an **axis** or **line of symmetry**. In this case either *axis* (plural *axes*) or *line* of symmetry can be used. There may be more than one axis of symmetry. In the shapes shown below the axes of symmetry are marked with dotted lines, and the number of axes of symmetry is written under each diagram.

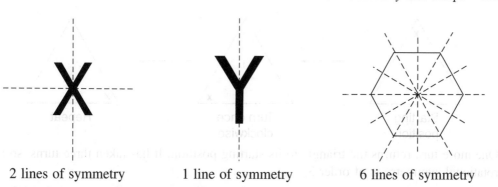

2 lines of symmetry   1 line of symmetry   6 lines of symmetry

Example 6

Copy the diagrams, draw the axes (if any) of symmetry. State the number of axes or lines of symmetry in each case.

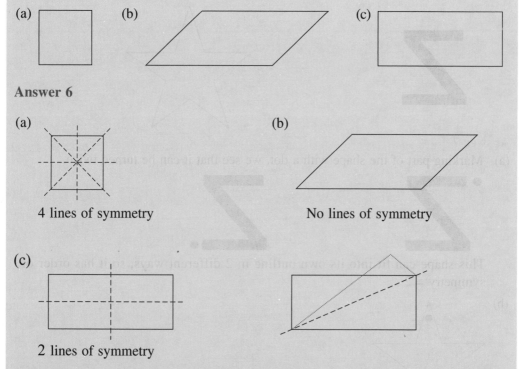

Answer 6

(a)

4 lines of symmetry

(b)

No lines of symmetry

(c)

2 lines of symmetry

NOTE: Check for yourself that there are not more than 2 by drawing and cutting out a rectangle and then folding along a diagonal line as shown here.

Rotational symmetry

A shape is said to have rotational symmetry if it can be picked up and rotated (but not turned over) through an angle less than 360° to fit again into its own outline. The number of ways it can be made to fit is called the **order of rotational symmetry**.

The diagram below shows an equilateral triangle with one corner marked with an **x** so that we can more easily see it turn and count the number of times it fits into its own outline.

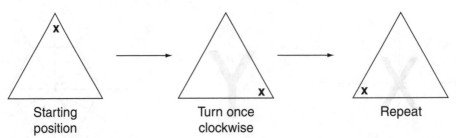

Starting position → Turn once clockwise → Repeat

One more turn returns the triangle to its starting position. It has taken three turns, so the rotational symmetry is of order 3.

Example 7

Find the order of rotational symmetry of each of these shapes.

(a)

(b)

Answer 7

(a)   Marking part of the shape with a dot, we see that it can be turned twice.

This shape can fit into its own outline in 2 different ways, so it has order of symmetry = 2

(b)

If necessary mark one of the points of the star with a dot, to help you see that it can be turned five times to reach its original position. It has order of rotational symmetry = 5.

Notice that the letter Z has no lines of symmetry, and the star has five lines of symmetry.

Exercise 6.6

State the number of lines of symmetry (if any) and the order of rotational symmetry for each of the shapes shown below.

1.

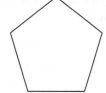

Regular pentagon

2.

3.

4.

5.

6.

7.

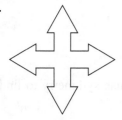

8.

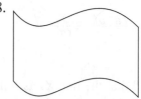

NOTE: Double check any lines of symmetry you might see. If in doubt copy and cut out the shape and fold it.

9.

NOTE: Take care ... the shaded triangles cannot be put on to the unshaded triangles!

Using symmetry

We have seen that a line of symmetry in a shape means that the shape can be folded along that line and one side will fit exactly over the other. This means that pairs of angles on each side are equal and lengths of corresponding sides are equal. In particular, if a line, *AB*, crosses a line of symmetry, *XY*, then the line of symmetry must be the perpendicular bisector of the line *AB*. This is shown in the diagram.

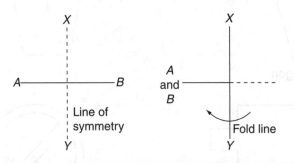

Example 8

The following shape has a line of symmetry, marked *XY* on the diagram.
Find the sides and angles marked by letters.

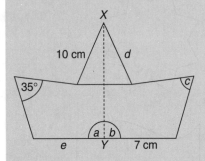

Answer 8

We are told that the line *XY* is a line of symmetry, so we can use symmetry to find the required sides and angles.

$a = b = 90$ $c = 35°$ $d = 10$ cm $e = 7$ cm

Quadrilaterals

A quadrilateral is a shape with 4 straight line sides. The sum of its four angles is 360°. The lines joining opposite angles are called diagonals. You need to know the names and properties of the following special quadrilaterals.

This is presented in the form of a table which you should copy and complete, and then check your answers at the back of the book before continuing.

Exercise 6.7

Copy and complete the table.

Shape	□	□	▱	◇	trapezium	kite
Name	square	rectangle	parallelogram	rhombus	trapezium	kite
Sides	all equal; opposite sides parallel	opposite sides equal in length; opposite sides parallel	opposite sides equal in length; opposite sides parallel	all sides equal; opposite sides parallel	one pair of opposite sides parallel	two pairs of equal sides; no parallel sides
Angles	all 90°	(a)	opposite angles equal	(b)	(c)	one pair of opposite angles equal
Diagonals	equal lengths; bisect each other at right angles	(d)	different lengths; bisect each other	(e)	different lengths; do not bisect	different lengths; one bisects the other at right angles
Lines symmetry	4	(g)	0	(h)	(i)	(j)
Order of rotational symmetry	(k)	2	(l)	(m)	no rotational symmetry	no rotational symmetry

Example 9

Find the angles marked with letters in the following quadrilaterals. In each case state your reasons.

(a)

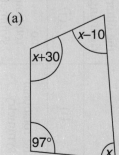

(b)

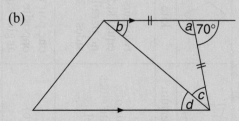

Answer 9

(a) $97 + x + (x - 10) + (x + 30) = 360°$ (the angle sum of a quadrilateral is 360°)

$97 + 3x - 10 + 30 = 360$

$\qquad 117 + 3x = 360$ (−117)

$\qquad\qquad 3x = 243$ (÷ 3)

$\qquad\qquad\quad x = 81$

The angles are: 81°, 71° and 111°

(b) $a = 180 - 70$ (angles on a straight line)

$\quad a = 110°$

$\quad b + c = 180 - 110$ (angle sum of a triangle)

$\quad b + c = 70$

$\quad b = c = 35°$ (isosceles triangle)

$\quad c + d = 70$ (alternate angles)

$\quad d = 70 - c = 35°$

Example 10

Using the diagram below explain how you know that *AB* is parallel to *DC*.

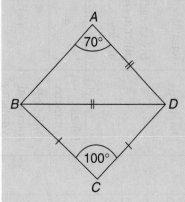

Answer 10

$\angle ABD = 70°$ (isosceles triangle)

$\angle ADB = 180 - 70 - 70 = 40°$ (angle sum of a triangle)

$\angle CBD = \angle DBC = \dfrac{180 - 100}{2} = 40°$ (angle sum of an isosceles triangle)

So $\angle ADB = \angle DBC = 40°$

and *AB* and *CD* are parallel with $\angle ADB$ and $\angle DBC$ alternate angles.

Exercise 6.8

Find the angles marked with letters in the following shapes.

1.

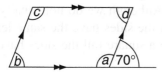

Parallelogram

2.

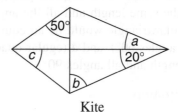

Kite

3.

Trapezium

4.

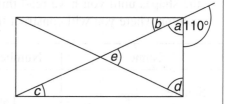

Rectangle

NOTE: Copy the diagram and extend the parallel lines.

5.

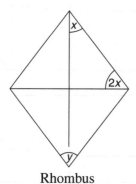

Rhombus

6.

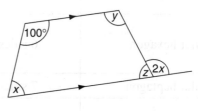

Trapezium

7. Explain how you know that *PQ* is parallel to *RS* in the diagram below.

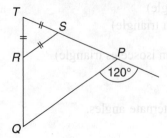

Polygons

Polygons are many-sided shapes. They include triangles and quadrilaterals. You need to know the names of the **regular polygons** shown below. Regular means that all the sides are of the same length and all the angles are the same. Although we do not usually say it a regular triangle would be an equilateral triangle (all the sides have the same length and all angles 60°), and a regular quadrilateral would be a square (all the sides have the same length and all angles 90°).

Investigation

1. Copy the table for use in parts 2 and 3 of the investigation. You do not need to draw the shapes until you have read through part 2, but leave plenty of space in the last column where you will construct the shapes.

Name	Number of sides	Interior angle	Shape
Square (regular quadrilateral)	4 sides	90°	
Regular pentagon	5 sides		
Regular hexagon	6 sides		
Regular heptagon	7 sides		
Regular octagon	8 sides		
Regular nonagon	9 sides		
Regular decagon	10 sides		

2. Drawing regular polygons.
   (a) Using angles at the centre.

      Regular polygons can be constructed from isosceles triangles as shown in the following diagram.

      The angle shown at the centre of the pentagon is obtained by dividing 360° (the complete turn) by 5 (the number of sides in the pentagon), to give 72°.

      The angles are drawn accurately, and the sides of the triangles are drawn all having the same length. An easy way to do this is to use your compasses, opening them to the length that you need to draw the pentagon. 2 centimetres is a convenient size. With the compass point on the centre, mark each line as shown.

      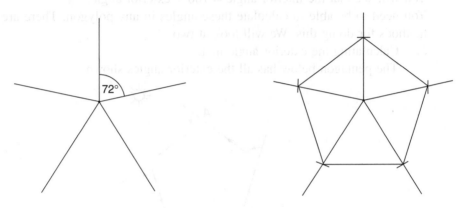

   (b) The hexagon is the only polygon that is made up from equilateral triangles. It can be drawn by the above method or by drawing a circle, then stepping round the circumference using the compasses as shown below. If you keep your compasses at exactly the same length as the radius of the circle you should be able to fit in 6 steps round the circumference. Join these to form the hexagon.

   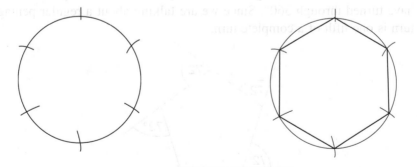

   Complete the right hand column of the table by drawing each of the polygons as accurately as possible.

3. Calculating the interior and exterior angles in regular polygons.
    The following diagram of part of a polygon shows an interior and an exterior angle.

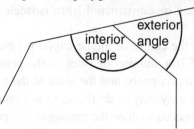

You will see that the interior angle = 180°– exterior angle.

You need to be able to calculate these angles in any polygon. There are several methods for doing this. We will look at two.

(a) Calculating the exterior angle first.

The pentagon below has all the exterior angles shown.

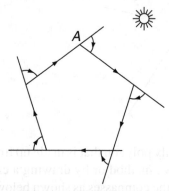

Imagine that you are walking round the outside of the pentagon, each time you come to a corner you turn through the angle shown. You start facing the sun at *A* and continue round, turning 5 times until you face the sun again, back at *A*. You have turned through 360°. Since we are talking about a regular pentagon each turn is one fifth of a complete turn.

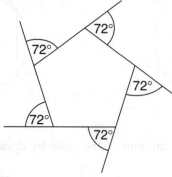

Each exterior angle = $\dfrac{360}{5}$ = 72°.

Each interior angle = 180 – 72 = 108° (angles on a straight line).

(b) Finding the total interior angle first.

The pentagon below has been divided into 3 triangles, with all of them drawn from one point on the pentagon. Each of these triangles has a total angle sum of 180°, so the total interior angle of the pentagon is 3 × 180 = 540°

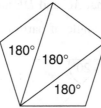

Each individual interior angle of the regular pentagon is thus $\dfrac{540}{5}$ = 108°.

(c) Using a formula.

If you prefer to learn formulae you could use the following, where *n* is the number of sides in the polygon.

Total interior angle of *any* polygon = 180 × (*n* − 2)

Interior angle of a *regular* polygon = $\dfrac{180 \times (n - 2)}{n}$

There are other methods. Look at your constructions of regular polygons and see if they suggest another method.

Complete the final column of the table by calculating the interior angles for each polygon. What do you notice about the size of the interior angle of a regular polygon as the number of sides increases?

Exercise 6.9

1. Calculate the total interior angle for a 15-sided polygon.
2. Calculate the exterior angle in a regular 12-sided polygon.
3. Calculate the angle *a* in this regular hexagon.

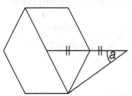

4. *ABCDE* is a regular pentagon, with centre *O*, and side *DE* extended to *F*.

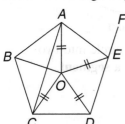

NOTE: **This is a long question. Make it easier by drawing a larger version of the diagram and marking in each angle as you** *calculate* **it.**

Calculate:

(a) ∠*AEF*   (b) ∠*AED*   (c) ∠*AEO*   (d) ∠*OAE*

(e) ∠*AOC*   (f) ∠*CAO*   (g) ∠*CAE*

Hence explain how you know that *AC* and *DE* are parallel.

5. The diagram shows the exterior angle of part of a regular polygon with *n* sides. Calculate *n*.

6. The total interior angles of a regular polygon add up to 2520°. Calculate:

(a) the number of sides,

(b) the interior angle of the polygon.

Circles

Once again there are words connected with circles that you need to know. The following diagrams show these names, many of which you will probably know already.

Circumference

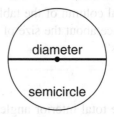

The diameter divides the circle into two halves. Each half is a **semicircle**.

Radius

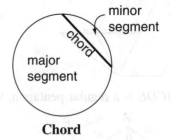

Chord

A chord divides a circle into two parts. Each part is a **segment**.

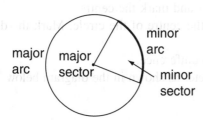

Arc and **Sector**

Two radii (plural of radius) divide
a circle into two parts. Each part is
a **sector**. Each part of the
circumference is an **arc**.

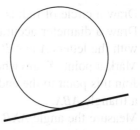

Tangent

A tangent is a line outside
the circle which *just*
touches the circle, making a
single point of contact.

Practical Work

Try these two constructions, making your drawings as accurate as possible (sharp pencil!)

1. Draw a circle with your compasses, radius 10 cm. Mark the centre accurately.
    Draw a tangent at any point, making sure it just touches the circle, as in the
    diagram above showing a tangent.
    From the point of contact of the tangent and the circle draw a radius to the centre
    of the circle.
    Measure the angle between the radius and the tangent.
    If you have been accurate you should find that the **angle between a tangent and
    the radius at the point of contact is 90°**.

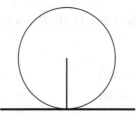

This also makes sense if you consider the symmetry of the diagram below.
The diameter is a line of symmetry of the circle. The tangent is a straight line. Folding
the diagram along the diameter makes one half of the diagram fit onto the other half,
as in the diagram below. In particular, the angles between the diameter and the tangent
must be the same. They must both be half a straight angle, and so must both be 90°.

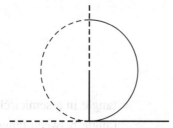

2. Draw a circle of radius 10 centimetres and mark the centre.
    Draw a diameter accurately through the centre of the circle. Mark the diameter
    with the letters *A* and *B*.
    Mark a point, *X*, anywhere on the circumference.
    Join this point to the ends of the diameter as shown in the diagram below, making
    a triangle *ABX*.
    Measure the angle *AXB*.
    Repeat with other positions of *X* on the same circle, and with circles with different
    radii.
    If you have been accurate you should find that the angle *AXB* is always 90°.
    This angle is referred to as the angle in a semicircle because the diameter divides
    the circle into two semicircles.
    You have shown that **the angle in a semicircle is a right angle**.

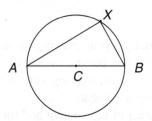

NOTE: There is one other point to notice about circles. In many of the questions about circles you will
find isosceles triangles, because all the radii are equal length, and a triangle made with two radii and
a chord will be an isosceles triangle. *Look out for isosceles triangles in questions on circles!*

Example 11

The diagram shows a circle with centre *O*. *AOB* is a diameter, *DBT* is a tangent, and
$\angle COB = 40°$.

Find, giving reasons,

(a) $\angle ACB$ (b) $\angle ABD$ (c) $\angle OCB$

(d) $\angle OCA$ (e) $\angle CAO$ (f) $\angle ODB$

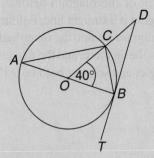

Answer 11

(a) $\angle ACB = 90°$ (angle in a semicircle)

(b) $\angle ABD = 90°$ (angle between tangent and radius)

(c) $\angle OCB = \dfrac{180 - 40}{2} = 70°$ (isosceles triangle)

(d) $\angle ACO = 90 - 70 = 20°$ ($\angle ACB$ is a right angle)

(e) $\angle CAO = 20°$ (isosceles triangle)

(f) $\angle ODB = 180 - 90 - 40 = 50°$ (triangle *ODB* is right-angled)

Exercise 6.10

Calculate the angles marked with letters in these diagrams. *O* is the centre of each circle.

1.
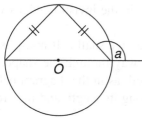

2.
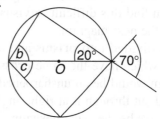

3.
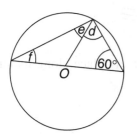

4.
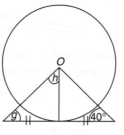

5.
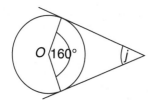

6.
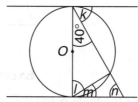

Solid Shapes

There are a few basic solid shapes that you need to know.

Solid figures are shapes in three dimensions, and are not easy to draw on two dimensional paper. They have **faces, edges** and **vertices** (the plural of **vertex**). These are shown in the diagram of a cube.

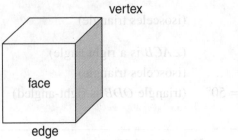

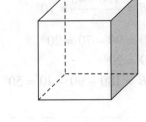

vertex

face

edge

An edge is the line where two faces meet, and a vertex is the point where more than two faces meet. If you look at the first diagram of the cube you should be able to see that it has 6 faces (only three are visible), 12 edges (only 9 are visible), and 8 vertices (only 7 are visible). If you find this difficult to visualise try drawing in the invisible edges with dotted lines as in the next diagram.

The next table is divided into **prisms** and other solids which we shall call non-prisms. Prisms are solids which have the same **cross-section** all the way through the shape. An example of a prism would be an unsharpened hexagonal pencil, as in the diagram below, because you could cut through it at right angles anywhere along its length and the cross-section would always be the same hexagon.

A cylinder is also a prism because you can cut through it at any place along its length and its cross-section will always be the same circle.

Exercise 6.11

Copy and complete the table below.

Prisms					
Name	Shape	Example	Number of faces	Number of edges	Number of vertices
Cube		dice	6	12	8

Prisms					
Name	Shape	Example	Number of faces	Number of edges	Number of vertices
Cuboid		box of matches	6	(a)	(b)
Cylinder			3	2	0
Triangular prism			(c)	(d)	(e)
Non-prisms					
Sphere		rubber ball	1	0	0
Tetrahedron (triangular based pyramid)			(f)	(g)	(h)
Square (or square-based) pyramid		the pyramids of Egypt	(i)	(j)	(k)
Cone			2	(l)	(m)

Nets

As we mentioned earlier, solids are difficult to draw on paper, so it is useful to be able to draw the **net** of a solid. This is a two-dimensional shape that can be cut and folded to make the three-dimensional shape. Two examples are shown below.

A cube

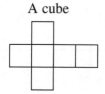

A square pyramid

Some nets are not as obvious as these. Copy these diagrams onto paper or card, cut them out and see if they can be folded into any of the shapes given in the table.

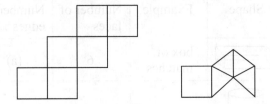

Not all shapes with six squares will make a cube. Try this one.

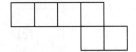

Example 12

Draw accurately a net for a tetrahedron (triangular-based pyramid) with all edges = 2 cm.

Answer 12

In the diagram *AB* was drawn first.

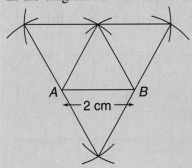

A ← 2 cm → B

Exercise 6.12

1. Draw accurately a net for the cuboid shown below.

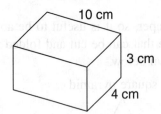

10 cm

3 cm

4 cm

2. What solid would this net make?

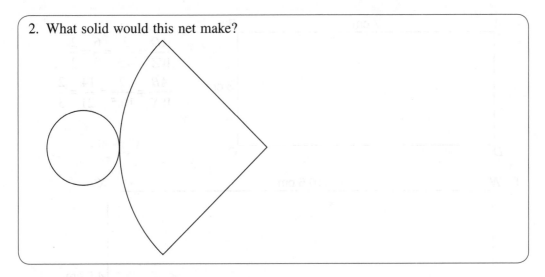

Similar and Congruent Shapes

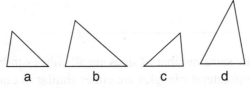

Look at the triangles above. They all have the same angles and all look the same shape, but some are different sizes.

Figures that have exactly the same shape *and* exactly the same size are called **congruent.** The pair of triangles **b** and **d** above are a set of congruent triangles. Triangles **a** and **c** are also another pair of congruent triangles, but they are not congruent with the first set of triangles.

Figures which have exactly the same shape, but different sizes are called **similar.** For example, triangles **c** and **d** are similar.

Similar triangles are easy to recognise because they have equal angles, but it is not so easy to recognise similar quadrilaterals. For example, look at the rectangles below.

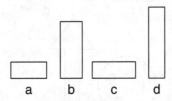

All the rectangles have angles of 90°, but only **a** and **b** are similar.

For quadrilaterals the sides have also to be in the same ratio. In the rectangles **e** and **f** below, rectangle *ABCD* has sides of 3 centimetres and 7 centimetres, and rectangle *WXYZ* has sides of 4.5 centimetres and 10.5 centimetres. The ratios of the corresponding sides are 2:3 as shown below. Therefore **e** and **f** are similar.

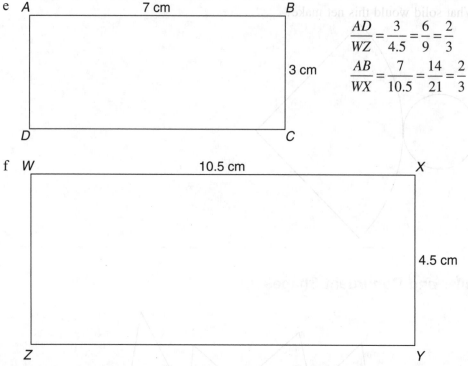

$$\frac{AD}{WZ} = \frac{3}{4.5} = \frac{6}{9} = \frac{2}{3}$$

$$\frac{AB}{WX} = \frac{7}{10.5} = \frac{14}{21} = \frac{2}{3}$$

Regular polygons with the same number of sides are all either similar or congruent to each other. For example, all equilateral triangles are either similar or congruent and all regular pentagons are either similar or congruent.

You will notice that congruent figures are also similar, but similar figures are not necessarily congruent.

Example 13

Match pairs of:

(a)   congruent   (b)   similar figures from the shapes below.

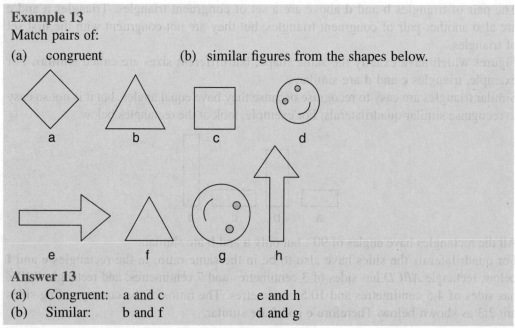

Answer 13

(a)   Congruent: a and c   e and h

(b)   Similar: b and f   d and g

Exercise 6.13

1. Match pairs of similar letters from the selection below.

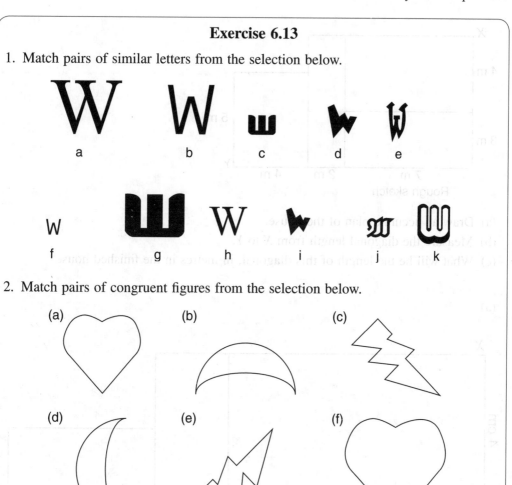

2. Match pairs of congruent figures from the selection below.

Scale Drawings

You have already met scales on maps and on diagrams and solid objects. We will now practise constructing some accurate scale drawings.

Example 14

An architect is drawing the plan of a proposed house. She uses a scale of 1 centimetre to represent 1 metre. She starts with a rough sketch on which she can mark the proposed measurements.

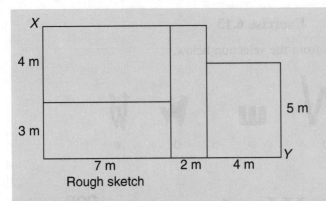

Rough sketch

(a) Draw an accurate plan of the house.
(b) Measure the diagonal length from *X* to *Y*.
(c) What will be the length of this diagonal, in metres in the finished house?

Answer 14
(a)

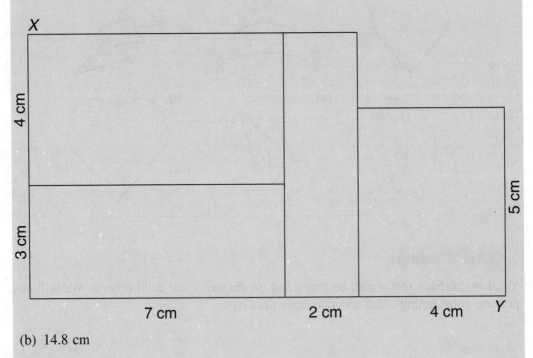

(b) 14.8 cm
(c) 14.8 m

Exercise 6.14

1. A ship sails 20 kilometres due East from port *P*. The captain then alters course to avoid a busy shipping lane and sails 22 kilometres due South. The ship then turns due East again and sails for 36 kilometres, to reach a point *Q*.
   The sketch shows this information.

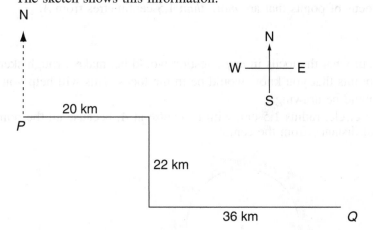

   (a) Make a scale drawing using a scale of 1 centimetre to 5 kilometres.
   (b) Join *P* and *Q*, and measure the length of the line in centimetres.
   (c) How far is the ship from port *P*?

2. The diagram shows the relationship of four towns, *A*, *B*, *C* and *D*.

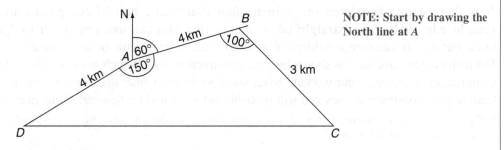

   NOTE: Start by drawing the North line at *A*

   (a) Make a scale drawing, using a suitable scale.
   (b) Use your drawing to find the distance, in kilometres, between *A* and *C*.

Locus

The Latin word *locus* means 'place'. Think of the words 'location' or 'locality'.
The plural of locus is loci. The loci we will be constructing are based on the shapes we have studied in this chapter.

Methods for constructing loci

Construct the following loci to ensure that you know how to do each one. The first locus is a circle.

1. (a) Construct the locus of points that are 1.5 centimetres from a point marked *A*.

   (b) Shade the locus of points that are *more than* 1.5 centimetres from *A*.

Method

If you are not sure about what the locus in any question would be, make a rough sketch showing some of the points that you know would be in the locus. This will help you to see what shape you should be drawing.

The required locus is a circle, radius 1.5 cm, with its centre at *A*, because all the points on a circle are an equal distance from the centre.

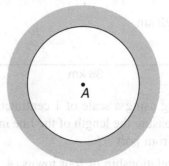

You will be expected to draw loci accurately, and sometimes you will be asked to draw them with a restricted set of instruments such as a pair of compasses and a straight edge only. The 'straight edge' means that you can use your ruler to draw lines, but not to measure anything. You cannot use a protractor or set square.

The method you have used is shown by your construction lines, which *must* be left in. This is equivalent to showing your working when asked to do so in other questions. If you do not leave in your construction lines you will probably get *no marks* for that part of the question.

NOTE: It is no good measuring with a ruler and then adding fake 'construction lines' afterwards. The examiners have seen it all before!

The examples show what is meant by construction lines.

NOTE: In all these constructions make your arcs reasonably large. The smaller they are the less accurate your results will be.

The next two loci are based on the shape of a rhombus.

Before you start drawing the next locus, follow the method to construct the rhombus shown below.

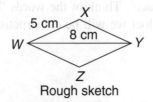

Rough sketch

Draw a line *WY* 8 cm long.

Open your compasses to 5 cm and, with the compass point at *W*, draw arcs above and below the line *WY*. Without altering the compasses move the compass point to *Y* and draw arcs above and below to intersect the first set of arcs at *X* and *Z*.

Join *X* and *Z* to *W* and *Y*.

Join *X* and *Z*.

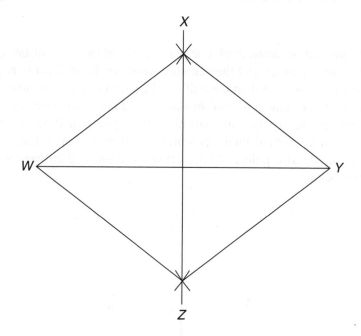

From our study of the symmetry of the rhombus earlier in the chapter we know that the diagonals of a rhombus bisect each other at right angles, so *XZ* is the perpendicular bisector of *WY*.

Look at the rhombus *PQRS* shown below. *Any* point on the diameter *QS* is the same distance from *P* as it is from *R*. Copy the figure, mark a few points on *QS* extended in either direction and measure the distances to check this fact.

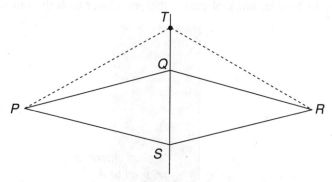

The dotted lines in the diagram show one example. Here *PT* = *TR*.

This is the fact that is used in the construction of the next locus.

2. (a) Construct the locus of points that are equidistant (equal distance) from two points *A* and *B*. Use a straight edge and compasses only.
    (b) Shade the region which is closer to *B* than to *A*.
    (c) Join *AB*. What is the name given to the line of the locus you drew in part (a) in relation to the line *AB*?

Method

If you are not sure what to draw, mark one point that is at the same distance from *A* as it is from *B*. Try to find another, and then another, until you know what is required.

Open your compasses to more than the half the distance between the two points. Put the compass point on *A* and draw a pair of arcs above and below the line between *A* and *B*. Keep the compasses exactly the same but place their point on *B*. Draw two more arcs above and below which should intersect with the first pair of arcs. Use your ruler as a straight edge to join the two points of intersection. The finished answer should look like the first diagram below.

(a)

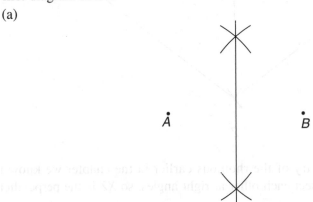

We have drawn a rhombus, but without joining two of the sides. As we have seen above, the symmetry of the rhombus means that we have drawn the perpendicular bisector of *AB*.

(b)   The diagram is repeated below for clarity, but you would not be expected to do this. As before, think of points that are closer to *B* than to *A*.

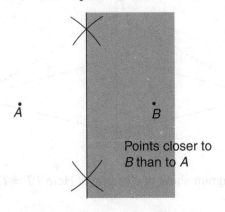

Points closer to *B* than to *A*

(c)   The locus we have drawn is the perpendicular bisector of *AB*.

The next locus is also based on a rhombus but with a slightly different method of construction. Before you go on to the next locus, draw accurately the rhombus shown below.

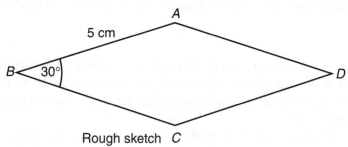

Rough sketch   *C*

Draw a line *BC*, longer than 5 cm and measure and draw the angle *ABC* = 30°.
Open your compasses to 5 cm and, with the compass point at *B*, draw an arc on both arms of the angle *ABC*. These intersections are *A* and *C*.
Keeping your compasses to the same measurement, put the point at *A* and then at *C*, drawing intersecting arcs as in the diagram.
The intersection of these two arcs is *D*. Join *BD*.

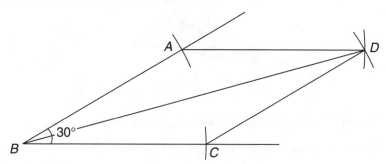

Once again we have constructed a rhombus. In this case we are going to use the fact that the symmetry of the rhombus means that *BD* is the bisector of angle *ABC*.
In particular, any point on *BD* is at the same distance from *BA* or *BA* extended as it is from *BC* or *BC* extended. As before, check a few points on *BD* to confirm this. The diagram below shows one such point with dotted lines to indicate the shortest distance to *BA* extended, and to *BC* extended. Remember that the shortest distance from a point to a line is the perpendicular from the point to the line.

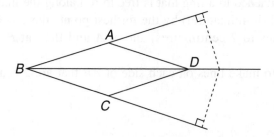

3. (a) Using a straight edge and compasses only, draw the locus of points that are equidistant from the lines *AB* and *BC* shown below.

   (b) Shade the locus of points *within the triangle ABC* which are closer to *AB* than to *BC*.

   (c) What is the name used to describe the line you drew in part (a)?

Method

As before, if you are not sure what the locus would look like always try to find a few points that would be on the locus until you see what has to be drawn.

If you worked through the last example you should be able to see from the diagram how this locus is constructed.

Start with your compass point on *B*, marking an arc on both arms of the angle.

Move your compasses to the points of intersection between each arc and line in turn and mark a pair of arcs as shown.

Join the point *B* to the intersection of the arcs.

We have again drawn part of a rhombus, which as we know from earlier work, has a line of symmetry which, in this case, divides ∠*ABC* into two equal parts.

(a)

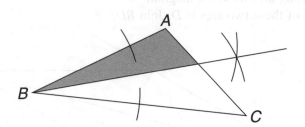

(b) The shading in part (a) shows the required locus.

(c) The locus is the bisector of angle *ABC*.

NOTE: However tempting it is to measure the angle, halve it, and then draw the bisector this way, remember that you will get no marks for this. However, if you then go on to complete another part of the same question you may obtain the marks for the next part.

4. Construct the locus of points that are 2 centimetres from the given line segment *AB*.

Method

Imagine that the line is a length of wire that is stretched tightly between two posts. A dog is on a leash that is attached to a ring that is free to run along the line. The locus of points would be represented (in miniature!) by the furthest points the dog could reach.

Open your compasses to 2 centimetres, and at *A* and then at *B* draw semicircles as shown.

Join the semicircles to make lines on each side of *AB* that are 2 centimetres from *AB*.

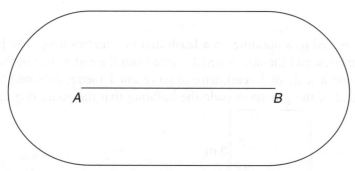

You should now have enough information to answer the examination questions on locus.

Example 15

(a) Draw a rectangle *ABCD* with *AB* = 8 centimetres and *BC* = 6 centimetres.
(b) Construct the angle bisector of angle *ABC*.
(c) Construct the perpendicular bisector of *BC*.
(d) Shade the region where the points are closer to *BC* than to *AB*, and closer to *AB* than to *DC*.

Answer 15

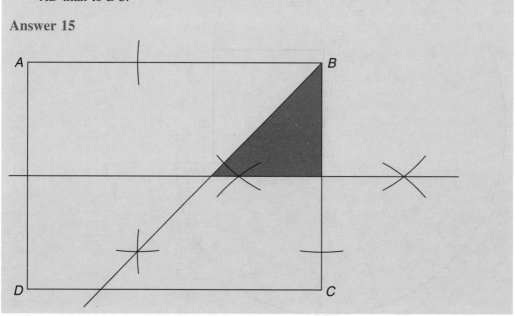

The next example is almost certainly more difficult than you would be asked to do in an examination, but it does usefully show how to construct loci.

Example 16

A guard dog is tied to a building on a leash that is 5 metres long. The building is 3 metres by 3 metres, and the dog is tied 1 metre from the end of one of the short sides as shown. Using a scale of 1 centimetre to represent 1 metre, construct a diagram to show the extent of the ground outside the building that the guard dog can patrol.

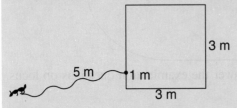

Answer 16

NOTE: Remember that each time the dog comes to a corner its leash is effectively shortened so the arc drawn needs to have the reduced radius.

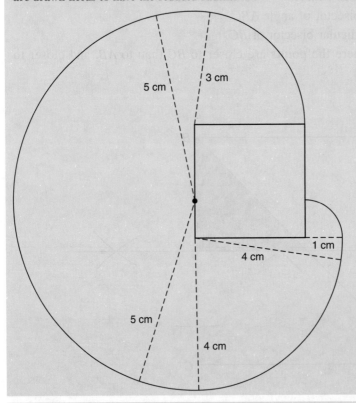

Example 17

Do not use a protractor in this construction.
(a) Construct an equilateral triangle with sides 3 cm.
(b) Construct the locus of points that are 2 cm outside the triangle.
(c) Shade the locus of points that are outside the triangle *and* less than 2 cm from the triangle.

Answer 17

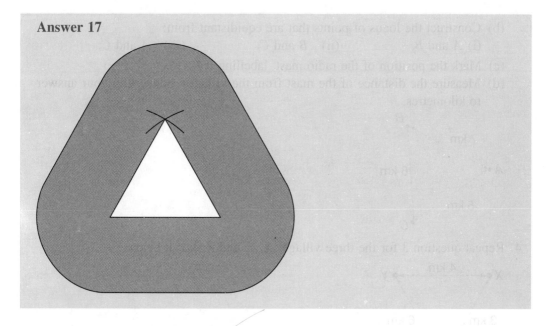

Exercise 6.15

1. Copy the diagram, and construct the bisector of angle *XYZ*, using a pair of compasses and a straight edge only.

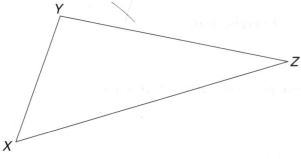

2. Copy the diagram and construct the loci of points that are equidistant from *AB* and *CD*. Use a pair of compasses and a straight edge only.

   NOTE: There are two lines of points that are equidistant from these two intersecting lines.

   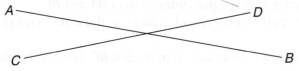

3. *A*, *B* and *C* are three villages as shown in the diagram.
   (a)   Draw an accurate diagram of the location of the three villages using a suitable scale.

   A radio mast is to be placed in a position that is at equal distance from each of the villages.

(b) Construct the locus of points that are equidistant from:
   (i) *A* and *B*,   (ii)   *B* and *C*,   (iii)   *A* and *C*.

(c) Mark the position of the radio mast, labelling it *R*.

(d) Measure the distance of the mast from the villages, converting your answer to kilometres.

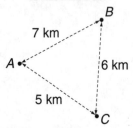

4. Repeat question 3 for the three villages *X*, *Y*, and *Z* shown below.

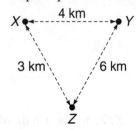

Exercise 6.16

Mixed Exercise

1. The diagram shows a triangular park with gates at *A*, *B*, and *C*.

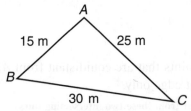

A fountain is to be placed so that it is at equal distance from *BA* and *BC*.

(a) Using a scale of 1 centimetre to represent 2 metres, draw the triangle accurately.

(b) Draw the locus of points that are equidistant from *BA* and *BC*, using a straight edge and compasses only.

The fountain is also equidistant from *A* and *B*.

(c) Draw the locus of points that are equidistant from *A* and *B*, using a straight edge and compasses only.

(d) Mark the position of the fountain and label it *F*.

The water from the fountain sprays over a circle of radius 2 metres.
(e) Using the given scale draw the locus of points that are 2 metres from *F*.
(f) Make appropriate measurements on your drawing to find (in metres):
   (i) How close the spray comes to the gate *A*,
   (ii) The distance of the fountain from gate *C*.

2. Describe the symmetry of each of the diagrams below.

(a)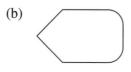

(b)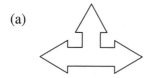

(c)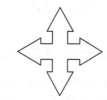

(d)

(e)
X X X X
X X O X
X O X X
X X X X

(f)

3. Calculate the angles and sides marked with letters in the following diagrams. In each case give reasons for your answers.

(a)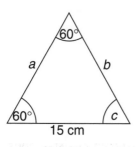

(b)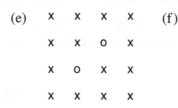

(c)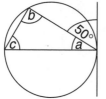

(d)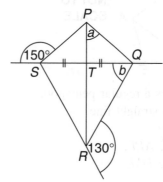

(e)

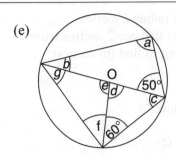

Examination Questions

Many of these examination questions are based on diagrams which will have to be copied into your own exercise book. Try using tracing paper.

4. Reflex   Right   Acute   Obtuse

   Use one of the above terms to describe each of the angles given.

   (a) 100°

   (b) 200°

   (0580/01 Oct/Nov 2004 q 7)

5.

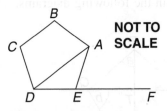

NOT TO SCALE

ABCDE is a regular pentagon.

DEF is a straight line.

Calculate:

   (a) angle *AEF*,

   (b) angle *DAE*.

   (0580/02 Oct/Nov 2005 q 17)

6.

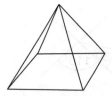

The diagram shows a pyramid with a square base.

All the sloping edges are of the same length.

Sketch a net of the pyramid.

   (0580/01 Oct/Nov 2004 q 13)

7. (a)
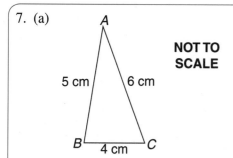

NOT TO
SCALE

5 cm

6 cm

A

B

4 cm

C

    (i) Using a ruler and compasses only, construct the above triangle accurately.

    (ii) Using the triangle you have drawn, measure and write down the size of angle *ACB*.

(b) In the diagram below, two points, *P* and *Q*, are joined by a straight line.

P

Q

    (i) Copy the diagram and draw the locus of all the points that are 4 centimetres from the line *PQ*.

    (ii) On the same diagram, using a straight edge and compasses only, construct the locus of the points that are equidistant from *P* and *Q*. **Show all your construction lines.**

    (iii) Shade the region which contains the points that are closer to *P* than to *Q* **and** are less than 4 centimetres from the line *PQ*.

(0580/03 Oct/Nov 2004 q 4)

8. Copy this diagram. (You can use tracing paper.)

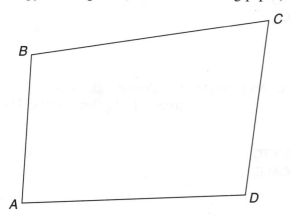

B

C

A

D

In this question show clearly all your construction arcs.

(a) Using a straight edge and compasses only, construct on the diagram,

    (i) the perpendicular bisector of *BD*,

    (ii) the bisector of angle *CDA*.

(b) Shade the region, inside the quadrilateral, which is nearer to *D* than *B* **and** nearer to *DC* than *DA*.

(0580/02 May/June 2004 q 17)

9. Copy this triangle.

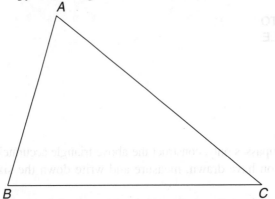

(a) In this part of the question use a straight edge and compasses only.
Leaving in your construction lines,
     (i)   construct the angle bisector of angle *ACB*,
     (ii) construct the perpendicular bisector of *AC*.
(b) Draw the locus of all the points inside the triangle *ABC* which are 6 cm from *C*.
(c) Shade the region inside the triangle which is nearer to *A* than *C*, nearer to *BC* than *AC* and less than 6 cm from *C*.

                       (adapted from 0580/02 Oct/Nov 2005 q 22)

10.

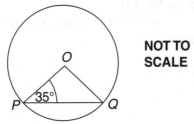

PQ is a chord of a circle, centre *O*. Angle *OPQ* = 35°. Calculate angle *POQ*.

                       (0580/01 May/June 2004 q 11)

11.

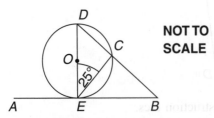

In the diagram, *DE* is a diameter of the circle, centre *O*.
AEB is the tangent at the point *E*. The line *DCB* cuts the circle at *C*.
Angle *DEC* = 25°.
(a) Write down the size of angle *DCE*.
(b) Calculate the size of angle *CDE*.
(c) Calculate the size of angle *DBE*.   (0580/01 May/June 2005 q 23)

12. The net of a solid is drawn below.

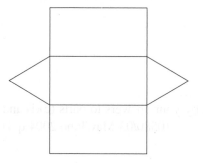

Write down the special name for:
(a) the triangles shown on the net,   (b) the solid.

(0580/01 May/June 2007 q 11)

13. The diagram shows a regular hexagon and a square.

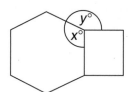

NOT TO SCALE

Calculate the values of *x* and *y*.   (0580/01 May/June 2007 q 15)

14. In this question the diagrams are not to scale.
   (a)   Calculate the value of *s*.   (b)   Calculate the value of *t*.

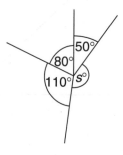

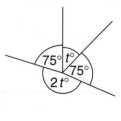

   (c)   (i)

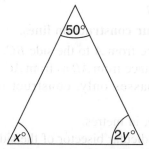

Complete the equation $x + 2y = $ ....

(ii)

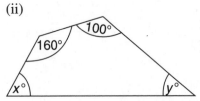

Complete the equation $x + y =$ ....

(iii)   Solve the simultaneous equations given by your answers to parts (c)(i) and
(c)(ii) to find the values of x and y.   (0580/03 May/June 2004 q 4)

15. A farmer owns a triangular field *ABC*.
A scale diagram of this field is drawn below.
1 centimetre represents 10 metres.

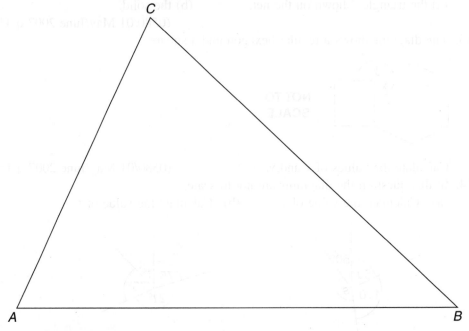

Copy the diagram.
(a) (i) Complete the following statement.
         The side of the field, *AC*, is .... metres long.
     (ii) Measure, in degrees, the angle *ACB*.
     In the following parts, leave in all your construction lines.
(b) The farmer divides the field with a fence from *A* to the side *BC*.
     Each point on the fence is the same distance from *AB* as from *AC*.
     (i) Using a straight edge and compasses only, construct the line
          representing the fence.
     (ii) Write down the length of this fence, in metres.
(c) He puts another fence along the perpendicular bisector of the side *AC*.
     Using a straight edge and compasses only, construct the line representing
     this fence.

(d) He decides to keep goats in the region of the field which is closer to *AC*
     than to *AB* and closer to *A* than to *C*.
     Label the region *G* in the field where he can keep goats.

(0580/03 May/June 2006 q 9)

16.

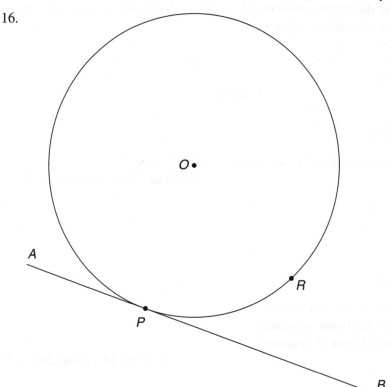

The diagram shows a circular garden with centre *O*. A straight path *AB* touches
the circle at *P*.

Copy the diagram by drawing a circle, radius 4 cm, the tangent at *P* and
∠*POR* = 60°.

(a) (i)   Draw on the diagram the diameter *PQ* and label the point *Q*.
     (ii) Without measuring, write down the size of angle *APQ*.
     (iii) The point *R* is marked on the circumference of the circle. Draw the
           lines *PR* and *QR*.
     (iv) Write down the reason why the angle *PRQ* is 90°.

(b) Showing all your construction lines, use a straight edge and compasses only
     to construct:
     (i)   the perpendicular bisector of *QR*,
     (ii) the bisector of angle *PRQ*.

(c) Shade the region of the garden between *PQ* and *QR* which is closer to *R*
     than to *Q* and closer to *RQ* than to *RP*.

(0580/03 May/June 2007 q 8)

17. In triangle *LMN*, *LM* = 120 mm, *LN* = 70 mm, and *MN* = 86 mm.
    (a) Calculate the perimeter of the triangle *LMN*.
    (b) Construct the triangle *LMN*, leaving in your construction arcs.
                                            (0580/01 May/June 2007 q 19a and b)

18. (a) Copy these diagrams and draw all the lines of symmetry on the shapes.
         (Shape *B* is a regular polygon.)

 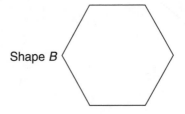

Shape *A*   Shape *B*

    (b) Write down the order of rotational symmetry of shape *A*.
                                            (0580/01 May/June 2006 q 12)

19.

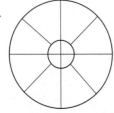

    For the diagram above write down:
    (a) the order of rotational symmetry,
    (b) the number of lines of symmetry.
                                            (0580/02 May/June 2007 q 1)

20.

    (a) Write down the order of rotational symmetry of the diagram.
    (b) Draw the lines of symmetry on a copy of the diagram.
                                            (0580/02 May/June 2005 q 5)

21. (a) Calculate the size of one exterior angle of a regular heptagon
         (seven-sided polygon). Give your answer correct to 1 decimal place.
    (b)

    D _____ *A* _____ *E*
                        $s°$ $t°$
                        $r°$

                                            NOT TO
                                            SCALE

    _____ $130°$ $p°$ _____ $q°$ _____
    F   *B*   *C*   *G*

    In the diagram above, *DAE* and *FBCG* are parallel lines.
    AC = *BC* and angle *FBA* = 130°.

(i) What is the special name given to triangle *ABC*?

(ii) Work out the values of *p*, *q*, *r*, *s* and *t*.

(c) *J*, *K*, and *L* lie on a circle with centre *O*.

KOL is a straight line and angle *JKL* = 65°.

Find the value of *y*.

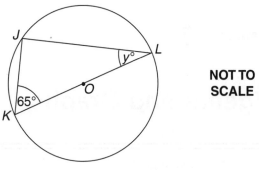

NOT TO SCALE

(0580/03 Oct/Nov 2005 q 9)

22.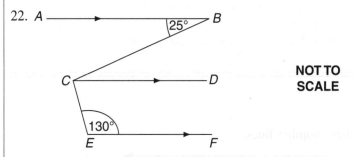

NOT TO SCALE

In the diagram, *AB*, *CD* and *EF* are parallel lines.

Angle *ABC* = 25° and angle *CEF* = 130°.

Calculate angle *BCE*.

(0580/01 May/June 2007 q 10)

23.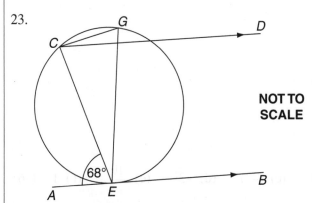

NOT TO SCALE

EG is a diameter of the circle through *E*, *C* and *G*.

The tangent *AEB* is parallel to *CD* and angle *AEC* = 68°

Calculate the size of the following angles and give a reason for each answer.

(a) angle *CEG*,   (b) angle *ECG*,   (c) angle *CGE*,   (d) angle *ECD*.

(0580/03 Oct/Nov 2008 q 4)

Chapter 7

Algebra and Graphs

You must have seen graphs in many places in everyday life. A graph is a good visual method for displaying the relationship between two quantities or measures. For example, a newspaper may have a graph to show how the price of oil has increased over a period of time. After working through this chapter you should have a better understanding of both commonly used graphs and those which display algebraic relationships.

Essential Skills

1. Copy and complete these number lines.

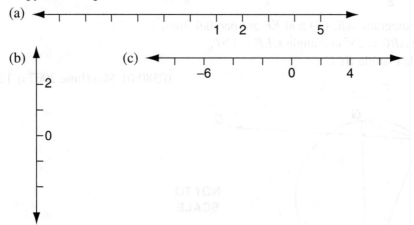

2. Calculate the following:

   (a) $-2 - 5$ (b) $2 \times (-3)$ (c) $(-3)^2$ (d) -3^2 (e) $\dfrac{10}{-5}$ (f) $1 - (-6)$

3. Simplify the following:

   (a) $\dfrac{3}{21}$ (b) $\dfrac{18}{6}$ (c) $\dfrac{2}{-10}$ (d) $\dfrac{-3}{-7}$ (e) $\dfrac{6}{-1}$

4. List the integer values that satisfy the following inequality:
   $-3 \leqslant x \leqslant 3$

Answers

1. (a)

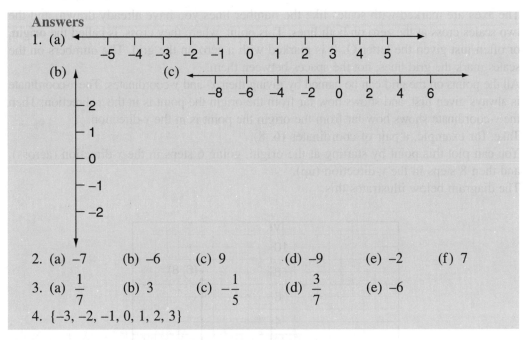

   (b)

   (c)

2. (a) –7   (b) –6   (c) 9   (d) –9   (e) –2   (f) 7

3. (a) $\dfrac{1}{7}$ (b) 3   (c) $-\dfrac{1}{5}$ (d) $\dfrac{3}{7}$ (e) –6

4. $\{-3, -2, -1, 0, 1, 2, 3\}$

Axes, Coordinates, Points and Lines

Most graphs you meet will be drawn on a grid with x- and y-axes as in the diagram.
The positive directions of the x- and y-axes are marked with arrows and the letters x and y.
The x-axis is always across the page, and the y-axis is up and down the page.

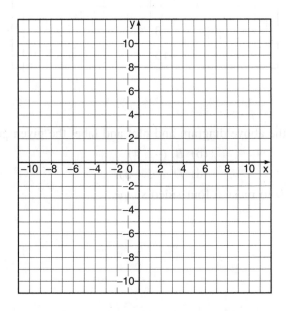

NOTE: You may find it helps you to remember each axis by thinking of the *y*-axis being parallel to the tail of a handwritten *y*. Or perhaps the *x*-axis (x is *a cross*) goes *across* the page.

The axes are marked with scales like the number lines you have already drawn, and the two scales cross at the zero on both lines. This point, where they cross, is called the **origin**, or often just given the letter **O**. It is marked with a zero on the grid. The numbers on the scales mark the grid lines, not the spaces between them.

All the points on the grid can be named by giving their *x*- and *y*-coordinates. The *x*-coordinate is always given first, and shows how far from the origin the point is in the *x*-direction. Then the *y*-coordinate shows how far from the origin the point is in the *y*-direction.

Take, for example, a pair of coordinates (6, 8).

You can plot this point by starting at the origin, going 6 steps in the *x*-direction (across), and then 8 steps in the *y*-direction (up).

The diagram below illustrates this.

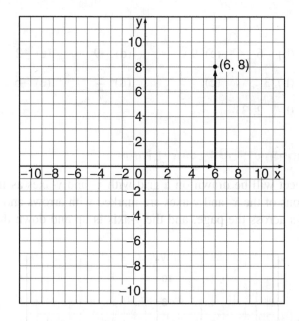

Example 1

Copy the grid in the above diagram and plot and label the given points.

(a) *A*, (5, 3)   (b) *B*, (−2, 4)

(c) *C*, (3, −2)   (d) *D*, (−3, −5)

Join *A*, *B*, *C* and *D* to make a four-sided shape.

Answer 1

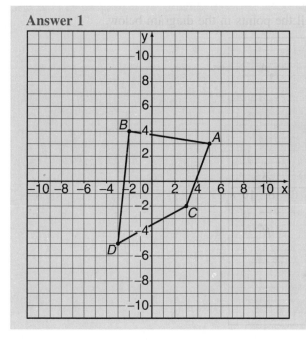

Exercise 7.1

1. (a) Draw a grid with both the axes running from −10 to +10, as in the diagram above.
   (b) Plot and label the following points.
      (i) A, (6, 1)
      (ii) B, (−5, −3)   **NOTE: Go back (left) 5 and down 3.**
      (iii) C, (4, −6)
      (iv) D, (−8, 5)

2. (a) Draw another grid as in question 1.
   (b) Plot and label the following points.
      (i)   A, (3, 2)
      (ii) B, (6, 2)
      (iii) C, (3, 7)
      (iv) D, (−6, 2)
      (v)   E, (−3, 2)
      (vi) F, (−3, 7)
   (c) Join A, B and C to form a triangle.
   (d) Join D, E and F to form another triangle.
   (e) In what way are the two triangles the same?
   (f) In what way are they different?

3. Write down the coordinates of all the points in the diagram below.

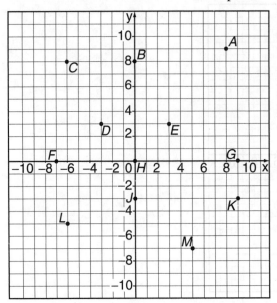

NOTE: To get to *B* from the origin you go along zero and up 8. Hence the *x*-coordinate is 0 and the *y*-coordinate is 8. *B* is the point (0, 8).

We can now see how points are specified, with their *x*- and *y*- coordinates given in that order, and written in brackets. The points can also be given capital letters to distinguish them.

Next we need to see how to specify lines on the grid.

Example 2

Draw three more grids as above.

(a) (i) On the first grid plot and join up all the points with *x*-coordinate 3. That is points from the set:
$\{(3, -10), (3, -9), (3, -8) \ldots (3, 10)\}$

(ii) What can you say about all the points on the line you have drawn?

(b) (i) On your second grid plot and join up all the points from the set:
$\{(-5, -10), (-5, -9), (-5, -8) \ldots (-5, 10)\}$

(ii) What do you think this line would be called?

(c) (i) On your third grid plot and join up all the points which have the *x*- and *y*-coordinates equal to each other.
That is all the points from the set:
$\{(-10, -10), (-9, -9), (-8, -8) \ldots (10, 10)\}$

(ii) What would you call this line?

Answer 2

(a) (i)

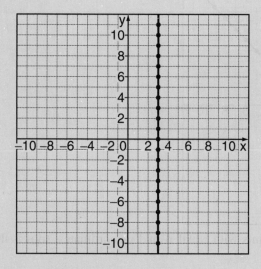

(ii) All the points have their x-coordinate equal to 3

The line is called $x = 3$, and its equation is $x = 3$

(b) (i)

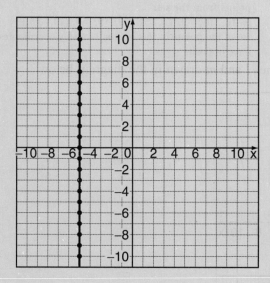

(ii) This is the line $x = -5$

(c) (i)

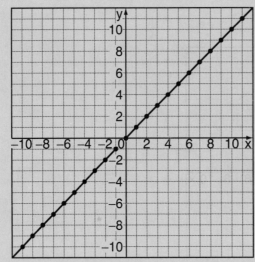

(ii) This line connects all the points with equal *x*- and *y*-coordinates, so it is the
line $y = x$

Exercise 7.2

1. On a grid draw the lines:
   (a) $y = 7$ (b) $x = -3$
   (c) $y = -x$

NOTE: The last line ($y = -x$) would join up all the
points from the set:
$\{(-10, 10), (-9, 9) \ldots (8, -8), (9, -9), (10, -10)\}$.
In all the points in this set the *x*- and *y*-coordinates
have opposite signs.

2. Write down the names for the lines *l* and *m* drawn in the diagram below.

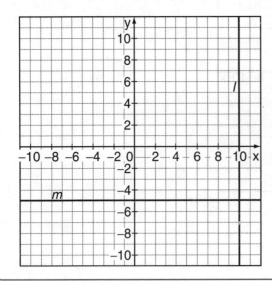

We have been referring to the *names* of lines, but from now on they will be referred to as the *equations* of the lines.

So for example, the line going up and down through $x = -1$ will be called the line with equation $x = -1$, or merely the line $x = -1$.

The lines do not have to go through integer values of x or y. The line $y = 0.75$ would join all the points whose y-coordinates were equal to 0.75.

Everyday Graphs

You will see graphs in newspapers, on television, in advertisements and so on. We will look at line graphs in this chapter, and in the chapters on statistics you will see examples of other kinds of graphs, for example, bar charts and pie charts.

The graph below is an example of the change of value of the US dollar compared with the UK pound over five days in one particular week. The points show the values at the close of business each evening.

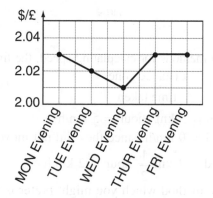

We can see that on Tuesday evening the financial markets closed with $\$2.02 = £1$. This is shown on the vertical axis as 2.02 \$/£, in other words for every 1 pound you get 2.02 dollars.

Example 3

The diagram shows a travel graph for a walking trip Abel and Gerry made. They both set out from Abel's home, and an hour later they were at a point B, 5 kilometres from Abel's home. They rested for a short time and then continued for one more kilometre to D before turning round and going home. The whole trip took 4 hours.

(a) Find their average speed from:
   (i) A to B,   (ii) B to C,   (iii) C to D,   (iv) D to E.
(b) For how long did they rest?
(c) What was their average speed from A to D?
(d) What was their average speed for the whole trip?

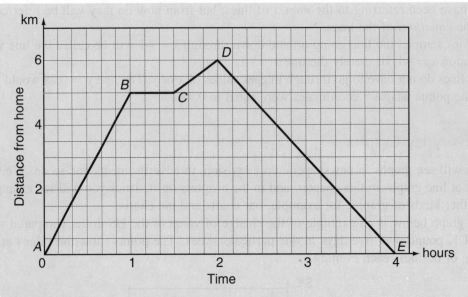

Answer 3

(a) (i) There are two methods for calculating speed, the first is to use the formula:

$$\text{average speed} = \frac{\text{distance gone}}{\text{time taken}} \qquad \text{(see chapter 4)}$$

Average speed = 5 km/1hour = 5 km/h

(ii) Average speed = 0 km/h, since they did not move

(iii) Average speed = 1 km/$\frac{1}{2}$ hour = 2 km/h

An alternative method which you might prefer is to work out how far they would have gone in one hour.

They travelled 1 kilometre in half an hour, so they would have gone 2 kilometres in 1 hour if they had continued at the same pace. So their average speed is 2 km/h.

(iv) From *D* to *E* they travelled six kilometres in 2 hours, so they would have travelled three kilometres in one hour.

Average speed is 3 km/h.

(b) From *B* to *C* the line is horizontal, the distance from home does not change, but the time changes by half an hour, so they rested for half an hour.

(c) From *A* to *D* is 6 kilometres, and they took altogether 2 hours. That is 3 kilometres in 1 hour, so again the average speed is 3 km/h.

(d) The total journey, there and back was 12 kilometres, and took 4 hours, so their average speed was 12 ÷ 4 = 3 km/h.

NOTE: Remember that average speed for a whole journey is:

$$\frac{\text{total distance gone}}{\text{total time taken}}$$

Example 4

(a) Draw a conversion graph to convert kilometres per hour into miles per hour for $0 \leqslant$ kilometres per hour $\leqslant 280$, using the fact that 5 miles = 8 kilometres. Use 2 mm graph paper.

(b) The speed restriction in Britain in built up areas is 30 mph. Use the graph to convert this to km/h.

(c) The fastest Formula One average speed recorded by a certain year was 243 km/h correct to 3 significant figures. Change this to miles per hour.

(d) Cricketers are fast bowlers if they bowl at speeds between 140 and 160 km/h. Find the difference between these two values expressed in miles per hour.

Answer 4

(a) 5 miles = 8 kilometres, so 50 miles = 80 kilometres.

Travelling 50 miles in one hour is the same as travelling 80 kilometres in one hour, so 50 mph = 80 km/h.

The line is drawn through (0, 0) and through (80, 50) as shown

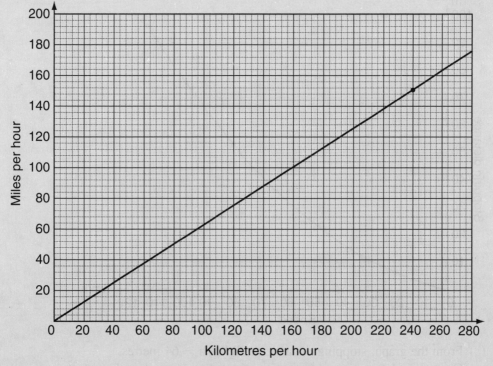

(b) The line goes through (48, 30), so 30mph = 48 km/h

(c) From the graph 243 km/h = 152 mph

(d) From the graph 140 km/h = 88 mph and 160 km/h = 100 mph

The difference between the two speeds is 12 mph

Example 5

Stopping distances for cars with good tyres, good brakes and on a dry road surface are shown in the table.

speed (kilometres/hour)	0	32	48	64	80	96	112
distance (metres)	0	12	23	36	53	73	96

(a) Use the table to plot a graph with distance on the vertical axis, and speed on the horizontal axis. Use 2 mm graph paper.

(b) If the average length of a car is 4 metres, find how many car lengths will be needed to stop if a car is travelling at 90 km/h.

(c) Find the minimum safe distance you should allow between your car, and the one in front if both cars are travelling at 60 km/h.

Answer 5

(a)

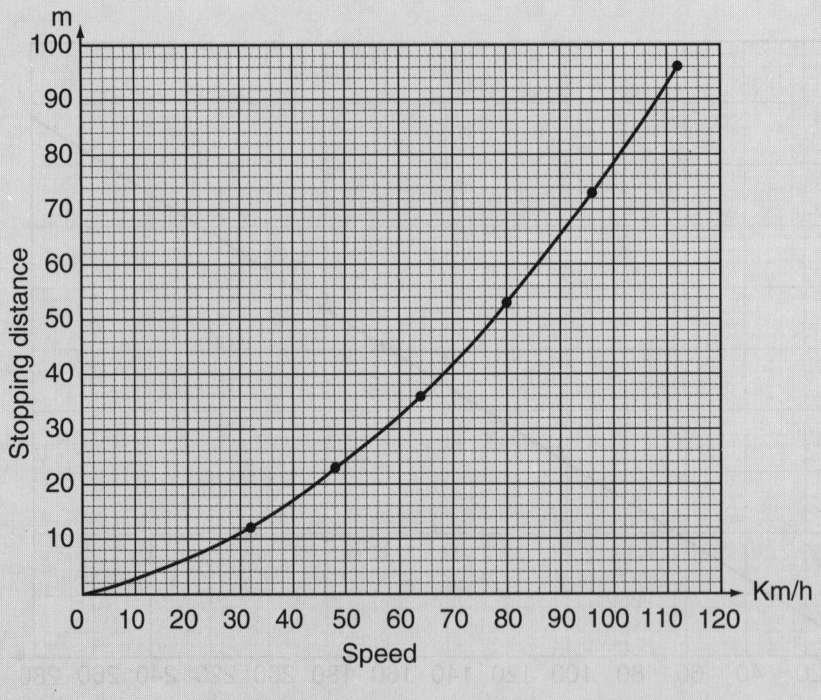

(b) From the graph, stopping distance at 90 km/h = 64 metres.
One car = 4 metres, so 64 metres = 64 ÷ 4 = 16 car lengths

(c) From the graph, stopping distance at 60 km/h = 32 metres, so the minimum safe distance is 32 metres

Exercise 7.3

1. The graph shows the number of daylight hours for a certain town in the northern hemisphere on the first day of every month.

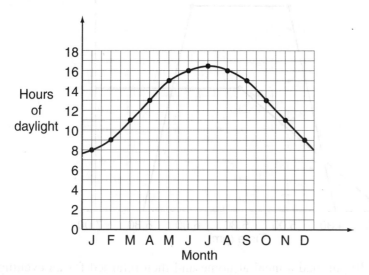

(a) Which month has the most daylight?

(b) For how many months are there more than 10 hours of daylight?

2. On a certain day the exchange rate for UK Pounds (£) to Indian Rupees (Rs.) was 1 pound: 80 rupees.

(a) Draw a graph to show this information, for $0 \leqslant$ Pounds (£) $\leqslant 100$. Use 5 mm squared paper.

(b) Use the graph to find:

(i) the value of £75 in Rupees (Rs.),

(ii) the number of pounds you would get in exchange for Rs. 3000.

3. An electrician charges $10 callout fee and $15 per hour worked. The callout fee is charged regardless of how much work is done, if any, and has to be added to every bill.

(a) Draw horizontal axes with $0 \leqslant$ time in hours $\leqslant 10$ and vertical axes with $0 \leqslant$ cost in dollars $\leqslant 180$.
Use 5 mm squared paper.

(b) Draw a graph showing charges for up to 10 hours work.

(c) From the graph how much would be charged for 5.5 hours?

(d) A bill came to $122.50. How many hours did the electrician work?

4. The graph shows two journeys, made by Anton and Bethany from school.

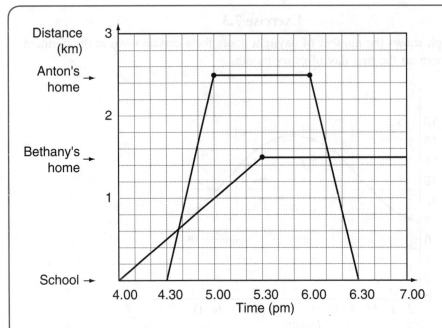

Anton left at 4.30 pm, had a meal at home and then returned for an evening games match. Bethany left school at 4.00 pm.
(a) How long did Anton spend at home?
(b) How long did it take Bethany to reach home?
(c) How far is Bethany's home from the school?
(d) What was Bethany's average speed on her journey home?
(e) Who travelled the fastest?

Graphs with Algebra

Graphs can also be drawn to illustrate algebraic equations which show the relationship between two variables, commonly x and y.

Look, for example, at the equation $y = 2x + 3$.

There is no single solution to this equation but there is a set of solutions that can be shown on an xy graph.

We can find a few members of this solution set by trying different values for x and calculating the corresponding values of y.

Trying $x = 1$,

then
$$y = 2 \times 1 + 3$$
$$y = 5.$$

This shows that $x = 1$, $y = 5$ is one solution to this equation.
This solution can be plotted on the graph as (1, 5).
We could try other values of x, for example, $x = 91.5$,

then
$$y = 2 \times 91.5 + 3 = 186.$$

So the point (91.5, 186) would also represent a solution to the equation and could be plotted on a graph.
Negative values of x also can be tried. For example,

when $\qquad\qquad x = -2, \quad y = 2 \times -2 + 3,$
so $\qquad\qquad\qquad\qquad y = -1,$
giving the point (−2, −1).

This is clearly a rather haphazard way of finding solutions, and it is more usual to decide a set of values for x and calculate the corresponding values of y, and put this solution set in a table.

Drawing straight line graphs

As an example of the method, using the equation $y = 2x + 3$ we will draw the graph for values of x: $-3 \leqslant x \leqslant 3$.

When $\qquad\qquad x = -3, \quad y = 2 \times -3 + 3,$
so $\qquad\qquad\qquad\qquad y = -3.$

The table shows the values of x and y for $x = -3$ to $x = 3$.

x	−3	−2	−1	0	1	2	3
y	−3	−1	1	3	5	7	9

Check these values for yourself to make sure you know how to find them.
Examination questions will often give you some of the values of y, but not all, and you will be asked to find the others.
The table shows that y varies from −3 to 9, so the y-axis must be drawn for values of y: $-3 \leqslant y \leqslant 9$.

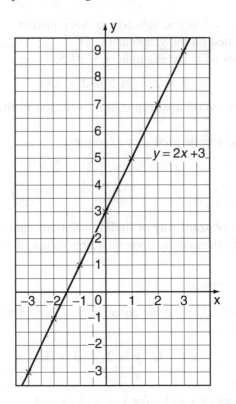

Note the following points.

- Straight lines should always be drawn with a ruler.
- Use the ruled, coordinate lines on the graph paper to find the x and y values.
- The line may be continued to the edge of the grid although in this case we have only been asked to plot points between $x = -3$ and $x = 3$.
- Write the equation for the line beside the line you have drawn on the graph.
- Make sure that whatever scale you choose is evenly spaced on both axes.
  For example, do *not* write:

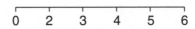

Example 6

Draw the graph of $y = -x + 5$ for values of x: $-3 \leqslant x \leqslant 3$.

Answer 6

x	-3	-2	-1	0	1	2	3
y	8	7	6	5	4	3	2

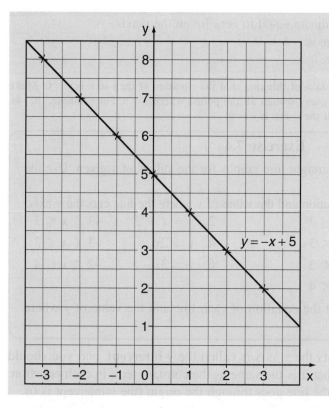

$y = -x + 5$

Example 7

(a) Draw and label each of the following lines on the same graph.
    (i) $x = 3$ (ii) $x = -2$ (iii) $y = -4$ (iv) $y = 1$

(b) (i) Where do all the points whose x-coordinate is 0 lie?
    (ii) What is the equation of the y-axis?

(c) What is another name for the line $y = 0$?

Answer 7

(a)

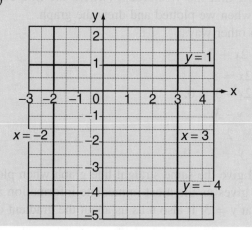

$y = 1$

$x = -2$

$x = 3$

$y = -4$

(b) (i) The points with x-coordinate equal to zero lie on the y-axis.
    (ii) The equation of the y-axis is $x = 0$.
(c) The line $y = 0$ is the x-axis.

NOTE: It is very easy to make the mistake of thinking that the equation of the y-axis is $y = 0$! Make sure that you LEARN that the y-axis goes through all the points where $x = 0$. For example, (0, –4), (0, –3), (0, 2) and so on. Also learn that the x-axis is $y = 0$!

Exercise 7.4

For questions 1 to 7 draw the straight line graphs for the values of x given. Use the same scale on both axes.

Beside each graph write its equation and the value of y where the line cuts the y-axis.

1. $y = x$ $-3 \leqslant x \leqslant 3$
2. $y = x + 2$ $-3 \leqslant x \leqslant 3$
3. $y = x - 1$ $-3 \leqslant x \leqslant 3$
4. $y = 2x$ $-3 \leqslant x \leqslant 3$
5. $y = 2x + 2$ $-3 \leqslant x \leqslant 3$
6. $y = 3x - 3$ $-2 \leqslant x \leqslant 4$
7. $y = 5$ $-2 \leqslant x \leqslant 4$

8. What did you notice about the equation of each line and the value of y where the line cuts the y-axis?

The value of y where the line cuts the y-axis is called the y-**intercept**, and you should have seen in the exercise above that it is the same as the constant term in each equation. Where there is no constant term the line goes through the origin (the y-intercept is 0).

Understanding straight line graphs

An equation will be a straight line when it is plotted on a graph if it has terms in x and/or y, and a constant term only. If the constant term is zero it will not be shown in the equation, and the line will go through the origin as you have seen.

The equations of straight lines do not have terms in y^2, y^3, x^2, xy or $\frac{1}{x}$, or any other terms like these.

Earlier we looked at the graph of $y = 2x + 3$.

When $x = 0$, $y = 3$. Therefore we know that the y-intercept (where the line cuts, or intercepts, the y-axis) is 3, as we found when we plotted and drew the graph.

The equation could have been written in other ways:

$$y = 2x + 3,$$
$$y - 2x = 3,$$
$$y - 2x - 3 = 0,$$
$$x = \frac{y - 3}{2},$$

or $$2x - y + 3 = 0.$$

These are all the same equation and will give the same straight line graph when plotted. The first form is the easiest to use, and gives us the most immediate information about the graph, such as that it cuts the y-axis at $y = 3$. It also tells us about the gradient of the graph as we shall see.

Gradients

For the rest of this section we are going to use equal scales on both axes so that we can count the squares to calculate the gradient.

Remember that gradient is the steepness of a line, and is measured by

$$\text{gradient} = \frac{\text{change in } y}{\text{change in } x}$$

You might find it helps you to remember which way up this fraction should be if you draw a triangle (like a hill) and fit in the words 'up' and 'along'. Hills have gradients and the words fit conveniently as shown in the diagram.

The gradient of a line that slopes *up* from left to right is positive, while one that slopes *down* from left to right is negative. These signs could be added to the hill as shown.

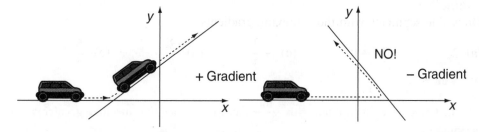

Example 8

Find the gradients of these line segments. Give your answer in its simplest terms. (A line *segment* is just part of a line. In general a line could go on forever!)

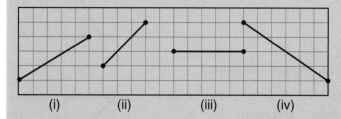

Answer 8

NOTE: Draw a right-angled triangle at any convenient point on the line and count the number of squares up and the number along. Remember to make the answer negative if the line slopes backwards.

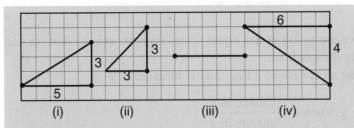

(i) $\dfrac{\text{up}}{\text{along}} = \dfrac{3}{5}$ (ii)   gradient $= \dfrac{3}{3}$ (iii) $\dfrac{\text{up}}{\text{along}} = \dfrac{0}{5}$

gradient $= \dfrac{3}{5}$ gradient $= 1$ gradient $= 0$

(iv)   line slopes backwards, so comparing with $\dfrac{\text{up}}{\text{along}}$.

gradient $= \dfrac{-4}{6}$ or $\dfrac{4}{-6}$, which simplifies to: gradient $= \dfrac{-2}{3}$ or $-\dfrac{2}{3}$

Example 9
Draw line segments with the following gradients:

(a) 2   (b) –1   (c) $-\dfrac{2}{3}$ (d) $\dfrac{2}{3}$ (e) $\dfrac{5}{2}$

Answer 9

NOTE: Compare each fraction with $\dfrac{\text{up}}{\text{along}}$, and remember that negative gradients slope backwards.

(a) $2 = \dfrac{2}{1}$. Therefore, along 1 and up 2.

(or, along 3 and up 6)

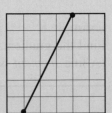

(b) Either $-1 = \dfrac{-1}{1}$. Therefore, along 1 and *down* 1.

Or $-1 = \dfrac{1}{-1}$. Therefore *back* 1 and up 1.

(or *back* 6 and up 6)

You should see that either way of looking at it gives a line sloping backwards

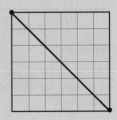

(c) Either $-\dfrac{2}{3} = \dfrac{-2}{3}$. Therefore along 3 and *down* 2.

Or $-\dfrac{2}{3} = \dfrac{2}{-3}$. Therefore *back* 3 and up 2.

Again, either way gives a line sloping backwards

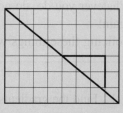

(d) $\dfrac{2}{3}$ This is 3 along and 2 up

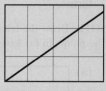

(e) $\dfrac{5}{2}$ This is 2 along and 5 up

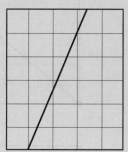

- Looking at these examples, you can see that it makes no difference to the gradient of the line if we count 5 mm squares or 1 cm squares, as long as we use the *same in both directions*.
- Have you noticed that a gradient of 1 slopes at 45° to the horizontal?
- What can you say about a gradient of zero?
- Gradients between 0 and 1 (such as $\dfrac{2}{3}$ in the example above) slope at *less* than 45°.
- Gradients bigger than 1 (such as $\dfrac{5}{2}$) are *steeper* than 45°.
- Gradients which are negative make the same angles with the horizontal, but slope backwards.
- Along 1 and up 2 is the same as along 2 and up 4 and so on. Always give your gradient in its simplest form.

As you have seen there are three ways which might help you to deal with negative gradients.

The first is to notice that negative gradients slope backwards, and deal with them accordingly.

The second is instead of *along* and *up* to count *along* and *down*. So $-\dfrac{2}{3}$ is $\dfrac{-2}{3}$, that is 3 along and 2 down.

The third is instead of *along* and *up* to count *back* and *up*. So $-\dfrac{2}{3}$ is $\dfrac{2}{-3}$, that is 3 back and 2 up.

Whatever you do, do not count *back* and *down* or you will end up with a positive gradient again!

Noticing these things will help you ensure that you get the correct answer.

Exercise 7.5

1. Find the gradients of the following line segments. Give each answer in its simplest terms.

   NOTE: Draw the triangle at the points where the line crosses the intersections of the grid lines. The points in (a) have been marked with two dots to show this.

   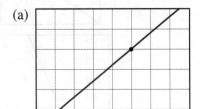

   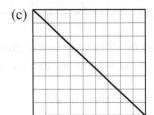

   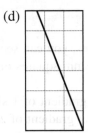

2. Draw line segments with the following gradients:

   (a) $\dfrac{1}{2}$ (b) 3   (c) $-\dfrac{1}{4}$ (d) –8

   (e) 0   (f) 1   (g) $\dfrac{6}{5}$

3. Match these line segments with their possible gradients.

   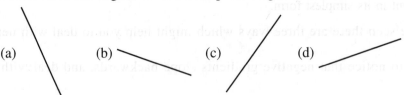

   (i) 2   (ii) –3   (iii) $\dfrac{2}{9}$ (iv) $-\dfrac{1}{2}$

4. Copy and complete the table for the gradients of lines given and the angles each line would make with the *x*-axis.
Two have been done for you.

(a) gradient $= \dfrac{1}{2}$ (b) gradient $= \dfrac{2}{2}$ (c) gradient $= \dfrac{3}{5}$

(d) gradient $= 5$ (e) gradient $= \dfrac{1}{5}$ (f) gradient $= \dfrac{4}{3}$

angle made with *x*-axis	less than 45°	exactly 45°	between 45° and 90°
gradient	$\dfrac{1}{2}, \dfrac{1}{5},$		

Drawing straight line graphs without using a table of values

We now know enough to be able to recognise and sketch straight line graphs very quickly. For example, looking at $y = 3x + 1$, we know that it crosses the *y*-axis at 1, and has a gradient of 3.

3 is the same as $\dfrac{3}{1}$, so comparing with $\dfrac{\text{up}}{\text{along}}$, we get 1 along and 3 up.

Go to the *y*-intercept (0, 1), and draw a line with a gradient of 3 by stepping 1 along and 3 up repeatedly as in the diagrams.

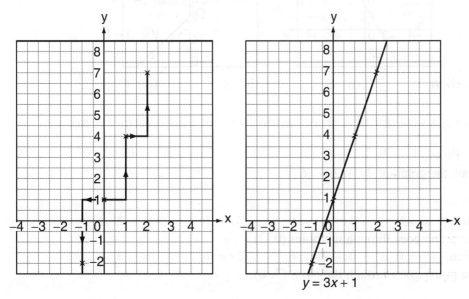

$y = 3x + 1$

Of course, 1 along and 3 up is the same as 1 *back* and 3 *down* as you can see.

Example 10
(a) Sketch the graphs of:
   (i) $y = 2x + 2$ (ii) $y = -x - 2$

(b) Match the following diagrams with their possible equations.

(i)
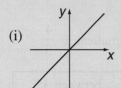

(ii)
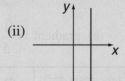

(iii)
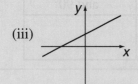

(iv)
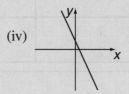

A. $y = \dfrac{1}{2}x + 1$ B. $y = x$ C. $y = -6x + 1$ D. $x = 4$

Answer 10

(a) (i)   (ii)

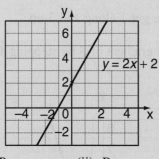

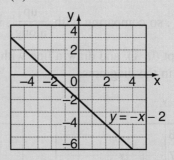

(b) (i) *B*   (ii) *D*   (iii) *A*   (iv) *C*

Drawing a straight line graph when it is not in the form y = mx + c

Example 11
Draw the graph of $3y + 2x = 6$.

Answer 11
When $x = 0$ $3y + 0 = 6$ $3y = 6$ $y = 2$
The graph goes through the point (0, 2).
When $y = 0$ $0 + 2x = 6$ $2x = 6$ $x = 3$
The graph goes through the point (3, 0).

Checking with one more point:

When $x = 2$ $3y + 2 \times 2 = 6$ $3y + 4 = 6$ $3y = 2$ $y = \dfrac{2}{3}$

The checkpoint is $(2, \dfrac{2}{3})$.

NOTE: Any value of x may be chosen for the checkpoint.

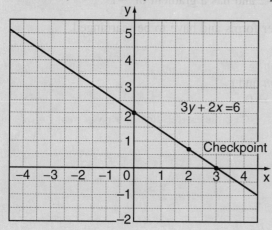

NOTE: It must be emphasised that all the work we have done with gradients has been on graphs with the same scale on both axes. If the scales are different the gradients do not look the same, and you must read the change in y and the change in x using the scales on the axes.

Exercise 7.6

Draw the following straight line graphs by any of the methods we have used.

1. $y = -2x - 3$ 2. $2y - 3x = 12$ 3. $y = \dfrac{1}{4}x + 2$

4. $5y + 2x = 10$ 5. $6y - 3x + 12 = 0$ 6. $y = x + 5$

The general form of a straight line equation

We have been working so far with equations in the form of $y =$ a term in x and a constant term, for example,

$$y = 5x + 3.$$

When the equation is arranged in this way it can be compared with the **general equation** of a straight line, which is,

$$y = mx + c.$$

In this general form m represents the gradient, and c represents the y-intercept.
For example, for the equation,

$$y = 5x + 3,$$
$$m = 5 \text{ and } c = 3.$$

It is important to notice that the y term has to be $1y$ for this to be true.
For example, if you have the equation,

$$3y = 5x - 6,$$

then you need to rearrange it first into the correct form.

$$3y = 5x - 6 \qquad (\div 3)$$
$$y = \frac{5}{3}x - 2$$

Now $m = \frac{5}{3}$ and $c = -2$.

The line $3y = 5x - 6$ cuts the y-axis at -2 and has a gradient of $\frac{5}{3}$.

The equations may also be given in other forms and have to be rearranged.

Example 12

For each equation find the values of m and c.

(a) $2y = 5x - 4$ (b) $3y - 3x - 5 = 0$

(c) $5x = 4y + 1$ NOTE: Be careful! Remember you want $y = \ldots$

Answer 12

(a) $2y = 5x - 4 \qquad (\div 2)$

    $y = \frac{5}{2}x - 2$

    $m = \frac{5}{2} \qquad c = -2$

(b) $3y - 3x - 5 = 0 \qquad (+\, 3x + 5)$

    $3y = 3x + 5 \qquad (\div 3)$

    $y = x + \frac{5}{3}$

    $m = 1 \qquad c = \frac{5}{3}$

(c) $5x = 4y + 1 \qquad$ (turn around)

    $4y + 1 = 5x \qquad (-1)$

    $4y = 5x - 1 \qquad (\div 4)$

    $y = \frac{5}{4}x - \frac{1}{4}$

    $m = \frac{5}{4} \qquad c = -\frac{1}{4}$

Exercise 7.7

For each equation find the value of m and the value of c.

1. $y = \frac{1}{2}x - 5$ 2. $2y + x = 3$ 3. $2x = 4 - 5y$ 4. $x + y + 1 = 0$

5. $5x - 4y = 20$ 6. $y = \dfrac{x + 1}{2}$ 7. $2x + 3y - 4 = 0$ 8. $y = \dfrac{x}{2} + 3$

9. $y = 6$ NOTE: This is the same as $y = 0x + 6$
10. $y = x$ NOTE: This is the same as $y = 1x + 0$
11. $y = -10$ 12. $y = 2x$ 13. $y = -x$ 14. $x + y = 4$

Finding gradients when the scales are not the same on both axes

Example 13
(a) Find the gradient of the line shown on the grid below.

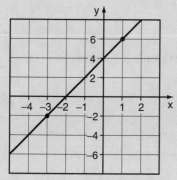

(b) Find the gradient of the line joining the points $(3, -4)$ and $(-2, -1)$.

Answer 13

(a) **NOTE: Choose two points on the graph, for example, (1, 6) and (−3, −2) as shown.
The change in y and the change in x have to be read from the axes. Counting the squares will
not work when the scales are different on both axes.**

For the change in y think "how do you get from -2 to 6?" These are the two
y-coordinates. The answer is you add 8.
So the change in $y = 8$
For the change in x think "how do you get from -3 to 1?" The answer is you
add 4.
So the change in $x = 4$

The gradient of the line is $\dfrac{\text{change in } y}{\text{change in } x} = \dfrac{8}{4} = 2$

**NOTE: In this particular case you could have chosen the points where the line crosses each axis
$(-2, 0)$ and $(0, 4)$ and read 4 units on the y-axis and 2 units on the x-axis, giving a gradient of**
$\dfrac{4}{2} = 2$. **However, it is important that you understand the first method because the line you are
given might not be shown cutting both axes.**

(b) The points are: $(3, -4)$ and $(-2, -1)$.
For the change in y: ask yourself "how do you get from -4 to -1?"
Change in $y = +3$.
For the change in x: "how do you get from 3 to -2?
Change in $x = -5$.

The gradient of the line is $\dfrac{\text{change in } y}{\text{change in } x} = \dfrac{3}{-5} = -\dfrac{3}{5}$

If you prefer you can plot the two points on a grid and find the gradient from the grid, taking account of the scale you have chosen.

Finding the equation of the straight line on a graph

To find the equation of a straight line, find its gradient (m) and the y-intercept (c) and use $y = mx + c$.

Example 14

Find the equation of the line shown on the grid in example 13 above.

Answer 14

Gradient, $m = 2$, and y-intercept, $c = 4$

So $y = mx + c$ becomes $y = 2x + 4$

Exercise 7.8

Find the equation for each of the lines shown in questions 1 to 4.

1.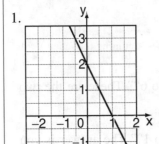

2.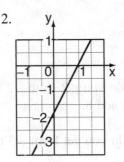

3.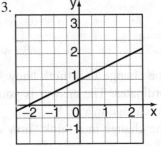

4.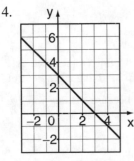

5. Find the gradient of the line joining the point (2, 4) to the point (10, 5).

6. Find the gradient of the line joining the points (1, −3) and (−2, 6).

Parallel lines

Parallel lines have the same gradient, but will cut the y-axis at different places.

For example, $y = 5x + 6$ and $y = 5x - 2$ both have a gradient of 5, but cut the y-axis at 6 and –2 respectively.

You may be asked to find parallel lines when the equations are not arranged conveniently in the $y = mx + c$ form, so you may have to rearrange the equations into this form first.

Example 15

(a) Find a pair of parallel lines from these equations.

    (i) $y - 3x + 5 = 0$ (ii) $y - 2x + 4 = 0$

    (iii) $2y - 3 = 4x$ (iv) $5y = 2x$

(b) Write down the equation of a line which is parallel to $y = -x + 4$.

Answer 15

(a) (i) $y - 3x + 5 = 0$ (ii) $y - 2x + 4 = 0$

        $y = 3x - 5$ $y = 2x - 4$

        $m = 3$ $m = 2$

    (iii) $2y - 3 = 4x$ (iv) $5y = 2x$

        $2y = 4x + 3$ $y = \dfrac{2}{5}x$

        $y = 2x + \dfrac{3}{2}$ $m = \dfrac{2}{5}$

        $m = 2$

  So the lines $y - 2x + 4 = 0$ and $2y - 3 = 4x$ both have the same gradient and are parallel

(b) Any line parallel to the given line will be acceptable.

    For example, $y = -x + 6$

Exercise 7.9

1. Find the pair of parallel lines.

    (a) $3y = x - 9$ (b) $2y - 6x = 9$ (c) $15y - 5x + 7 = 0$

2. Write down the equation of a line parallel to $2y = \dfrac{1}{2}x - 3$.

3. Find the gradients of each of these lines to find a pair that are parallel.

    (a) 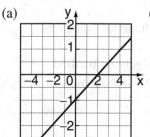   (b) 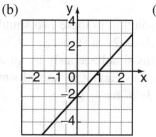   (c)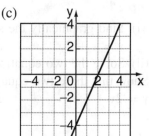

> NOTE: Look carefully at the scales and remember that gradient = $\dfrac{\text{change in } y}{\text{change in } x}$

Using graphs to solve simultaneous equations

We have already solved simultaneous equations using algebra in Chapter 5. We shall now look at a graphical method.

The graphs of straight lines link every single point whose x- and y-coordinates obey the equation for the line. This includes all the rational numbers as well as the integers. No other point on the grid has x- and y-coordinates which obey this equation.

Look at the first diagram, which is the graph of $y = 2x + 1$.

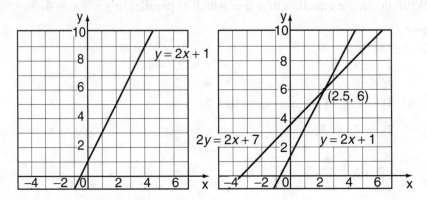

You can see that when, for example, $x = 2.5$, then $y = 6$.

But this point also lies on the line $2y = 2x + 7$.

(Check: when $x = 2.5$, $2y = 5 + 7$, $2y = 12$, $y = 6$.)

If this equation is plotted on the graph you will see that it crosses the $y = 2x + 1$ at the point (2.5, 6) as in the second diagram above.

Simultaneous equations are equations which share a point in common, so that the x- and y-coordinates obey *both* equations. If two simultaneous equations are plotted on a graph then the point where they cross, is a solution to *both* equations. When you are asked to solve two equations simultaneously, you are looking for the point which satisfies both of the equations. Two straight lines will always cross unless they are parallel, in which case they never meet.

Example 16

(a) (i) Find graphically the solution to these two simultaneous equations.
   $y = x + 2$ $3x + 2y = 9$
   (ii) Solve the two equations simultaneously by an algebraic method.
(b) Show that these two equations cannot be solved simultaneously.
   $y = 2x + 1$ $3y = 6x + 1$

Answer 16

(a) (i) $y = x + 2$

$m = 1, c = 2$

$3x + 2y = 9$

when $x = 0, y = 4.5$

when $y = 0, x = 3$

Checkpoint:

when $x = 2, y = 1.5$

Solution: $x = 1, y = 3$

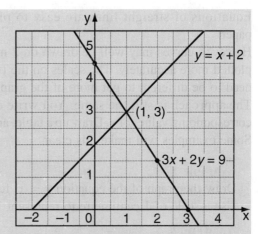

(ii) $y = x + 2 \quad\longrightarrow\quad y - x = 2 \quad\xrightarrow{\times 2}\quad 2y - 2x = 4$

$3x + 2y = 9 \quad\longrightarrow\quad 2y + 3x = 9 \quad\longrightarrow\quad \underline{2y + 3x = 9} \qquad$ subtract

$-5x = -5$

$x = 1, \quad y = 3$

(b) $y = 2x + 1, m = 2$

$3y = 6x + 1 \qquad (\div 3)$

$y = 2x + \dfrac{1}{3}, \qquad m = 2$

The lines are parallel and will never meet

Exercise 7.10

1. $y + x = 7$

$y - 2x = 1$

Solve this pair of simultaneous equations:

(a) by a graphical method, (b) by an algebraic method.

2. Solve these pairs of simultaneous equations by a graphical method.

(a) $x + y = 2$ (b) $y = 3x + 8$

$y = x + 4$ $2x + y = 3$

NOTE: In your examination you may be asked to solve two simultaneous equations using algebra, or by using a graph. It is important to read the question carefully to ensure that you are using the required method to find the solution. If you have been asked to calculate the solution, you will get no marks for a graphical method.

Drawing Curves

So far we have looked at equations that produce straight line graphs, but other equations produce curves which you need to be able to draw.

Equations of straight lines are easy to plot, and are often drawn on 5 mm squared paper.

However, curves may well be drawn on 2 mm squared graph paper to allow points to be plotted more accurately. The scales on the two axes need not be the same, and they often need to be different to get more of the graph on a reasonable amount of paper.

The approach is still the same. You write down a set of values for x, and calculate the corresponding values of y. Draw up a table, and then plot the curve.

Some examples will make this clear.

Example 17

Draw the graph of the equation $y = x^2$ for values of x: $-3 \leqslant x \leqslant 3$.

Use a scale of 1 centimetre to represent 1 unit on each axis.

Answer 17

$$y = x^2$$

x	-3	-2	-1	0	1	2	3
y	9	4	1	0	1	4	9

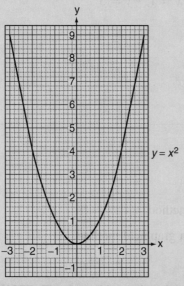

NOTE: The shape this graph makes is called a *parabola*. It may be this way up, or upside down. It may be symmetrical about the y-axis as this is, or about some other line parallel to the y-axis.

Example 18

The table shows values of x and y for x: $-2 \leqslant x \leqslant 2$ for the equation $y = 2x^2 - 3$.

x	-2	-1	0	1	2
y	5	-1	-3	-1	5

(a) Use the table to draw the graph of $y = 2x^2 - 3$.

Use a scale of 1 centimetre to represent 1 unit on the y-axis and 2 centimetres to represent 1 unit on the x-axis.

(b) Use your graph to solve the equation $2x^2 - 3 = 0$.
Give your answers correct to 1 decimal place.

Answer 18

(a)

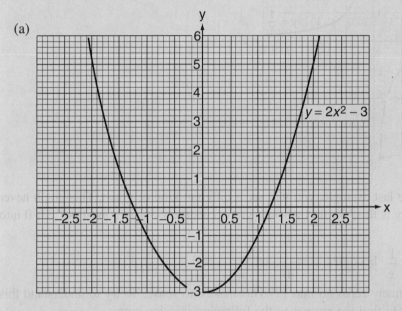

$y = 2x^2 - 3$

(b) $2x^2 - 3 = 0$ is found on the graph of $y = 2x^2 - 3$ when $y = 0$. The line $y = 0$ is the x-axis, so the solutions to the equation $2x^2 - 3 = 0$ are the points where the curve cuts the x-axis. $x = -1.2$ or $x = 1.2$.

Example 19

x	-4	-2	-1	$-\dfrac{1}{2}$	$-\dfrac{1}{4}$	$\dfrac{1}{4}$	$\dfrac{1}{2}$	1	2	4
y	$-\dfrac{1}{4}$	$-\dfrac{1}{2}$	-1	-2	-4	4	2	1	$\dfrac{1}{2}$	$\dfrac{1}{4}$

Use this table of values of x and y to draw the graph of $y = \dfrac{1}{x}$ for $x: -4 \leqslant x \leqslant 4$.

Use a scale of 1 centimetre to represent 1 unit on both axes.

Answer 19

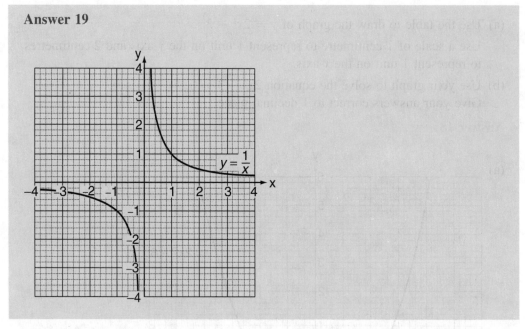

In the answer to the last example, you will see a different kind of graph. This graph never cuts the *y*-axis. Why is this? To find where a graph cuts the *y*-axis we must put $x = 0$ into the equation.

When $x = 0$, $y = \dfrac{1}{x}$ becomes $y = \dfrac{1}{0}$, or $y = 1 \div 0$.

$1 \div 0$ means 'how many zeros go into 1?' This makes no sense. To try to understand this use your calculator to find the answer to the following division sums:

$$1 \div 0.1 =$$
$$1 \div 0.01 =$$
$$1 \div 0.001 =$$
$$1 \div 0.0001 =$$
$$1 \div 0.00001 =$$
$$1 \div 0.00000001 =$$

You should see that as you divide by a smaller and smaller number, the answer gets larger and larger. You could say that zero is infinitely small so $1 \div 0$ must be infinitely large.

In the graph of $y = \dfrac{1}{x}$ you see the curve getting closer and closer to the *y*-axis, but we could never draw a graph large enough to find where it might touch the *y*-axis.

Similarly, see if you can work out an argument to show why *y* can never be zero when $y = \dfrac{1}{x}$. Try different values of *x* to see how close you could get to $y = 0$.

NOTE: For the purposes of your course you should remember that you must never divide by zero.

Example 20

The figure shows the graph of $y = x^3 - 4x$ for $x: -2.5 \leqslant x \leqslant 2.5$. Use the graph to find the solutions to $x^3 - 4x = 1$.

NOTE: For this question you do not need to copy the graph; laying your ruler across the diagram in the correct place will help you find the answers.

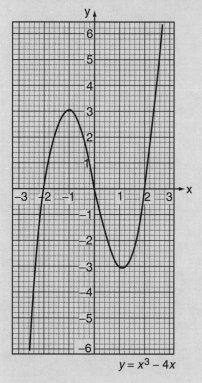

$y = x^3 - 4x$

Answer 20

When $y = 1$, $1 = x^3 - 4x$.

Laying a ruler along the line $y = 1$ we see that where this line cuts the curve
$x = -1.9$ or $x = -0.3$ or $x = 2.1$

Completing a table of values for a curve

Example 21

Find the missing values from the table for the equation
$y = 2x^2 - 3x - 5$ for values of $x: -3 \leqslant x \leqslant 3$.

x	-3	-2	-1	0	1	2	3
y	a	b	c	d	e	f	g

Answer 21

(a) $y = 2x^2 - 3x - 5$
$y = 2 \times (-3)^2 - 3 \times (-3) - 5$
$y = 2 \times 9 + 9 - 5$
$y = 18 + 9 - 5 = 22$
$a = 22$

NOTE: $(-3)^2$ means -3×-3 which means $+9$, and $-3 \times (-3)$ is also $+9$.

(b) $y = 2 \times (-2)^2 - 3 \times (-2) - 5$
$y = 8 + 6 - 5 = 9$
$b = 9$

(c) $y = 2 \times (-1)^2 - 3 \times (-1) - 5$
$y = 2 + 3 - 5 = 0$
$c = 0$

(d) $y = 2 \times (0)^2 - 3 \times 0 - 5$
$y = 0 - 0 - 5 = -5$
$d = -5$

(e) $y = 2 \times 1^2 - 3 \times 1 - 5$
$y = 2 - 3 - 5 = -6$
$e = -6$

(f) $y = 2 \times 2^2 - 3 \times 2 - 5$
$y = 8 - 6 - 5 = -3$
$f = -3$

(g) $y = 2 \times (3)^2 - 3 \times (3) - 5$
$y = 18 - 9 - 5 = 4$
$g = 4$

Exercise 7.11

Copy and complete the table for each of the equations given.

x	-3	-2	-1	0	1	2	3
y	a	b	c	d	e	f	g

1. $y = x^2$
2. $y = x^3$
3. $y = \dfrac{3}{x}$ ($x \neq 0$, so you cannot find d in this case)
4. $y = x^2 + x$
5. $y = x^2 - x$
6. $y = x^3 + 2$
7. $y = -x^2$ NOTE: $-(-3)^2 = -(9) = -9$

8. $y = -x^3$ NOTE: $-(-3)^3 = -(-3) \times (-3) \times (-3) = -(-27) = 27$
9. $y = x^2 - x - 5$
10. $y = -x^2 - 2x + 1$

When you are satisfied that you have the correct answers to the above exercise then carry on to the next exercise, taking careful note of the following points.

- The scales do not have to be the same on both axes.
- Curves should never be drawn with a ruler.
- Curves should be smooth, without angles.
- If you have to suddenly change direction to include one point you have probably made an error, either in the calculation or in the plotting of the point. Go back and check.
- Write the equation beside the curve and see if you can make connections between the equation and the shape of the graph.

Exercise 7.12

Using the values you found in the above exercise, draw the following curves for $x: -3 \leqslant x \leqslant 3$.

1. $y = x^2$, $0 \leqslant y \leqslant 9$

2. $y = x^3$, $-27 \leqslant y \leqslant 27$

3. $y = \dfrac{3}{x}$ $x \neq 0$, $-3 \leqslant y \leqslant 3$ $y \neq 0$.   NOTE: See Example 19 above.

4. $y = x^2 + x$, $-1 \leqslant y \leqslant 12$

5. $y = x^2 - x$, $-1 \leqslant y \leqslant 12$

6. $y = x^3 + 2$, $-25 \leqslant y \leqslant 29$

7. $y = -x^2$, $-9 \leqslant y \leqslant 0$

8. $y = -x^3$, $-27 \leqslant y \leqslant 27$

9. $y = x^2 - x - 5$, $-6 \leqslant y \leqslant 7$

10. $y = -x^2 - 2x + 1$, $-14 \leqslant y \leqslant 2$

Exercise 7.13

Mixed Exercise

1. The graph shows the flight of a ball thrown from a height of 1 metre above the ground.
   (a) How far away from the person who threw it does it land on the ground?
   (b) What is its maximum height?
   (c) Will the ball clear a wall 1.5 metres high 7.2 metres away?

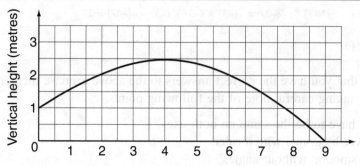

2. What is the equation of the line going through each of the following sets of points?

 NOTE: For each line plot the points on squared paper first.

  (a) (1, 1), (0, 0), (3, 3), (−2, −2)
  (b) (1, −1), (0, 0), (3, −3), (−2, 2)
  (c) (5, 0), (5, 1), (5, 3), (5, −4)
  (d) (1, −2), (3, −2), (−1, −2), (0, −2)

3. For values x: $-3 \leqslant x \leqslant 3$ draw each of the following lines.

  (a) $y = x$ (b) $y = x + 2$ (c) $y = x - 4$
  (d) $y = 2x$ (e) $y = -2x$

4. For values of x: $-3 \leqslant x \leqslant 3$ draw each of the following curves.

  (a) $y = x^2$ (b) $y = x^2 + 2$
  (c) $y = -x^2$ (d) $y = x^2 - 3$

5. An examiner is marking the following answers to questions on curves. All of them lose marks. Can you see why?

  (a)

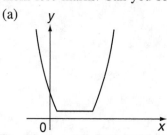

  (b)

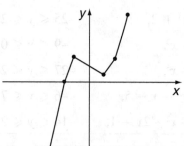

  (c)

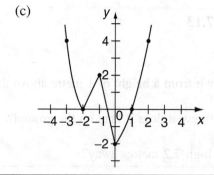

  $y = x^2 + x - 2$, $-3 \leqslant x \leqslant 2$

6. From your knowledge of the graphs in this chapter, try to match each graph to a possible equation.

(a)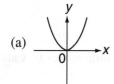

(b)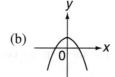

(c)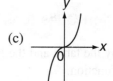

(d)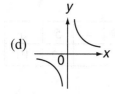

(e)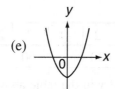

(f)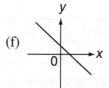

(i) $y = x^3$

(iii) $y = 1 - x$

(v) $y = x^2 - 2$

(ii) $y = x^2$

(iv) $y = -x^2 + 1$

(vi) $y = \dfrac{1}{x}$

7. Find the gradients of each of these lines.

(a)

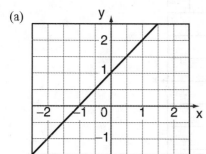

(b)

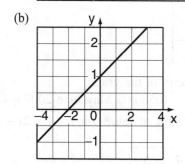

(c)

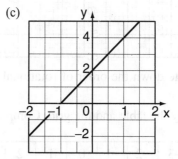

8. (a) Draw a graph to show the conversion of Singapore Dollars to Swiss Francs. The exchange rate is:

1 Singapore Dollar = 0.8 Swiss Francs.

The horizontal axis should show:

0 ⩽ Singapore Dollars ⩽ 100.

(b) Use your graph to convert 45 Swiss Francs to Singapore Dollars.

Examination Questions

9. (a) Copy the table and the grid. The table shows corresponding values of x and y for the function:

$$y = \frac{60}{x} \quad (x \neq 0).$$

x	−6	−5	−4	−3	−2	−1	1	2	3	4	5	6
y		−12	−15		−30		60				12	10

(i) Fill in the missing values of y in the table above.

(ii) Plot the points on the grid below and draw the graph for −6 ⩽ x ⩽ −1 and 1 ⩽ x ⩽ 6.

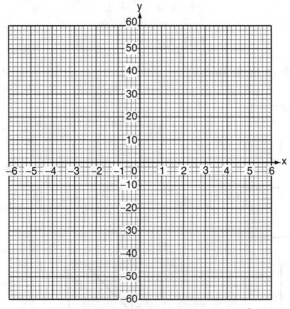

(b) Write down the order of rotational symmetry of the graph.

(0580/03 May/June 2007 q 4a)

10. (a) Copy the table and the grid. Complete the table of values for $y = 1 + 2x - x^2$.

x	−3	−2	−1	0	1	2	3	4	5
y	−14	−7				1	−2		−14

(b) Draw the graph of $y = 1 + 2x - x^2$ on the grid below.

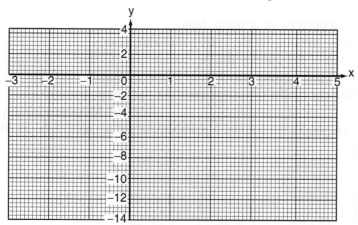

(c) Use your graph to find the solutions to the equation:
$1 + 2x - x^2 = 0$.

(d) (i) On the grid, draw the line of symmetry of the graph.
   (ii) Write down the equation of this line of symmetry.

(0580/03 May/June 2005 q 2)

11.

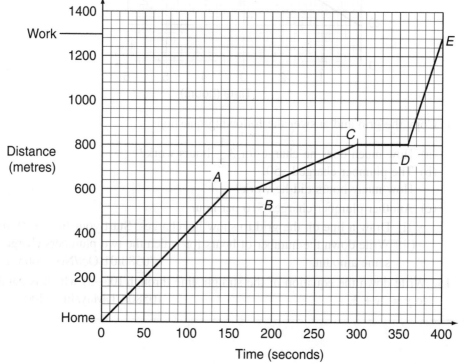

The graph shows the distance travelled by a cyclist on a journey from Home to Work.
(a) The cyclist stopped twice at traffic lights.
    For how many seconds did the cyclist wait altogether?
(b) For which part of the journey did the cyclist travel fastest?
(c) (i) How far did the cyclist travel from Home to Work?
    (ii) Calculate the cyclist's average speed for the whole journey.
                                            (0580/01 May/June 2002 q 21)

12. The graph below shows the amount a plumber charges for up to 6 hours work.

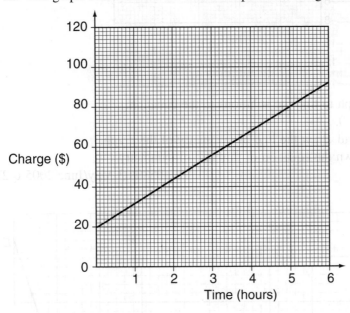

Copy the grid and answer the following questions.

(a) How much does he charge for $3\frac{1}{2}$ hours work?
(b) The plumber charged $50.
    How many hours did he work?
(c) Another plumber charges $16 per hour.
    (i) Draw a line on the grid to show his charges. Start your line at (0, 0).
    (ii) Write down the number of hours for which the two plumbers charge the
        same amount.   (0580/01 Oct/Nov 2005 q 21)

13. Write down the equation of the straight line through (0, −3) which is parallel
    to $y = 2x + 3$.   (0580/01 May/June 2007 q 12)

14.

	Monday	Tuesday	Wednesday	Thursday	Friday	Saturday	Sunday
Minimum temperature °C	4	6	0	−2	−4	2	
Maximum temperature °C	8	10	5	7	2	7	

The table shows the minimum and maximum temperatures on six days of a week. Copy the table and the grid.

(a) (i) On Sunday the minimum temperature was 5°C lower than on Saturday. The maximum temperature was 2°C higher than on Saturday. Use this information to complete the table.

(ii) Find the difference between the minimum and maximum temperatures on Thursday.

(b) Use the table to complete the graphs below for all seven days.

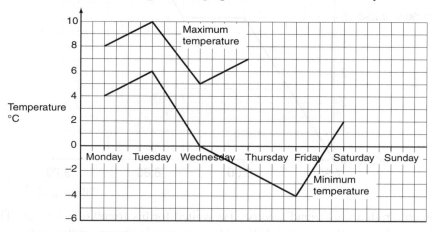

(c) Use your graphs to find:
   (i) on how many days the temperature fell below −1°C,
   (ii) which day had the largest difference between minimum and maximum temperatures.

(d) The formula for changing degrees Celsius (*C*) to degrees Fahrenheit (*F*) is

$$F = \frac{9C}{5} + 32.$$

Use the formula to change 6 degrees Celsius to degrees Fahrenheit. Show all your working.

(0580/03 May/June 2005 q 3)

15.

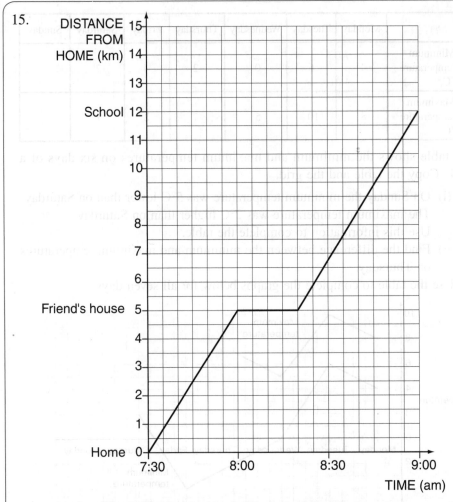

Ricardo rode to his friend's house. He waited for his friend to get ready. Then they cycled together to school. Ricardo's journey is shown on the grid.

(a) Work out the speed at which Ricardo cycled to his friend's house.

(b) How long did he wait for his friend?

(c) Ricardo's brother left home at 8:00 am.
He cycled directly to school at a constant speed of 15 kilometres per hour. Draw his journey.

(d) How many minutes earlier than Ricardo did his brother arrive at school?

(0580/01 May/June 2005 q 22)

16. (a) Copy and complete the table of values for the equation $y = x^2 + x - 3$.

x	-4	-3	-2	-1	0	1	2	3
y	9		-1	-3		-1		9

(b) On a copy of the grid, draw the graph of $y = x^2 + x - 3$.

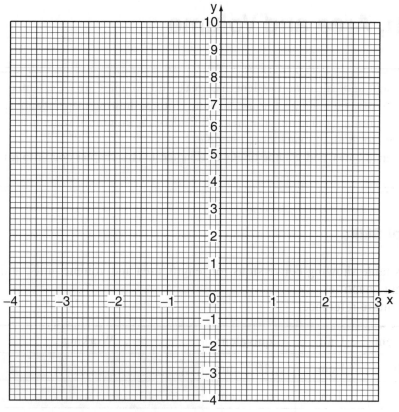

(c) Write down the coordinates of the lowest point of the curve.

(d) (i) Draw the line of symmetry of the graph.

(ii) Write down the equation of the line of symmetry.

(0580/03 Oct/Nov 2008 q 7)

Chapter 8

Length, Area and Volume

The measurement and calculation of lengths, areas, and volumes are an essential part of modern life. We may be building a house, making clothes, drawing maps, planning routes from one city to another, calculating the area of a field, or the volume of water in a reservoir. You will be able to think of many more examples.

We looked at length, area, and volume briefly in Chapter 4. This chapter goes into it in more detail.

Essential Skills

1. Copy this table showing conversions for length, area and volume units, and fill in the missing quantities.
   Keep for future reference.

Measurement	Unit	Equivalent	Working
Length	1 cm	...10....mm	
	1 m	.........cm	
Area	1 cm²	.........mm²	10 mm × 10 mm
	1 m²	.........cm²	100 cm × 100 cm
Volume	1 cm³	.........mm³	......mm × ......mm × ......mm
	1 m³	.........cm³	......cm × ....cm × ....cm
Capacity	1 millilitre	1 cm³	
	1 litre	1000 ml	
	1 litre	1000 cm³	
	1 m³	1000 litres	

2. (a) How do you 'undo' squaring?
   (b) What is the inverse of cubing?
   (c) Use your calculator to find the following. If the answers are not exact, give to 3 significant figures.
      (i) 13^2 (ii) $\sqrt{13}$ (iii) 13^3 (iv) $\sqrt[3]{13}$

Answers

1.

Measurement	Unit	Equivalent	Working
Length	1 cm	10 mm	
	1 m	100 cm	
Area	1 cm²	100 mm²	10 mm × 10 mm
	1 m²	10000 cm²	100 cm × 100 cm
Volume	1 cm³	1000 mm³	10 mm × 10 mm × 10 mm
	1 m³	1000000 cm³	100 cm × 100 cm × 100 cm
Capacity	1 millilitre	1 cm³	
	1 litre	1000 ml	
	1 litre	1000 cm³	
	1 m³	1000 litres	

2. (a) Square root
   (b) Finding the cube root
   (c) (i) 169   (ii) 3.61   (iii) 2197   (iv) 2.35

NOTE: In this chapter none of the diagrams are drawn to scale.

Length

A length is a measurement in one dimension. It may be the length of a straight line, or of a curved line *measured along the curve*. Adding lengths together makes a longer length, and multiplying by a number greater than one likewise makes a longer length. For example, 3 cm + 4 cm = 7 cm, and 2 cm × 4 = 8 cm. However, multiplying by a number between zero and one makes a shorter length. For example, 0.5 × 2 cm = 1 cm (or $\frac{1}{2}$ × 2 cm = 1 cm).

Lengths do not have direction, so are always positive.

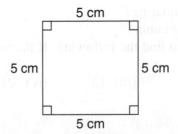

The diagram shows a square with each side having length of 5 centimetres. The distance (or length) all the way round the square is called the **perimeter**. The perimeter can either be calculated by adding all the lengths together:

$$5 \text{ cm} + 5 \text{ cm} + 5 \text{ cm} + 5 \text{ cm} = 20 \text{ cm},$$

or it may be calculated more simply in this case by multiplying one length by 4 since the lengths are all the same:

$$4 \times 5 \text{ cm} = 20 \text{ cm}.$$

You will encounter formulae in this chapter which will help you to calculate lengths, areas and volumes, and it is important that you understand a little about them.

As you can see from the above, if the side of the square was of length a centimetres, then the formula for the perimeter could be either:

$$\text{perimeter} = a \text{ cm} + a \text{ cm} + a \text{ cm} + a \text{ cm},$$

or $\text{perimeter} = 4 \times a \text{ cm}.$

The lengths have either been added together, or since they are all the same, they have been multiplied by 4.

It is not usual to write the units *in* the calculation, but it is done here to make the ideas clearer. However, you must state the units in the answer, and all the measurements in a single calculation must be in the same units. So for example, you cannot mix centimetres with metres in a single calculation.

Example 1

Find the perimeters of the following shapes.

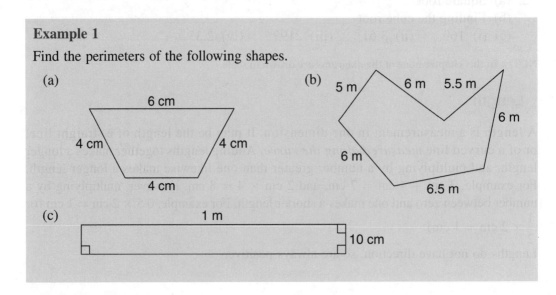

Answer 1

(a) Perimeter $= 6 + 3 \times 4 = 6 + 12$

Perimeter $= 18$ cm

NOTE: This has been calculated by multiplication of a common length by a number *and* by addition of another length.

(b) Perimeter $= 5 + 6 + 5.5 + 6 + 6.5 + 6$ (or $5 + 5.5 + 6.5 + 3 \times 6$)

$= 35$ m

(c) The units of measurement must all be made the same before addition, so either

Perimeter $= 100 + 10 + 100 + 10$

$= 220$ cm

or

Perimeter $= 1 + 0.1 + 1 + 0.1$

$= 2.2$ m

Practical Investigation

As you already know, the perimeter of a circle is called the circumference. For this experiment you need a cylindrical object such as a tin. If possible find one without a lid, or widening, at the end. A piece of drainpipe is ideal, or some other piece of tube. You will also need a strip of tracing paper long enough to wrap round the cylinder.

Rule a pencil mark across the strip of paper, and then wrap it tightly round the cylinder, so that the end overlaps the pencil mark. Trace the pencil mark onto the overlap.

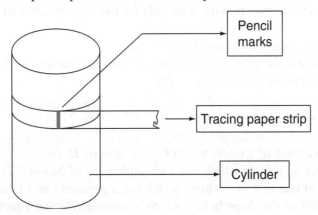

The pencil marks should coincide exactly.

Now take off the paper strip and measure between the pencil marks. This is the circumference of the circle which is the shape of the cross-section of the cylinder.

Also measure the diameter of the cylinder, by taking the largest measurement you can find across the circular end.

Copy and enter your measurements into this table. Two sets have already been entered.

circumference	26.3 cm	14.5 cm			
diameter	8.2 cm	4.5 cm			
circumference / diameter (to 1 dp)	3.2	3.2			

If you can, make the same measurements on one or two other cylinders and include these results.

Possible Experimental Errors

- Not holding the paper strip tightly round the cylinder before tracing the mark.
- Not measuring the diameter correctly (remember that it is the widest part of the circle).
- The cylinder curves out (like the rim of a tin) or in (like the base of a bottle) making it hard to get an accurate measurement.

If you have managed to avoid all these errors you should get a close similarity between the numbers along the bottom row.

The number you have found is an approximation to π (3.141 ...), a number you met in Chapter 1. Pi (π) is an irrational number. However, it is the ratio of the circumference of a circle to its diameter. How can we say it is an irrational number *and* that it is the ratio of the circumference to the diameter of a circle? The reason is that to be rational a number must be capable of being expressed as the ratio of two *integers*. It is impossible to find a circle which has integer measurements for both diameter and circumference.

We now have the information to write a formula for the circumference of a circle in terms of its diameter.

If the diameter is d, then the circumference $= \pi d$.

You already know that the diameter of a circle is twice the radius, so if the radius is r, then the circumference $= \pi \times 2r = 2\pi r$.

Example 2

Calculate:

(a) the circumference of a circle which has a radius of 15 cm,

(b) the diameter of a circle which has a circumference of 34 cm,

(c) the length of one side of a square which has a perimeter of 13 cm,

(d) the perimeter of the shape below, which is a semicircle and a rectangle.

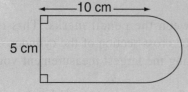

Answer 2

(a) Circumference $= 2\pi r$

$\qquad\qquad\quad = 2 \times \pi \times 15$ NOTE: Use the π button on your calculator.

$\qquad\qquad\quad = 94.247779\ ...$ NOTE: Check whether this seems reasonable:
radius $= 15$ cm, so diameter $= 30$ cm, and
circumference is approximately 3 times the diameter,
so it is approximately 90 cm. The answer is reasonable.

Circumference $= 94.2$ cm

(b) Circumference $= \pi \times$ diameter

$\qquad\quad 34 = \pi \times$ diameter $(\div\ \pi\ )$

$\qquad$ diameter $= \dfrac{34}{\pi}$

$\qquad\qquad\quad = 10.822536\ ...$

$\qquad$ diameter $= 10.8$ cm   NOTE: Is this reasonable? The circumference is
approximately 3 times the diameter and 3 times 10 cm
is 30 cm, so 10.8 cm is reasonable.

(c) The perimeter of a square is the four sides added together, or 4 times
the length of one side.

$\qquad$ Perimeter $= 4 \times$ side

$\qquad\qquad\ 13 = 4 \times$ side $(\div\ 4)$

$\qquad\quad$ side $= \dfrac{13}{4}$

The length of one side of the square $= 3.35$ cm

(d) Draw a dotted line to show the semicircle and the rectangle.

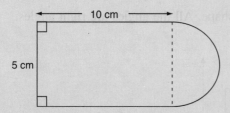

The semicircular end of the shape has a diameter of 5 cm (the same as
the other side of the rectangle).

$\qquad$ Curved end $= \dfrac{1}{2} \times \pi \times 5$

$\qquad$ Perimeter $= 10 + 5 + 10 + \dfrac{1}{2} \times \pi \times 5$

$\qquad\qquad\qquad\ = 32.85398\ ...$

$\qquad$ Perimeter $= 32.9$ cm   NOTE: Is this reasonable? The circumference of the whole
circle would be approximately $3 \times 5 = 15$ cm, so half the
circle would be approximately 7.5 cm. $10 + 5 + 10 + 7.5$
$= 32.5$ cm, so 32.9 cm is reasonable.

Example 3

A triangle has sides of length x, $x + 2$ and $2x - 1$ cm. The perimeter of the triangle is 30 cm. Calculate the lengths of the sides of the triangle.

Answer 3

The perimeter $= x + (x + 2) + (2x - 1)$

$$= 4x + 1$$

The perimeter $= 30$ cm

$4x + 1 = 30$

$\qquad 4x = 29 \qquad\qquad\qquad (-1)$

$\qquad x = 29 \div 4 = 7.25$ cm $\qquad (\div 4)$

The sides of the triangle are 7.25 cm, $(7.25 + 2)$ cm, and $(2 \times 7.25 - 1)$ cm.
The sides are 7.25 cm, 9.25 cm and, 13.5 cm.

NOTE: In all the following exercises the dotted lines are only there to help you; they are not part of the perimeter, area or volume of the shape. Arrows on sides indicate parallel lines, and single or double marks on sides indicate equal lengths. Diagrams are not drawn to scale, so you cannot assume that the sides are in proportion to those shown in the diagrams.

NOTE: In all these exercises it is wise to copy the diagram and mark in the lengths of all the sides before you try to calculate the answers. For example, if you are asked to find the perimeter of a shape which has pairs of equal sides, then mark the lengths on all the sides. See the next example.

Example 4

Calculate the perimeter of the following shape. All the angles are right angles.

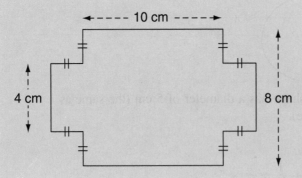

Answer 4

Copying the diagram and enlarging to allow room for the extra numbers:

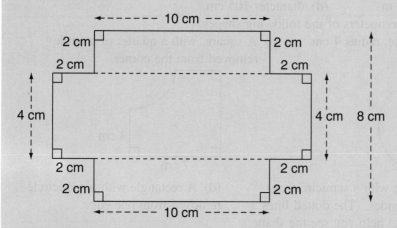

It is now easy to find the perimeter.

Perimeter $= (8 \times 2) + 2 \times 4 + 2 \times 10 = 16 + 8 + 20 = 44$ cm

NOTE: You will not have to redraw the diagram if you are expected to answer the questions on the printed examination question paper. In this case you can write on the printed diagram.

NOTE: Can you explain why, in the diagram drawn above, the perimeter is the same as that of a simple rectangle measuring 14 cm by 8 cm?
If you were to make this shape with, say, a length of string it would require exactly the same length as for the simple rectangle!

Exercise 8.1

1. Calculate the perimeters of the following shapes.

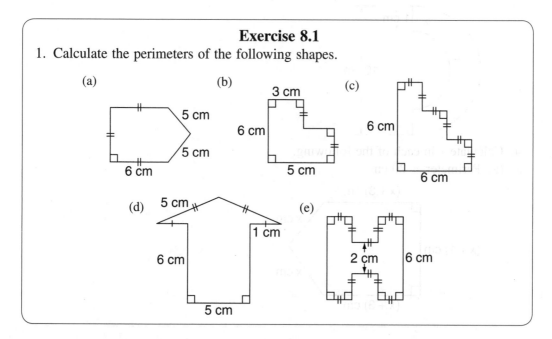

2. Calculate the circumferences of the following circles:
   (a) radius 5 cm,   (b) diameter 5 cm,
   (c) radius 7.15 m,   (d) diameter 105 cm.
3. Calculate the perimeters of the following shapes.
   (a) A semicircle, radius 4 cm.   (b) A square, with a quarter of a circle
                                           removed from the corner.

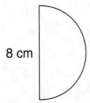

8 cm

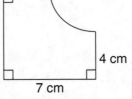

4 cm

7 cm

   (c) A rectangle with a semicircle on   (d) A rectangle with a semicircle
       two of the sides. (The dotted lines   removed from one edge.
       are drawn to help you see the shape;
       they are not part of the perimeter.)

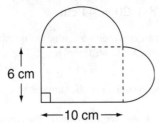

6 cm

←10 cm→

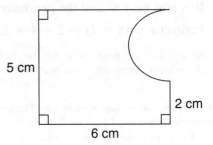

5 cm

2 cm

6 cm

   (e) Two semicircles joined by two straight lines each 1 cm long.

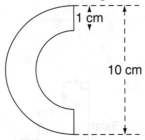

1 cm

10 cm

4. Calculate x in each of the following.
   (a) Perimeter = 20 cm

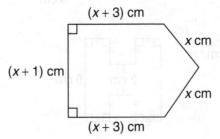

$(x + 3)$ cm

x cm

$(x + 1)$ cm

x cm

$(x + 3)$ cm

(b) A circle with circumference = 15 cm and diameter = x cm.

(c) A circle with radius x m and circumference = 11 m.

(d) A rectangle with perimeter = 24 cm.

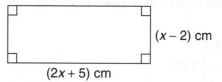

$(x-2)$ cm

$(2x+5)$ cm

5. Find, by calculation:

   (a) the length of the shortest side,

   (b) the length of the longest side in the diagram below.

All the measurements are in centimetres and the perimeter is 37 cm.

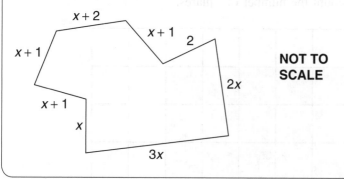

NOT TO SCALE

Area

An area is a measurement in two dimensions. For example, it could be a measurement of the amount of surface on one page of this book. If you think of the amount of surface on a postage stamp compared with the amount of surface on the envelope you will get some idea of area. You could make a reasonable guess at how many stamps would be needed to cover the whole surface of the envelope. Areas of rectangles are reasonably easy to visualise, it is a little harder when the shape is irregular or even has curved edges.

Because the area of a shape is a measurement in two dimensions it needs two length measurements to define it. For example, the area of a rectangle is calculated by multiplying its length by its breadth.

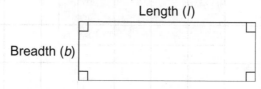

Length (l)

Breadth (b)

Area of rectangle = lb

The units of area are square units. Suppose the length and breadth are measured in centimetres:

area $= l$ cm $\times b$ cm $= lb$ cm $\times$ cm $= lb$ cm^2

This means that the two measurements must be in the same units (that is, both in centimetres, or both in metres, and so on).

The unit of area, 1 cm^2, is the amount of surface covered by a square measuring 1 cm by 1 cm, as shown below.

The area of a rectangle is the number of 1 centimetre squares that it covers. For example, looking at the rectangle drawn on the one centimetre squared paper shown below, you will see that we can actually count the number of squares.

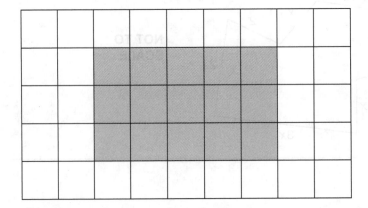

The rectangle is 5 centimetres long and 3 centimetres wide, and covers $5 \times 3 = 15$ centimetre squares. The area of the rectangle is 15 cm^2.

The length and breadth may not necessarily be whole numbers.

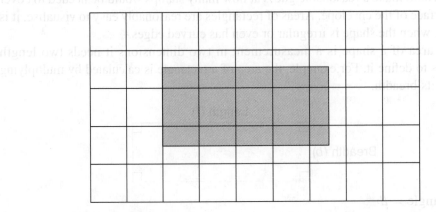

The rectangle above covers 4 whole squares, 9 half squares and 2 quarter squares.

This is $4 + 9 \times \dfrac{1}{2} + 2 \times \dfrac{1}{4} = 4 + 4.5 + 0.5 = 9$ centimetre squares.

The rectangle measures 4.5 cm by 2 centimetres, so its area is $4.5 \times 2 = 9$ cm².

- **Area of a rectangle = length × breadth**

NOTE: The lengths of the sides may be called length, breadth, width or height. Whatever they are called they must be measurements at right angles to each other.

Area of a triangle

Other shapes that you know can be related to the rectangle by various means. For example, the right-angled triangle is half a rectangle as shown.

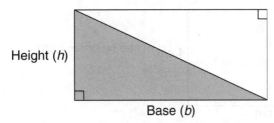

Height (*h*)

Base (*b*)

The area of the rectangle is base × height, so the area of the triangle is half the base × height.

The formula for the area of this triangle is $\dfrac{1}{2}bh$.

A rectangle can be drawn round any triangle as shown below.

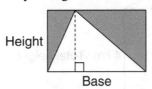

Height

Base

The area of the rectangle is base × height, and you should be able to see that the area of the triangle is half that of the rectangle, so the area of the triangle is half the base times the height. However, it is essential to see that the height is a line drawn from one vertex of the triangle *perpendicular* to the opposite side. The 'base' and the 'height' of the triangle are *always* perpendicular to each other. The 'base' does not always have to be at the bottom of the triangle as long as the two measurements are at right angles as you can see from the following examples of triangles.

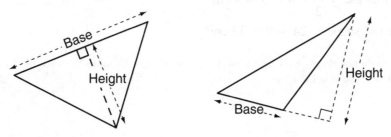

Base

Height

Height

Base

In each case the base and the height are at right angles to each other.

- **The area of a triangle $= \dfrac{1}{2} \times$ base $\times$ perpendicular height**

Example 5

Calculate the area of this shape.

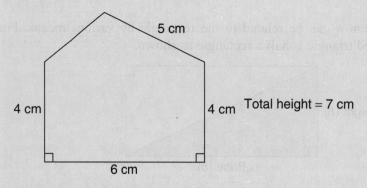

Answer 5

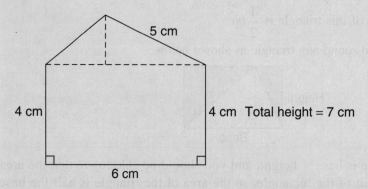

The shape is a rectangle with a triangle on top. The height of the triangle is 7 cm − 4 cm = 3 cm, and the base of the triangle is 6 cm.

Area of rectangle $= 4 \times 6 = 24$ cm^2

Area of triangle $= \dfrac{1}{2} \times 6 \times 3 = 9$ cm^2

Total area of shape $= 24 + 9 = 33$ cm^2

NOTE: You do not always have to use all the measurements given, in this case the 5 cm on one side of the triangle is not used.

Example 6

Calculate the area of the shape in Example 4.

Answer 6

Redrawing the shape with all its measurements marked on it:

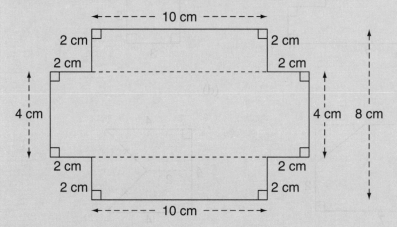

There are different ways to work through this question.

You could divide it into rectangles in different ways. Take, for example, the dotted lines shown above. There are now 2 rectangles measuring 10 cm by 2 cm, and one measuring 14 cm by 4 cm.

Area = $2 \times 10 \times 2 + 14 \times 4 = 40 + 56 = 96$ cm².

Or, you could calculate the area of the rectangle measuring 14 cm by 8 cm, and take away 4 squares each measuring 2 cm by 2 cm.

Area = $14 \times 8 - 4 \times 2 \times 2 = 112 - 16 = 96$ cm².

In the question about the perimeter of this diagram, we saw that the perimeter was, in fact, the same as that of the 14 cm by 8 cm rectangle. In the case of the area, however, it is not the same.

This shows that two different shapes with the same perimeter will not necessarily have the same area.

Exercise 8.2

1. Calculate the areas of the following shapes by dividing them up into rectangles and triangles when necessary.
   Remember:
   The double marks on the sides of the diagrams indicate equal lengths.
   The dotted lines are there to help you; the areas to be found are enclosed by solid lines.

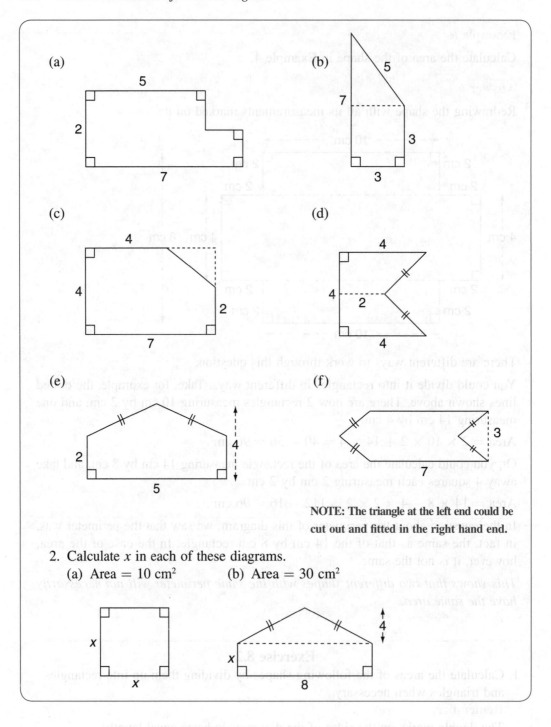

NOTE: The triangle at the left end could be
cut out and fitted in the right hand end.

2. Calculate *x* in each of these diagrams.
   (a) Area = 10 cm²   (b) Area = 30 cm²

The area of a circle

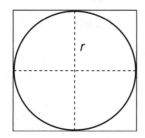
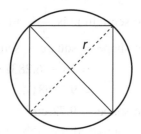

In the first diagram above, you can see that the area of the circle is less than the total area of the four squares. The area of each square is r^2, where r is the radius of the circle. So the area of the circle is less than $4r^2$.

In the second of the diagrams you can see that the area of the circle is more than the area of the inside square. This square is made up from two triangles; each with a base $2r$ in length, and each with a height of r. The area of one triangle is:

$$\frac{1}{2} \times \text{base} \times \text{height} = \frac{1}{2} \times 2r \times r = r^2$$

So the area of the circle is more than $2r^2$.

$$2r^2 < \text{area of circle} < 4r^2$$

It seems possible that the area is about 3 times the radius squared, and you can check this by drawing circles on graph paper and counting the squares.

In fact the magic number is once again π.

- **The area of a circle is πr^2.**

We now have two formulae connected with circles:

- **Circumference $= \pi d$ or $2\pi r$**
- **Area $= \pi r^2$**

  NOTE: Remember that an area needs two length measurements multiplied together, in this case the radius $\times$ the radius (r^2).

NOTE: Remember that the circumference is a length and has one length measurement multiplied only by numbers. (π is just a number, approximately 3.14 ...)

Example 7

Calculate the area of this shape. It is a square with side 4 cm with a semicircle removed from it.

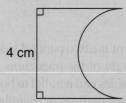

4 cm

Answer 7

The area of the square $= 4 \times 4 = 16$ cm^2

The radius of the circle is 2 cm ($\frac{1}{2} \times 4$)

The area of the whole circle is $\pi r^2 = \pi \times 2^2 = 4\pi$ cm²

The area of the semicircle is $\dfrac{1}{2} \times 4\pi = 2 \times \pi = 6.28318 \ldots$ cm²

The area of the shape = square − semicircle

$$= 16 - 6.28318 \ldots$$
$$= 9.71681 \ldots$$

The area of the shape = 9.72 cm²

NOTE: Check for yourself whether this seems reasonable.

Exercise 8.3

1. Calculate the areas of the following:
    (a) a circle with radius 6 cm
    (b) a circle with radius 3.2 metres
    (c) a circle with diameter 8 cm
    (d) a semicircle with radius 6 cm
    (e) a semicircle with diameter 8.1 cm
2. Calculate the areas of the shapes shown in Exercise 1, question 3.
3. Calculate x in each of the following:
    (a) a circle, area 17 cm², radius x cm
    (b) a circle, area 12 cm², diameter x cm
    (c) a semicircle, area 5 cm², radius x cm
    (d) a semicircle, area 6 cm², diameter x cm

    NOTE: What would be the area of the whole circle if the semicircle is 5 cm²?

4. Calculate the shaded area. The large circle has a diameter = 7.5 cm, and the small circle has diameter = 3 cm.

The area of a trapezium

In the diagram below the dark line shows a trapezium. The lengths of the two parallel sides are a cm and b cm. The distance between the parallel sides, or height of the trapezium, is h cm. A dotted line, *XY*, is drawn half way between the two parallel sides and parallel to both of them. The two shaded triangles can each be rotated to fill the dotted outlines of triangles. The result of moving these two triangles is to create a rectangle whose length is the length of the dotted line, *XY*, and whose width is the perpendicular height, h, of the trapezium.

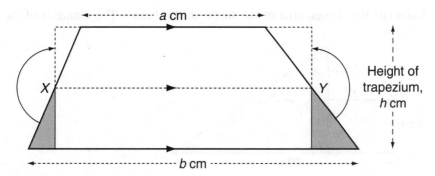

The area of the rectangle $= XY \times h$

$$XY = \frac{a+b}{2}$$

So the area of the rectangle $= \frac{a+b}{2} \times h.$

The area of the trapezium is the same as the area of the rectangle.

The area of the trapezium is usually written as:

- **Area of trapezium** $= \frac{1}{2}(a + b) \times h$

You may prefer to remember in words:

- **Area of trapezium** $= \frac{1}{2} \times$ **the sum of the parallel sides** $\times$ **the distance between them.**

NOTE: Remember that it is the *perpendicular* distance between the parallel sides.

Example 8

Calculate the areas of the following shapes.

(a) *ABCD* is a parallelogram.   (b) *PQRS* is a trapezium.

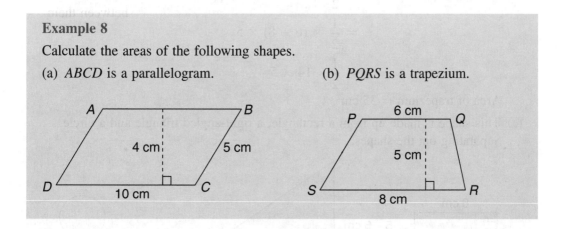

(c) Calculate the shaded area of the shape shown below. The diameter of the circle is 3 cm.

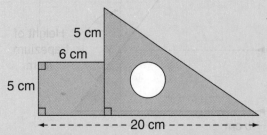

Answer 8

(a) Draw a line from *C* perpendicular to *AB*. The resulting right-angled triangle could be moved so that its hypotenuse *BC* fits along *AD*, forming a rectangle.

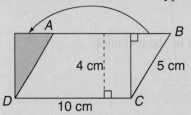

The area of the rectangle is $10 \times 4 = 40$ cm².

Area of *ABCD* = 40 cm²

NOTE: The length of the side *BC* (5 cm) was not required for the calculation.

(b) The area of a trapezium $= \dfrac{1}{2} \times$ the sum of the parallel sides $\times$ the distance between them

$$= \frac{1}{2} \times (6 + 8) \times 5$$

$$= \frac{1}{2} \times 14 \times 5$$

Area of trapezium = 35 cm²

(c) This shape is made up from a rectangle, a right-angled triangle and a circle. Separating out the shapes:

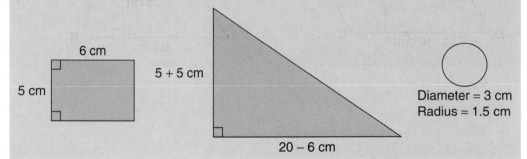

Area of rectangle $= 6 \times 5 = 30$ cm^2

Area of triangle $= \dfrac{1}{2} \times (20 - 6) \times (5 + 5)$

$= \dfrac{1}{2} \times 14 \times 10 = 70$ cm^2

Area of circle $= \pi r^2 = \pi \times 1.5^2 = 7.0685 \ldots$

Total area of shape $=$ rectangle $+$ triangle $-$ circle

$= 30 + 70 - 7.0685 \ldots$

$= 92.931 \ldots$ cm^2

Total area $= 92.9$ cm^2

In the answer to Example 8 (a), we calculated the area of the parallelogram by transforming it into a rectangle. This will always be possible.

However, if you prefer to learn a formula, then you can use the following:

- **the area of a parallelogram = the length of one of the parallel sides × the perpendicular distance between them.**

Exercise 8.4

Calculate the shaded areas of these shapes.

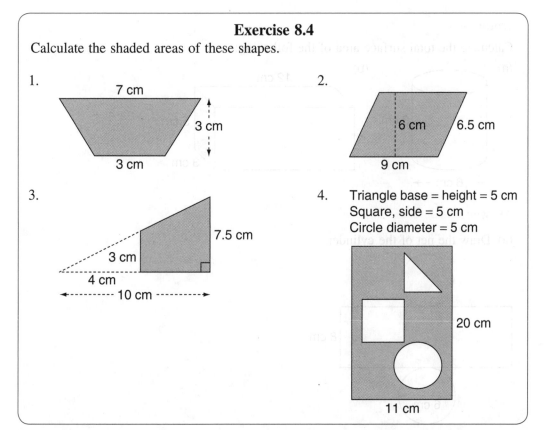

1.
7 cm
3 cm
3 cm

2.
6 cm   6.5 cm
9 cm

3.
7.5 cm
3 cm
4 cm
10 cm

4.   Triangle base = height = 5 cm
Square, side = 5 cm
Circle diameter = 5 cm
20 cm
11 cm

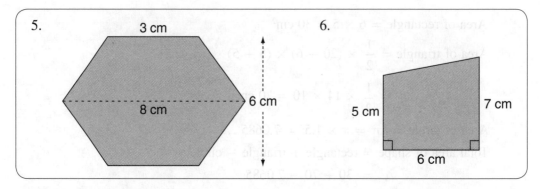

Total surface area

You may be asked to calculate the total surface area of a three dimensional or solid shape. This is the areas of all the faces of the solid added together. It can be useful to draw the net of the solid first (see Chapter 6).

Imagine you are going to make the solid out of card. The total surface area is the amount of card you would use.

Remember that a solid cuboid would be made of six rectangles or squares, while a box without a lid would have only five rectangle or square faces, with the sixth side being open.

Example 9

Calculate the total surface area of the following:

Answer 9

(a) Draw the net of the cylinder.

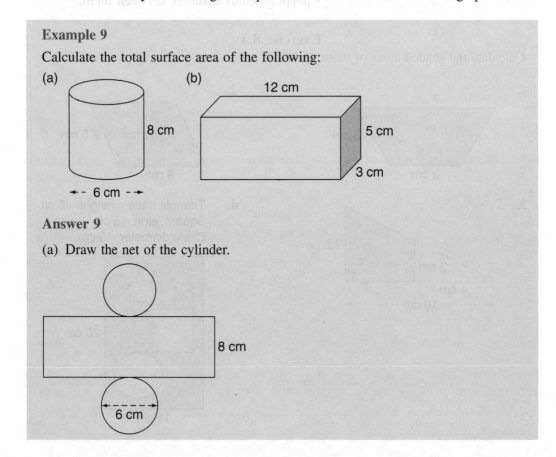

The net consists of a rectangle and two circles.

NOTE: If you find this difficult to visualise, think of a tin with a label all round it. Take off the label which will be a rectangle. The two ends of the tin are the circles.

The length of the rectangle is the same as the circumference of the circular ends.

Length of rectangle $= \pi \times$ diameter $= \pi \times 6$ cm $= 18.849 \ldots$ cm

Area of rectangle $= 8 \times 18.849 \ldots = 150.796 \ldots$ cm^2

Radius of circular ends $= 3$ cm

Area of one circle $= \pi r^2 = \pi \times 3^2 = 28.274 \ldots$ cm^2

Total surface area $=$ rectangle $+ 2$ circles

$$= 150.796 \ldots + 2 \times 28.274 \ldots$$

$$= 207.34 \ldots \text{ cm}^2$$

Total surface area $= 207$ cm^2

The surface area of a cylinder can always be calculated by drawing the net as you can see in this example.

However, if you prefer to learn a formula you can use the following:

- **The total surface area of a cylinder of height h and radius r is given by:**
  Total surface area $= 2\pi r^2 + 2\pi rh$

(b) The cuboid has 6 rectangular surfaces.

There are 2 rectangles 12×5, 2 rectangles 3×5 and 2 rectangles 12×3.

Total surface area $= 2 \times 12 \times 5 + 2 \times 3 \times 5 + 2 \times 12 \times 3$

$$= 120 + 30 + 72 \text{ cm}^2$$

Total surface area of cuboid $= 222$ cm^2

NOTE: If you find it difficult to visualise the rectangles, sketch a simple net of the cuboid.

Exercise 8.5

Calculate the total surface area of the following:

1. a box with no lid

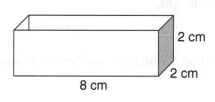

8 cm, 2 cm, 2 cm

2. a food can with its lid still on

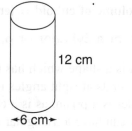

12 cm, 6 cm

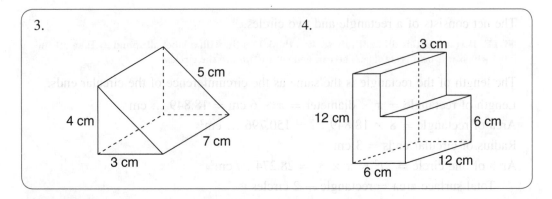

3.   4.

Volume

A volume is a measurement in three dimensions. For example, the amount of water that a bottle can hold, or the amount of space a brick takes up. As you will realise by now 3 dimensions means that 3 length measurements are needed to calculate a volume. You can also use an area and a length to calculate a volume because the area is already made up from two length measurements multiplied together.

A volume can either be measured in cubic units, for example, cm^3, or in units of *capacity*. Capacity is, for example, the amount of a liquid, which a container can hold.

Units of capacity are, for example, litres and millilitres. The calculations are the same, it is only the units used which are different. (See the table in Essential Skills at the beginning of the chapter for conversions.)

The unit of volume in the diagram below is a cube with sides measuring 1 cm, and its volume is 1 cm $\times$ 1 cm $\times$ 1 cm = 1 cm^3.

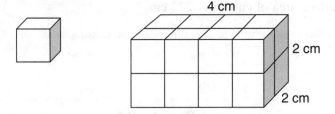

The cuboid measures 4 cm $\times$ 2 cm $\times$ 2 cm. It is made up from 16 one centimetre cubes, so its volume is $4 \times 2 \times 2$ cm^3.

The formula for the volume of a cuboid is:

- **volume of cuboid = length $\times$ breadth $\times$ height.**

Volume of a cylinder or prism

A prism is a shape which has the same cross-section throughout, and for our purposes, it has its ends at right angles to its length.

A cylinder is a prism, as is a cube or a cuboid.

A prism can have a triangular cross section, or a pentagonal or hexagonal cross section. Many pencils have hexagonal cross sections, and before it is sharpened such a pencil would be a hexagonal prism. Some containers for sweets are triangular prisms. Beams made from steel for the construction of buildings can have complicated cross sections, such as H shapes, but they are still prisms.

The volume of each of these prisms is calculated by multiplying the area of the cross section by the length (or height if it is standing up!) of the prism.

- **Volume of a prism = area of cross section × length.**
- **Volume of a cylinder = area of circular end × height**
$$= \pi r^2 h$$

Example 10

Calculate the following:

(a) The volume of a triangular prism which has a cross section which is a triangle with a height of 3 cm and a base of 4 cm. The length of the prism is 15 cm.

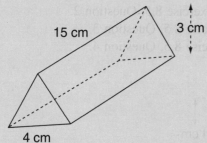

(b) A pipe has a cross section which is a circle with diameter 2.5 cm. The length of the pipe is 105 m. How much water can it hold? Give your answer in litres.

Answer 10

(a) The area of the triangular end (or cross section) of the prism is:

$\dfrac{1}{2} \times$ base $\times$ height

Area of end $= \dfrac{1}{2} \times 4 \times 3 = 6$ cm²

Volume = area of end × length

$= 6 \times 15 = 90$ cm³

Volume of prism = 90 cm³

(b) The area of the circular end of the pipe $= \pi r^2$

The radius of the pipe $= \dfrac{1}{2} \times 2.5 = 1.25$ cm

The units must be the same for the radius and the length, so we can choose either centimetres or metres.

Working in metres:

the radius of the pipe = 0.0125 m.

Area of end = $\pi \times 0.0125^2$ m^2

Volume of pipe = area of end × length

$$= \pi \times 0.0125^2 \times 105 \text{ m}^3$$

$$= 0.051541 \dots \text{ m}^3$$

Volume = 51.541 litres   (1 m^3 = 1000 litres)

Volume of pipe = 51.5 litres

Exercise 8.6

1. Calculate the capacity of the cuboid in Exercise 8.5, Question 1.
   (1 millilitre = 1 cm^3)
2. Calculate the capacity of the cylinder in Exercise 8.5, Question 2.
3. Calculate the volume of the prism in Exercise 8.5, Question 3.
4. Calculate the volume of the prism in Exercise 8.5, Question 4.
5. Calculate the volumes of these prisms:

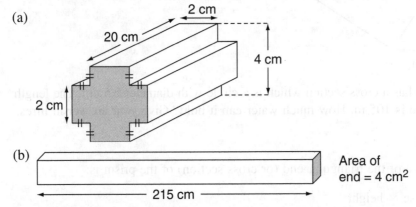

(a)

(b)

6. A cuboid has a volume of 25 cm^3. Its length is 5 cm, and its breadth is 2.5 cm.
   Calculate the height of the cuboid.
7. The length of a triangular prism is 16 cm, and its volume is 36 cm^3. Calculate
   the area of the cross section of the prism.
8. The volume of a cylinder is 18 cm^3, and its height is 4 cm. Calculate the radius
   of the circular cross section of the cylinder.
9. A plastic pipe has a cross section as shown in the diagram.
   The outer circle has a radius of 20 cm, and the inner circle
   has a diameter of 10 cm.
   (a) Calculate the shaded area.
   (b) The pipe is 1 metre long. Calculate the volume of plastic
       used to make the pipe.
   (c) Calculate the capacity of the pipe, giving your answer in millilitres.

Exercise 8.7

Mixed Exercise

1. A farmer has three steel drinking troughs.
   A has a semicircular cross section, B has a triangular cross section, and C is a cuboid. All are open at the top.
   (a) Calculate the area of steel required to manufacture each trough (that is: the total surface area of each trough).
   (b) Calculate the capacity of each trough.

   (c) Find the ratio $\dfrac{\text{volume}}{\text{surface area}}$ for each trough.

   (d) Which is the most economical shape? (Maximum volume for minimum steel)?

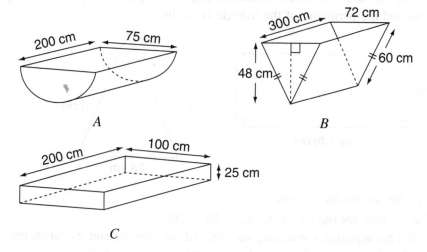

A

B

C

2. A laboratory measuring cylinder has a capacity of 250 millilitres when filled to the top.
   It has a diameter of 5 cm and a height of h cm.
   Calculate h.

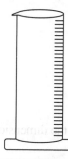

3. Another measuring cylinder is 20 centimetres tall and has a capacity of 100 millilitres when filled to the top with water.
   (a) Calculate the radius of this cylinder.
   (b) The contents of the measuring cylinder are poured into a container which is a cylindrical beaker with a radius of 2.5 cm. How deep is the water in the beaker?

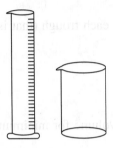

4. The diagram shows a rectangle and a triangle. The perimeter of the rectangle is 20 cm, and the perimeter of the triangle is 16 cm.

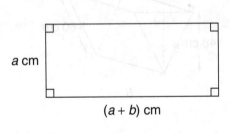
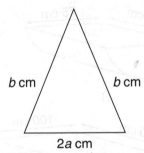

Using the information given:
   (a) show that, for the rectangle, $4a + 2b = 20$,
   (b) find the equation connecting the sides of the triangle and its perimeter,
   (c) solve the two equations simultaneously to find a and b,
   (d) write down the length and breadth of the rectangle.

5. The diagram shows a parallelogram and a rectangle. The perimeter of the parallelogram is 22 cm, and the perimeter of the rectangle is 14 cm.

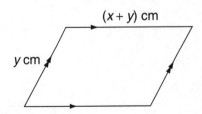
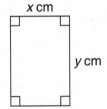

Form two equations in x and y, and solve them simultaneously to find the dimensions of the parallelogram.

Examination Questions

6.

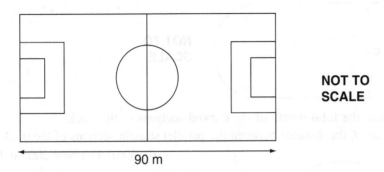

NOT TO
SCALE

90 m

(a) The diagram shows the plan for a new soccer field.
The length of the pitch is 90 metres.
The ratio length : width is 5 : 3.
Calculate the width of the pitch.

(b) The centre circle has a circumference of 57.5 metres.
Calculate the radius.

(0580/01 May/June 2006 q 17)

7.

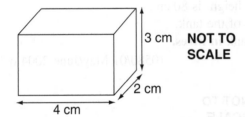

3 cm   **NOT TO
SCALE**

2 cm

4 cm

The solid shown is a cuboid with length 4 cm, width 2 cm and height 3 cm.

(a) Draw an accurate net of the cuboid on 1 centimetre squared paper.
(b) Using your net, calculate the total surface area of the cuboid.

(0580/01 May/June 2006 q 18)

8.

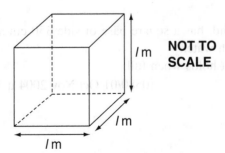

l m   **NOT TO
SCALE**

l m

l m

A cube of side *l* metres has a volume of 20 cubic metres.
Calculate the value of *l*.

(0580/01 May/June 2006 q 5)

9. A 400 metre running track has two straight sections, each of length 120 metres, and two semicircular ends.

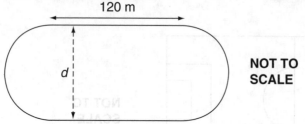

120 m

d

NOT TO SCALE

(a) Calculate the **total** length of the **curved** sections of the track.
(b) Calculate *d*, the distance between the parallel straight sections of the track.

(0580/01 Oct/Nov 2005 q 18)

10.

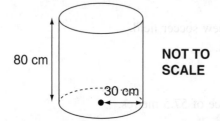

80 cm

NOT TO SCALE

30 cm

The diagram shows a cylindrical tank.
The radius is 30 cm and the height is 80 cm.
(a) Calculate the area of the base of the tank.
(b) Calculate the volume of the tank **in litres.**

(0580/01 May/June 2004 q 19)

11.

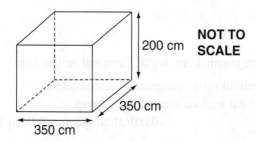

200 cm

NOT TO SCALE

350 cm

350 cm

A large tank, in the shape of a cuboid, has a square base of side 350 cm and height 200 cm. The tank is filled with water.
Find, in **litres**, the volume of water it holds when full.

(0580/01 Oct/Nov 2004 q 15)

12.

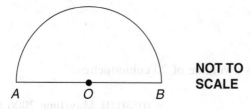

NOT TO SCALE

A   O   B

The diagram shows half of a circle, centre *O*.

(a) What is the special name of the line *AB*?

(b) *AB* = 12 cm.

    (i) Calculate the perimeter of the shape.

    (ii) Calculate the area of the shape.

(0580/01 Oct/Nov 2004 q 22)

13. The diagram shows a swimming pool with cross-section *ABCDE*.
The pool is 6 metres long and 3 metres wide.
AB = 2 m, *ED* = 1 m and *BC* = 3.6 m.

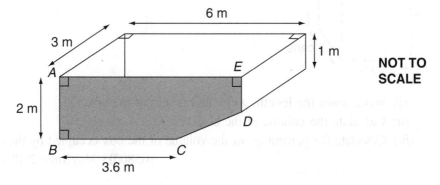

(a) (i) Calculate the area of the cross-section *ABCDE*. **Show your working**.

    (ii) Calculate the volume of the water in the pool when it is full.
Give your answer in **litres**. [1 cubic metre is 1000 litres.]

    (iii) One litre of water evaporates every hour for each square metre of the water surface. How many litres of water will evaporate in 2 hours?

(b) **Another pool** holds 61 500 litres of water.
Jon uses a hosepipe to fill this pool.
Water flows through the hosepipe at 1000 litres per hour.

    (i) Calculate how long it takes to fill the pool.
Give your answer in hours and minutes.

    (ii) Change 61 500 litres to gallons.
[4.55 litres = 1 gallon.]

    (iii) Every 10 000 **gallons** of water needs 2.5 litres of purifier.
How many litres of purifier does Jon use for this pool?

    (iv) The purifier is sold in 1 litre bottles.
How many **bottles** of purifier must Jon buy for this pool?

(0580/03 Oct/Nov 2005 q 6)

14. The area of a square is 42.25 cm².
Work out the length of one side of the square.

(0580/01 Oct/Nov 2008 q 4)

15. (a)   (i) Calculate the area of a circle with radius 3.7 centimetres.

(ii) A can of tomatoes is a cylinder with radius 3.7 centimetres and height *h* centimetres. The volume of the cylinder is 430 cubic centimetres. Calculate *h*.

(b) Twelve cans fix exactly inside a box 3 cans long, 2 cans wide and 2 cans high.

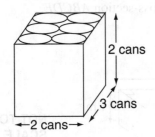

(i) Write down the length, width and height of the box.

(ii) Calculate the volume of the box.

(iii) Calculate the percentage of the volume of the box occupied by the cans.

(0580/03 May/June 2008 q 5)

Chapter 9

Trigonometry

Trigonometry involves the calculations of the angles and the lengths of sides in triangles. Your Core trigonometry work will be entirely with right-angled triangles, and you will find out how to calculate the remaining angles and sides of triangles which have some known measurements.

You have seen that triangles need three measurements to define them exactly. In this chapter, one of these measurements will be a right angle. The other two measurements may be two sides or one side and another angle.

You will need graph paper or squared paper, protractor, ruler and a scientific calculator for this chapter.

Please note that if you are asked to **calculate** measurements in triangles you will get **no marks** for scale drawings! This does not mean that you cannot occasionally make a scale drawing to check your own work, or to try to understand a problem better. Do **not**, however, give it as your working in your examination!

Essential Skills

1. Round to 3 significant figures.
   (a) 56.1935   (b) 7.9514   (c) 6.1445
2. Round to 1 decimal place.
   (a) 60.199   (b) 72.954   (c) 14.57801
3. Find x, giving to 3 significant figures if not exact.
   (a) $x^2 = 4$ (b) $x^2 = 5$ (c) $x^2 = 10.34$
4. Calculate x, giving your answers to 3 significant figures if necessary.
   (a) $6.9x = 5.1$ (b) $7.3 = \dfrac{x}{4.6}$ (c) $9.2 = \dfrac{7.9}{x}$
5. Make x the subject:
   (a) $ax = b$ (b) $a = \dfrac{x}{b}$ (c) $a = \dfrac{b}{x}$
6. (a) Find the area of a square of side 3.9 centimetres.
   (b) Find the length of the sides of a square with area 5.7 cm², giving your answer to 3 significant figures.

7.

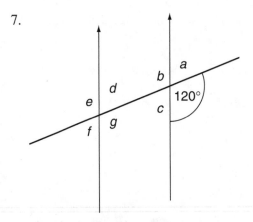

(a) Using the diagram above name a pair of:
   (i) corresponding angles,
   (ii) alternate angles,
   (iii) vertically opposite angles.
(b) Find the sizes of all the angles in the diagram.

Answers

1. (a) 56.2   (b) 7.95   (c) 6.14
2. (a) 60.2   (b) 73.0   (c) 14.6
3. (a) $x = 2$ (b) $x = 2.24$ (c) $x = 3.22$
4. (a) $x = 0.739$ (b) $x = 33.58$ (c) $x = 0.859$
5. (a) $x = \dfrac{b}{a}$ (b) $x = ab$ (c) $x = \dfrac{b}{a}$
6. (a) Area $= 15.21 \text{ cm}^2$ (b) Length of side $= 2.39$ cm
7. (a) (i) for example, d and a (ii) for example, d and c
   (iii) for example, d and f
   (b) $a = 60°$, $b = 120°$, $c = 60°$, $d = 60°$, $e = 120°$, $f = 60°$, $g = 120°$

NOTE: In your answers to the questions in this chapter give all angles to 1 decimal place, and the lengths of all sides to 3 significant figures unless otherwise stated or the answers are exact. Always give the units of the lengths as part of your answer.

The Tangent Ratio

Practical investigation

To get the most out of this work you need to be as accurate as possible, so have a sharp pencil and work carefully.

Before you start work make sure that your calculator is in *degree mode*.

Try this to check: press ⬚tan ⬚45 ⬚= . If your calculator has a different logic you might have to press 45 tan. Either way you should get 1 *exactly*. If you get either 1.6 … or 0.85 …, your calculator is in the wrong mode and you may have to refer to your instruction book to change it to degrees. Most calculators will show either a D or *deg* on the display if they are in degree mode, but it can be very small and difficult to read.

It is very important that you get to know your own calculator, its logic (the order in which things must be entered) and its settings.

- Take a sheet of 1 mm or 2 mm squared graph paper and draw *x*- and *y*-axes with scales of 1 cm to represent 1 unit on each axis.
- Draw an angle with its vertex at the origin and one side along the *x*-axis as in the first diagram below. Make it a different angle from the one shown here and make the lines at least 12 cm long.
- At convenient places (where the line crosses the intersection of two grid lines) draw some perpendiculars from the top line down to the base line as in the second diagram, thus making a set of similar triangles.
- Starting with the smallest triangle copy and complete the table below for your own drawing. The *y*-measurement is the height of the triangle, and the *x*-measurement is the length of the base of the same triangle.
- Calculate $\dfrac{y}{x}$ to 2 decimal places.

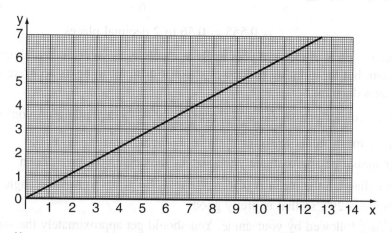

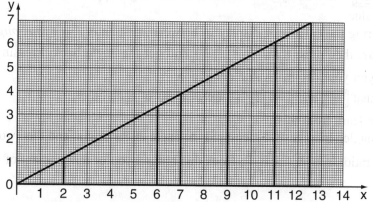

y-measurement	1.1	3.3	3.9	5	6.1	7
x-measurement	2	6	7	9	11	12.6
$\dfrac{y}{x}$	0.55	0.55	0.56	0.56	0.55	0.56

The table has been filled in for the angle shown above, but yours will not be the same, because you will have drawn a different angle and different perpendiculars.
You should notice that the bottom row of your table has nearly the same value all the way along. The values may not be exactly the same because of experimental error (even the thickness of your pencil line can make the work less accurate), but they should be very similar. If one is very different have another look at it and make sure that you have not made a mistake.
Find the mean of all your values. Calculating the mean involves adding up all the values along the bottom row and dividing by the number of values you have entered in that row.
For the table above:

the mean value of the ratio $\dfrac{y}{x} = \dfrac{0.55+0.55+0.56+0.56+0.55+0.56}{6} = \dfrac{3.33}{6}$

$$= 0.555 = 0.56 \text{ to 2 decimal places}$$

The next thing to do is to measure your angle as carefully as you can. It is about 29° in the given diagram, but the printing processes make it difficult to be as accurate as you can be in your own diagram.

Using your calculator press [shift] [tan] (or [2nd function] [tan] depending on your calculator) and enter your mean value. Press [=].
The answer *should* be approximately the size of the angle you have drawn.

For the given diagram the mean is 0.56, and [shift] [tan] [0.56] gives the angle 29.2°. The measured angle is 29°, so that is pretty good!
Now press [tan] followed by your angle. You should get approximately the same as your ratio mean value.

Without clearing the calculator display in between, first enter [tan] *your angle*, note the decimal value, then press [shift] [tan] [=] and get the angle back. Pressing [tan] [=] again gives you your decimal value once more.

This shows that [tan] and [shift] [tan] are *inverses* of each other.

What can we learn from this?

In our right-angled triangles we measured the angle at the origin, and then worked out the value of the ratio $\dfrac{y\text{-measurement}}{x\text{-measurement}}$. Does this remind you of the gradient of a line?

Can you see that the *y-measurement* is the length of the side **opposite** the angle, and the *x-measurement* is the length of the *side next to* the angle?

The mathematical name for something which is next to (or next door to) is **adjacent**.

The triangle has one more side, the one opposite the right angle, which is the longest side and is called the **hypotenuse**.

The names of the sides are summarised in the diagram below.

If you have trouble seeing which side is adjacent to and which is opposite your angle, draw a straight line arrow from within the angle, as in the diagram. The side it points to is the *opposite* side. The longest side is the hypotenuse, so the one remaining is the adjacent side.

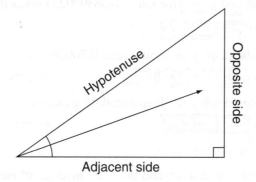

It is important that you are sure you can work out and remember which side is which before you go on.

Also you need to note that although the hypotenuse is always the longest side, and always opposite the right angle, the other two sides can change places according to which of the other two angles you are using. If you are interested in the other angle, as you can see in the diagram below, the opposite and adjacent change places.

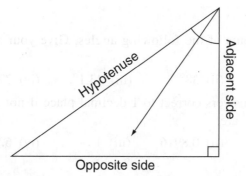

Looking back at our investigation work, you may have wondered why we pressed [shift] [tan] on the calculator.

The reason is that the ratio $\dfrac{\text{opposite side}}{\text{adjacent side}}$ for any angle in a right-angled triangle has been given the name **tangent**, and tan is short for tangent. Here, the word 'tangent' does not refer to a line touching a curve, which is also called a tangent.

We say that the tangent of an angle in a right-angled triangle is the ratio $\dfrac{\text{opposite side}}{\text{adjacent side}}$.

A note about calculators

If your calculator has a key labelled [Ans], you can press it to enter the answer to the previous calculation. If it does not have this key you may be able to use the [=] key. You can see this in the example below.

Your calculator probably shows the [tan] key with *tan*$^{-1}$ written above it. This is a very convenient notation for inverse tan, or shift tan. Pressing [tan] [20] finds the tangent of 20° (0.363970234 …), and [shift] [tan] (or tan^{-1}) 0.363970234 gives the angle back again.

Try this: enter [tan] [35] [=]

read the display (it should read 0.70020 …)

press [shift] [tan] [Ans] [=]   or press [shift] [tan] [=]

read the display (it should read 35 exactly)

press [tan] [Ans] [=]   or press [tan] [=]

read the display …

This is just to convince you that tan and tan^{-1} are inverses of each other!

NOTE: You need plenty of practice with your own calculator. Make sure that it is the same as the one you will be using in your examination!

Sometimes you might see the term *arctan* used to mean inverse tan (or tan^{-1}), but in this book we will use tan^{-1}.

From now on, tan^{-1} will mean [shift] [tan], or [second] [function] [tan].

Example 1

(a) Find the tangents of the following angles. Give your answers to 4 decimal places.

  (i) 20°   (ii) 38°   (iii) 43.1°   (iv) 77.2°

(b) Giving your answers correct to 1 decimal place if not exact, find the angles whose tangents are:

  (i) 0.3249   (ii) 0.8916   (iii) 1   (iv) 6.5184

Answer 1

(a) (i) tan 20° = 0.3640   (ii) tan 38° = 0.7813

  (iii) tan 43.1° = 0.9358   (iv) tan 77.2° = 4.4015

(b) (i) tan^{-1} 0.3249 = 18.0°   (ii) tan^{-1} 0.8916 = 41.7°

  (iii) tan^{-1} 1 = 45°   (iv) tan^{-1} 6.5184 = 81.3°

Exercise 9.1

1. Find the tangents of the following angles. Give your answers to 4 decimal places.

   (a) 56°   (b) 75°   (c) 27.12°   (d) 30°   (e) 60°   (f) 49.4°

2. Giving your answers correct to 1 decimal place if they are not exact, find the angles whose tangents are:

   (a) 0.1651   (b) 0.8013   (c) 1.6571   (d) 5.9503   (e) 14.5710   (f) 0.5

We now are able to make use of the tangent ratio, and first we will find an angle in a right-angled triangle where two of the sides are known.

Here is a good routine to follow in order to get the best marks in an examination question.

- Remember that the tangent ratio is $\dfrac{\text{opposite side}}{\text{adjacent side}}$, which is often abbreviated to $\dfrac{\text{OPP}}{\text{ADJ}}$.
- Check that the triangle has a right angle.
- If necessary sketch the triangle.
- Label the sides in the triangle OPP and ADJ, in relation to the angle you are calculating.
- Calculate the tangent ratio as a decimal.
- Write down the decimal, but *do not clear your calculator*.
- Press ⌜shift⌟ ⌜tan⌟ ⌜Ans⌟ (or ⌜shift⌟ ⌜tan⌟ ⌜=⌟) to get the angle.
- *Without rounding* write down the angle to a few decimal places.
- Give the answer correct to 1 decimal place.

NOTE: Think 'the angle whose tangent is ...' when you see tan⁻¹...

Example 2

In the diagram, triangle *ABC* has angle *BAC* = 90°, *BA* = 5 cm, and *CA* = 7 cm.

Find angle *BCA*. Give your answer correct to 1 decimal place.

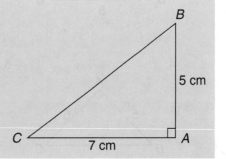

Answer 2

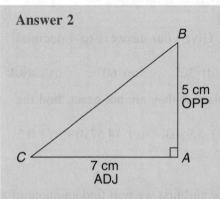

$\tan$ angle $BCA = \dfrac{\text{OPP}}{\text{ADJ}} = \dfrac{5}{7} = 0.71428 \dots$

$\tan^{-1}$ Ans $= 35.537 \dots$

angle $BCA = 35.5°$ correct to 1 decimal place.

NOTE: For maximum marks in an examination follow the above method. DO NOT ROUND your tangent ratio before finding tan⁻¹ or you will lose accuracy marks. This is referred to as premature approximation, and is an error that can so easily be avoided. Never round until the end of the question or part question.

Exercise 9.2

Calculate the angle B in each of these triangles. Measurements are all in centimetres. Give your answers correct to 1 decimal place unless exact.

1.

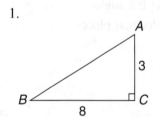

2.

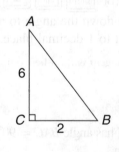

3.

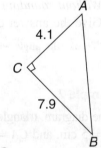

4.

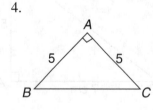

5.

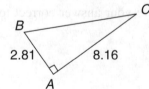

6. In triangle *ABC*, ∠*BAC* = 90°, *AB* = 10 cm and *AC* = 7 cm. Calculate ∠*ABC*.

NOTE: Sketch the triangle *ABC*. Make sure that you label it correctly with ∠*BAC* = 90°, and make it *look* like a right angle, so that you are less likely to make mistakes!

7. In triangle *ABC*, ∠*BCA* = 90°, *AC* = 7.96 cm and *BC* = 3.52 cm. Calculate ∠*ABC*.

NOTE: Note that ∠*BCA* is the right angle.

Finding a side

If you are given an angle and the length of the opposite side you can use the tangent ratio to find the length of the adjacent side. Alternatively, if you are given the length of the adjacent side and an angle you can find the length of the opposite side.

Finding the opposite side

We will start with finding the length of an opposite side. Using the triangle *PQR* below as an example, we will calculate the length of *QR*.
The routine is as follows.

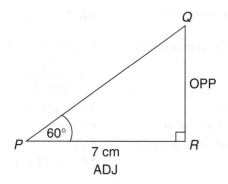

- Check the right angle.
- Label the sides.
- Write down: $\quad\quad\quad\quad\quad\quad\quad\quad$ tan angle $QPR = \dfrac{\text{OPP}}{\text{ADJ}}$

- Fill in the known measurements: $\quad\quad$ $\tan 60 = \dfrac{\text{OPP}}{7}$
- Using normal algebra methods
  rearrange: $\quad\quad\quad\quad\quad\quad\quad$ $7 \times \tan 60 = \text{OPP}$ $\quad$ (× 7)
- Enter into calculator: $\quad\quad\quad\quad$ $7 \times \tan 60 =$
- Write down calculator display: $\quad\;$ 12.12435 ...
- Round to 3 significant figures: $\quad$ 12.1
- Answer: $\quad\quad\quad\quad\quad\quad\quad\quad$ $QR = 12.1$ cm

Finding the adjacent side

Using triangle *XYZ* as an example, calculate side *YZ*.
This is a good time to start using a convenient new notation. If a length of a side is unknown it may be referred to by using the lower case letter of the opposite angle, as shown in the diagram. We use *x* to represent the length of the side opposite *X*.

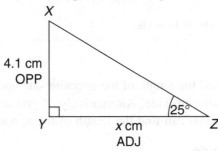

- Check the right angle.
- Label the sides.
- Write down: $\qquad$ tan angle $YZX = \dfrac{\text{OPP}}{\text{ADJ}}$

- Fill in the known measurements: $\qquad$ $\tan 25 = \dfrac{4.1}{x}$

- Use algebra methods to rearrange: $\qquad$ $x \times \tan 25 = 4.1 \qquad (\times \, x)$

$$x = \frac{4.1}{\tan 25} \qquad (\div \tan 25)$$

- Enter into calculator: $\qquad$ $4.1 \div \tan 25 =$
- Write down calculator display: $\qquad$ 8.79247...
- Give answer to 3 significant figures: $\qquad$ 8.79
- Answer: $\qquad$ $YZ = 8.79$ cm

Example 3

In triangle *DEF*, angle *DEF* is a right angle, angle *FDE* is 56°, and *EF* is 6.3 centimetres. Calculate the length of *DE*.

Answer 3

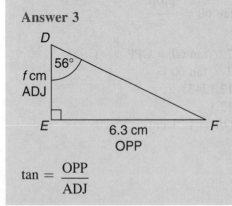

$$\tan = \frac{\text{OPP}}{\text{ADJ}}$$

$$\tan 56 = \frac{6.3}{f}$$

$f \times \tan 56 = 6.3 \qquad (\times f)$

$f = \dfrac{6.3}{\tan 56} \qquad (\div \tan 56)$

$f = 4.24940 \ldots$

$f = 4.25$

Answer: $DE = 4.25$ cm

Exercise 9.3

1. Calculate the side marked with a letter in each of the following triangles.
   The lengths of the sides are all in centimetres.

   (a)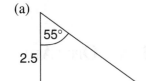

   (b)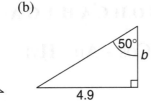

   (c)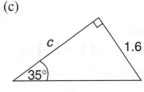

   (d)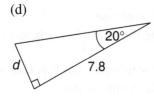

   (e)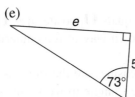

   (f)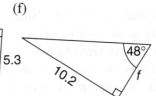

2. Calculate the angles or lengths of sides marked with letters. All lengths are in centimetres.

   (a)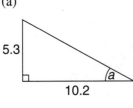

   (b)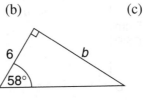

   (c)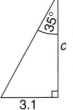

   (d)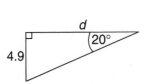

   (e)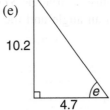

The Sine and Cosine Ratios

If you were to repeat the practical work on page 278, but this time measure the length of the hypotenuse in each triangle, you would find that, not only is the tangent ratio $\left(\dfrac{y}{x}\right.$, or

$\left.\dfrac{\text{OPP}}{\text{ADJ}}\right)$ constant for each angle, but so are the ratios $\dfrac{\text{OPP}}{\text{HYP}} \left(\dfrac{y}{\text{HYP}}\right)$ and $\dfrac{\text{ADJ}}{\text{HYP}} \left(\dfrac{x}{\text{HYP}}\right)$.

The ratio $\dfrac{\text{OPP}}{\text{HYP}}$ is called the SINE of the angle, normally abbreviated to SIN.

The ratio $\dfrac{\text{ADJ}}{\text{HYP}}$ is called the COSINE of the angle, normally abbreviated to COS.

You will need to learn the three ratios.
A made up 'word' is often used to remember these ratios:

S O H C A H T O A

This stands for:

SIN = **O**PP / **H**YP,   **C**OS = **A**DJ / **H**YP,   **T**AN = **O**PP / **A**DJ

It may be helpful to write:

O   **A**   **O**
S   H   **C   H**   **T   A'**

to show that, for example, **S**ine equals **O**pposite *over* **H**ypotenuse.
You may want to make up your own way of remembering the ratios, but, whatever you do, you need to learn them!
The routines for using sines and cosines are the same as for tangents, but it is now more important to label the sides of the triangle to work out which ratio to use.
We will work through one example to show this step.

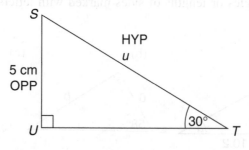

Find the length of the hypotenuse in triangle *STU*.
In this triangle, we are given an angle and the opposite side and are asked to find the hypotenuse.

- Label the sides we are using (in this case OPP and HYP).
- Choose appropriately sine, cosine or tangent. Using SOHCAHTOA we can see that it is SINE, which links OPP and HYP.

- Write down: $\quad\quad\quad\quad\quad\quad\quad\quad \sin = \dfrac{OPP}{HYP}$

- Fill in the measurements. $\quad\quad\quad \sin 30 = \dfrac{5}{u}$

- Use algebra to rearrange: $\quad\quad u \times \sin 30 = 5 \quad\quad (\times u)$

$$u = \dfrac{5}{\sin 30} \quad\quad\quad (\div \sin 30)$$

- Enter $5 \div \sin 30$ into calculator.
- Write down display: $\quad\quad\quad 10$
- Answer is exact so $\quad\quad\quad u = 10$
- Answer: $\quad\quad\quad\quad\quad\quad\quad ST = 10$ cm

Example 4

Calculate the measurements represented by letters in the following triangles.

(a) 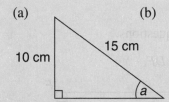   (b) 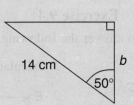   (c)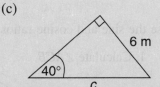

Answer 4

(a)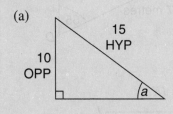

$$\sin = \dfrac{OPP}{HYP}$$

$$\sin a = \dfrac{10}{15} = 0.66666\,...$$

$$\sin^{-1} \text{Ans} = 41.813\,...$$

$$a = 41.8°$$

(b)

$$\cos = \dfrac{ADJ}{HYP}$$

$$\cos 50 = \dfrac{b}{14}$$

$$14 \times \cos 50 = b \quad\quad (\times 14)$$

$$b = 8.999020\,...$$

$$b = 9.00 \text{ cm}$$

(c)

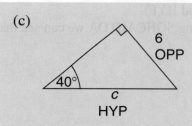

$$\sin = \frac{\text{OPP}}{\text{HYP}}$$

$$\sin 40 = \frac{6}{c}$$

$$c \times \sin 40 = 6 \qquad (\times c)$$

$$c = \frac{6}{\sin 40} \qquad (\div \sin 40)$$

$$c = 9.3343\ ...$$

$$c = 9.33 \text{ m}$$

NOTE: Remember, we are always referring to the ratios of the lengths of sides. The lengths must always be expressed in the same units. (For example, in cm or m or km, but not in mixtures of any two!)

NOTE: The maximum value of the sine or cosine of any angle is 1. If you enter, for example, $\sin^{-1} 1.21$ into your calculator you will get an error message and, if this is the result of calculation you must have made a mistake in the calculation. However, the tangent of an angle can be any number. But, note that the tangent of 90° does not exist!

Exercise 9.4

Use the sine and cosine ratios to answer the following questions.

1. Calculate ∠*ACB*.

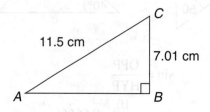

2. Calculate *DF*.

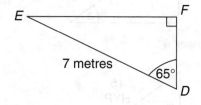

3. Calculate ∠*IGH*.

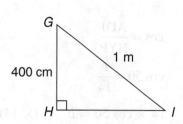

4. Calculate *JK*.

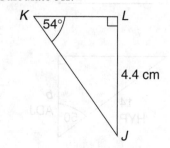

NOTE: In questions 5 to 10 draw the triangles and label the sides before trying to do the calculations. In each question just sketch a right-angled triangle and make sure that you write the correct letter beside the right angle, otherwise your answer will be wrong!

5. In triangle *MNP*, $\angle P = 90°$, $\angle M = 31°$, *NP* = 10.6 cm. Calculate the length of *NM*.

6. In triangle *QRS*, $\angle Q = 90°$, *RS* = 9.46 cm, *SQ* = 5.12 cm. Calculate $\angle S$.

7. In triangle *TVW* , $\angle V = 90°$, *VW* = 11.5 cm, $\angle T = 80°$. Calculate the length of *WT*.

8. In triangle *XYZ*, $\angle Y = 90°$, *ZX* = 51 metres, $\angle X = 57°$. Calculate the length of *ZY*.

9. In triangle *ABC*, $\angle C = 90°$, *AB* = 12.7 cm, *AC* = 11.3 cm. Calculate $\angle B$.

10. In triangle *DEF*, $\angle F = 90°$, $\angle D = 37°$, *DF* = 7.23 metres. Calculate the length of *DE*.

Pythagoras' Theorem

There is one more useful relationship that you need to know, this time connecting the lengths of all the sides of a right-angled triangle. The theorem actually connects the *areas* of the *squares* drawn on each of the sides, and from these we can work out the lengths of the sides.

Pythagoras was a Greek philosopher who lived 2500 years ago. He found that the 'square on the hypotenuse of a right-angled triangle is equal to the sum of the squares on the other two sides'.

This has been very important ever since and is used every day in, for example, architecture, engineering, surveying, science and so on. It was especially important for the Greeks in their architecture.

The diagram illustrates Pythagoras' theorem.

You will see that the 'square on the hypotenuse' and the 'squares on the other two sides' have been drawn.

If you trace this diagram and cut the two smaller squares up into smaller pieces you will be able to fit these pieces exactly onto the larger square, showing that the sum of the *areas* of the two smaller squares is indeed equal to the *area* of the largest square.

The diagrams show an easy method for cutting one of the two smaller squares into pieces which can be rearranged with the smallest square to fit exactly on the largest square. You can use this method for any right-angled triangle if you want to check this theorem.

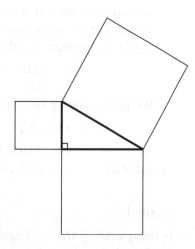

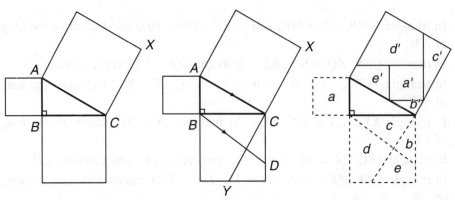

Draw a right-angled triangle. Draw *accurately* the squares on each of the sides.

Draw *BD* parallel to *AC*. Produce *XC* to *Y* so that *XY* is a straight line.

Cut out *a,b,c,d* and *e* and arrange on to *a',b',c',d',* and *e'* to exactly cover the square on the hypotenuse.

In practice the use of this theorem means that we can find the lengths of any of the three sides of a right-angled triangle if we know those of the other two sides.

For example, suppose we need to find the third side when the hypotenuse is 10 cm, and one of the other sides is 4 cm. We will call the unknown side *x*. The lengths of the sides are in centimetres.

- Sketch the triangle.
- Draw the squares *as recognisable squares*.

  NOTE: If you draw too carelessly you will not benefit from the visual information which helps you to see which two squares add to make the third.

- The areas of the two smaller squares add to make the area of the largest square, so the area of the square we want must be the *difference* between the other two.

$$x^2 = 100 - 16$$
$$x^2 = 84$$

- We undo squaring by square rooting.

$$x = \sqrt{84}$$
$$x = 9.16515 \ldots$$

- Answer: $x = 9.17$ cm

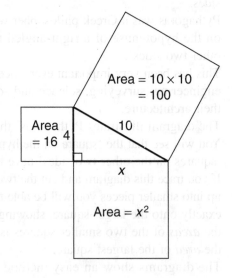

Area = 10 × 10
     = 100

Area = 16   10   4

x

Area = x^2

Example 5

In triangle *ABC*, *AB* = 6.5 centimetres, *BC* = 2.3 centimetres, and angle *ABC* = 90°.
(a) Sketch the triangle, showing all the given measurements.
(b) Calculate the length of *AC*.

Answer 5

(a)

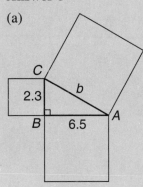

(b) The area of the square on the side $CB = 2.3^2$
The area of the square on the side $AB = 6.5^2$
The area of the square on the side $AC = b^2$
From the figure we can see that $b^2 = 2.3^2 + 6.5^2$

$$b^2 = 47.54$$
$$b = \sqrt{47.54}$$
$$b = 6.89492 \ldots$$

Answer $AC = 6.89$ cm

Work through the following exercise using the same routine, drawing the triangle and the squares on each side.

Exercise 9.5

1. Calculate *BC*.   2. Calculate *EF*.

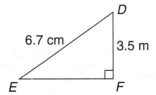

3. In triangle *PQR*, $\angle PRQ = 90°$, $PR = 6$ m, $QR = 11$ m. Calculate *PQ*.
4. In triangle *XYZ*, $\angle ZYX = 90°$, $ZY = 4.2$ cm, $ZX = 5.7$ cm. Calculate *XY*.

Once you are confident that you can see which two squares on which sides add to give the square on the third it is no longer necessary to draw the squares, but remember that the two smaller squares add to give the largest.

Example 6

Calculate the side marked x in the following diagram.

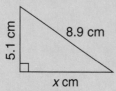

Answer 6

The hypotenuse is 8.9 cm.

So $x^2 = 8.9^2 - 5.1^2$

$\quad x^2 = 53.2$

$\quad x = \sqrt{53.2}$

$\quad x = 7.29383 \dots$

$\quad x = 7.29$ cm

NOTE: **Always finish by checking whether the answer is reasonable. If you find the side x to be longer than the hypotenuse you have made a mistake! Remember, however, that the diagrams will not be drawn to scale.**

Exercise 9.6

Calculate the lengths of sides marked with letters in the following diagrams.

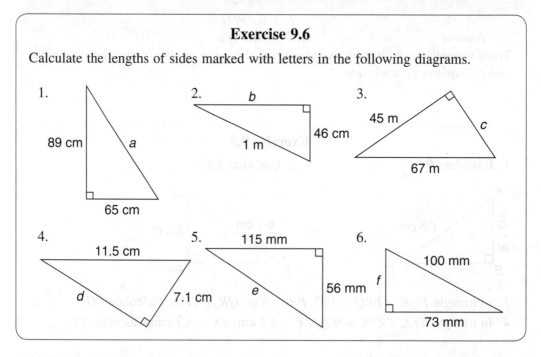

We will now practise choosing Pythagoras' Theorem or the sine, cosine and tangent ratios to solve right-angled triangles before going on to use them to solve problems.

- Remember that these only apply to right-angled triangles.
- Draw the diagram and write in the given measurements.
- Write a small letter beside each measurement you are asked to find.

- You now need to look carefully at the diagram to decide how to calculate the information.
- If you are given two sides and asked to find the third you will need to use Pythagoras' Theorem.
- If you are given two sides and asked to find an angle, or given an angle and a side you will need to use the sine, cosine or tangent ratios. In this case label the sides OPP, ADJ and HYP to help you decide which ratio to use.
- Remember SOHCAHTOA.

Exercise 9.7

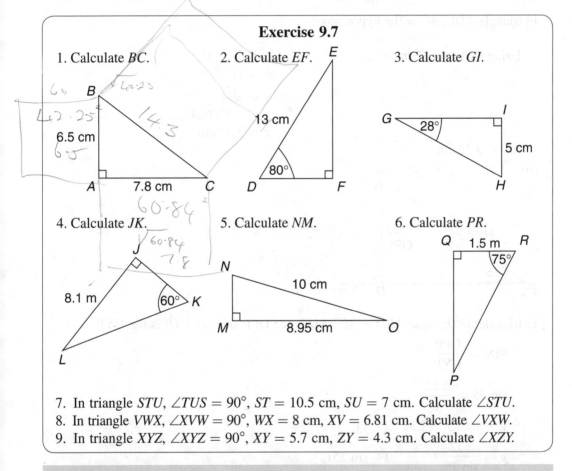

1. Calculate *BC*.

2. Calculate *EF*.

3. Calculate *GI*.

4. Calculate *JK*.

5. Calculate *NM*.

6. Calculate *PR*.

7. In triangle *STU*, ∠*TUS* = 90°, *ST* = 10.5 cm, *SU* = 7 cm. Calculate ∠*STU*.
8. In triangle *VWX*, ∠*XVW* = 90°, *WX* = 8 cm, *XV* = 6.81 cm. Calculate ∠*VXW*.
9. In triangle *XYZ*, ∠*XYZ* = 90°, *XY* = 5.7 cm, *ZY* = 4.3 cm. Calculate ∠*XZY*.

Example 7

The diagram shows the triangle *ABC* and the perpendicular *AD* from *A* to the side *BC*.

Using the information on the diagram calculate:

(a) *AD*

(b) *AB*

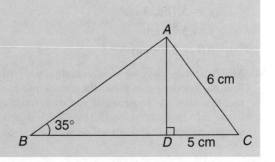

Answer 7

(a)

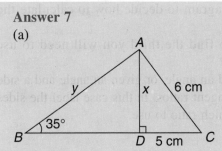

In triangle *ADC*, *AC* is the hypotenuse.

Using Pythagoras' theorem,

$$x^2 = 6^2 - 5^2$$
$$x^2 = 11$$
$$x = \sqrt{11}$$
$$x = 3.316624 \ldots$$
$$x = 3.32 \text{ cm}$$

$AD = 3.32$ cm

(b)

In triangle *ABD*, angle *ABD* = 35°, *AD* is the OPP side, and *AB* is the HYP.

$$\text{SIN} = \frac{\text{OPP}}{\text{HYP}}$$

$$\sin 35 = \frac{3.32}{y}$$

$$y \times \sin 35 = 3.32 \quad (\times y)$$

$$y = \frac{3.32}{\sin 35} \quad (\div \sin 35)$$

$$y = 5.78824 \ldots$$
$$y = 5.79 \text{ cm}$$
$$AB = 5.79 \text{ cm}$$

NOTE: In this case, for your examinations, it is acceptable to use the rounded answer (3.32) from the previous part (a) of the question to continue with part (b), but you should understand that this does lose accuracy. If you were working in a real life situation you would be expected to use the calculator value (3.316624 ...), which in this case actually gives the answer 5.78 to 3 significant figures. Either method will be accepted in your examination.

There are other shapes which have right-angled triangles, although they are not necessarily immediately obvious.

For example, the line of symmetry of an isosceles triangle, or the diagonal of a rectangle both divide the diagram into two equal right-angled triangles as is shown in the diagrams below.

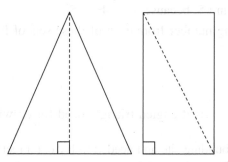

Example 8

ABC is an equilateral triangle with side 10 centimetres. Calculate the height of the triangle.

Answer 8

Sketch the triangle and draw the perpendicular height, *AD*. Let the height be *h* cm.

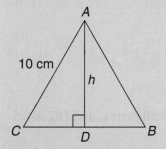

D is the midpoint of *CB* (symmetry of an equilateral triangle).
In triangle *ACD*, *AC* = 10 cm, *CD* = 5 cm, and $\angle ADC = 90°$.
Using Pythagoras:

$h^2 = 10^2 - 5^2$
$h^2 = 75$
$h = \sqrt{75} = 8.66025\ldots$

$h = 8.66$ cm

The height of the triangle is 8.66 centimetres.

NOTE: This problem could also be solved using the Sine, Cosine or Tangent ratios. Remember that the angles in an equilateral triangle are all 60°, and that the symmetry of the figure means that angle *CAB* = 30°.

Exercise 9.8

NOTE: When necessary draw sketches before answering the questions.

1. Pythagorean Triples are sets of whole numbers which, when used as the lengths of the sides of a triangle, make it a right-angled triangle.
   For example, 3, 4 and 5, because $3^2 + 4^2 = 5^2$.

   (a) Find the missing number from each of these sets of Pythagorean triples.

       (i) 5, ..., 13
       (ii) 6, 8 ...
       (iii) ..., 24, 25

   (b) A set square is a right-angled triangle used for drawing or setting out right angles.
   Lena was on a building site and needed to lay out a rectangle for the foundation of a building. She had one strip of wood 5 metres long and another 7 metres long, a saw and a tape measure. Describe how she could make a set square.

2. Calculate the length of the diagonal of a square which has sides of 4 cm.

3. Calculate the height of an equilateral triangle which has sides of length 10 cm.

4. The diagram shows the side of a house with a pitched roof. *BC* and *CD* are 4.5 metres each. *DE* is 6 metres and *AE* is 8 metres.

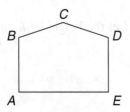

   Calculate the total height of the house.

5. *ABCD* is a trapezium. *AB* is 5 centimetres, *AD* is 4 centimetres, and *DC* is 8.5 centimetres.

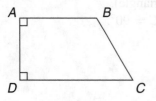

   NOTE: Copy the diagram and draw a perpendicular from *B* down to the side *CD* to make a triangle and a rectangle.

   Calculate (a) *BC*,   (b) $\angle BCD$,   (c) $\angle ABC$.

6. Using the diagram, calculate the length *AE*.

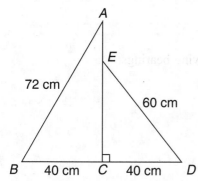

7. In the diagram below, $AB = 16$ centimetres, $AD = 20$ centimetres, $CD = 15$ centimetres.

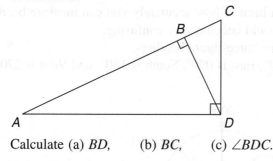

Calculate (a) *BD*,   (b) *BC*,   (c) $\angle BDC$.

Bearings

A bearing is the direction of one place from another, using due North as the reference direction.

The **mariner's compass**, shown in the diagram below uses the four main directions, North, South, East and West. These can then be divided into four more directions, as shown in the diagram. The order of the letters for each bearing is always given with the main bearing first, with North and South before East and West. For example, we would say North-West, not West-North.

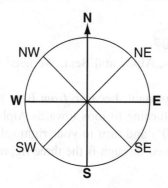

Half way between North and North-East is North-North-East or NNE, and in between North-East and East is East-North-East or ENE.

Example 9

Name, and give the abbreviations for, the following bearings.

(a) Half way between South-West and West

(b) Half way between South-East and South

Answer 9

(a) West-South-West, or WSW

(b) South-South-East, or SSE

With the mariner's compass there is a limit to how accurately you can measure bearings because the naming of the bearings would become very confusing.

The bearings used more often today are three-figure bearings.

For three figure bearings, North is 000°, East is 090°, South is 180°, and West is 270°. This is shown in the diagram below.

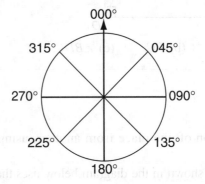

As you will see, the angles are measured in a clockwise direction, starting from North, 000°, and going all the way round to 360°, which is North again. This allows many more directions to be specified clearly, for example, 021° which would be difficult to express with the mariner's compass. Angles less than 100° normally have a zero in front to make them three-figure.

Using three figure bearings

The diagram shows two towns, Alpha and Beta. We need to know the bearing of Alpha *from* Beta.

The way this is written is important, because *from* Beta means that you must imagine yourself standing at Beta and turning to look towards Alpha.

Stand at Beta, face North (000°) and turn to your right (clockwise) until you are facing Alpha. The angle you have turned through is the three figure bearing of Alpha from Beta.

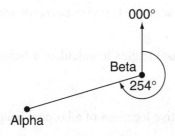

It is very convenient to have a 360° protractor for measuring bearings. If you do not have a 360° protractor, you need to either measure the obtuse angle in the above diagram and subtract it from 360°, or draw the South (180°) line in the diagram, measure the acute angle and add 180°.

Check with the diagram above that you can get the correct bearing by measuring. It should be about 254° allowing for slight errors in the printing process.

The next thing to be aware of is that the bearing of Beta from Alpha is not the same as the bearing of Alpha from Beta. For this you need to draw in another North line at Alpha, making sure that it is parallel to the north line at Beta.

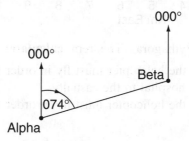

Now imagine you are standing at Alpha looking North, and then turn to your right (clockwise) until you are facing Beta. The angle you have turned through will give you the bearing of Beta from Alpha. It is 074° as shown in the diagram.

You should be aware that angles in parallel lines frequently come into bearings questions because North lines are always parallel. Drawing in the South lines will often help with calculations. The next diagram shows how the South lines help to show the relationship between the bearings of Alpha from Beta (254°) with the bearing of Beta from Alpha (74°) above. In the diagram 254° is split into 180° and 74°.

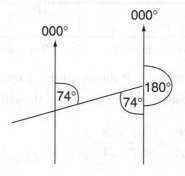

If you are not sure about these angles, revise *corresponding*, *alternate*, and *vertically opposite* angles.

The next example uses alternate angles to calculate a bearing.

Example 10

The diagram shows the relative locations of a hospital helipad (H) and a casualty (C).

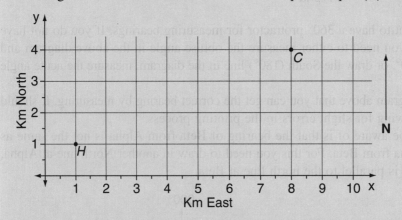

Using trigonometry and Pythagoras' Theorem, **calculate:**

(a) the bearing on which the helicopter must fly in order to reach the casualty,
(b) the distance from the hospital to the casualty,
(c) the bearing on which the helicopter must fly in order to return to the hospital.

Answer 10

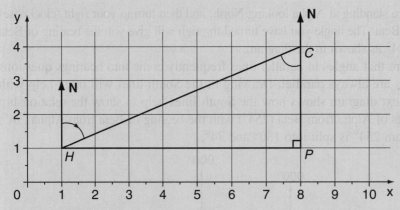

Using the right-angled triangle HCP shown on the diagram:

(a) the bearing of the casualty from the hospital is the same as angle HCP (alternate angles).

$$\tan \text{ angle } HCP = \frac{\text{OPP}}{\text{ADJ}} = \frac{HP}{CP} = \frac{8-1}{4-1}$$

$$\tan \angle HCP = \frac{7}{3} = 2.33333 \ldots$$

$$\tan^{-1} 2.33333 \ldots = 66.801 \ldots$$

$$\angle HCP = 66.8°$$

The bearing of the casualty from the hospital is 066.8°

(b) Using Pythagoras' Theorem,

HC is the hypotenuse.

$$HC^2 = 3^2 + 7^2 = 58$$

$$HC = \sqrt{58}$$

$$HC = 7.61577 \ldots$$

The distance of the casualty from the hospital is 7.62 km.

(c) The bearing of the hospital from the casualty is $180° + \angle HCP = 180° + 66.8° = 246.8°$

Example 11

The diagram shows three ships, *A*, *B* and *C*.

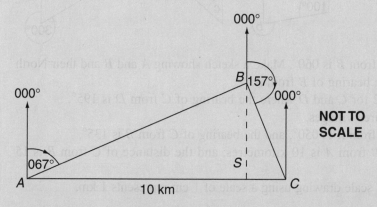

The bearing of *C* from *A* is 083°.

(a) Using the bearings given show that triangle *ABC* has a right angle at *B*.
The distance between *A* and *C* is 10 kilometres.

(b) Calculate the distance between *A* and *B*.

Answer 11

(a) *BS* is the South (180°) line from *B*.

$\angle ABS = 67°$ (alternate angles)

$\angle CBS = 180 - 157 = 23°$ (angles on a straight line)

So $\angle ABC = 67° + 23° = 90°$

(b) $\angle BAC = 83 - 67 = 16°$

$$\cos = \frac{ADJ}{HYP}$$

$$\cos 16° = \frac{AB}{10}$$

$AB = 10 \times \cos 16°$ $\qquad\qquad$ $(\times 10)$

$AB = 9.6126 \dots$

The distance between A and B is 9.61 km

Exercise 9.9

1. Find the angle marked with a letter in each of the diagrams below.

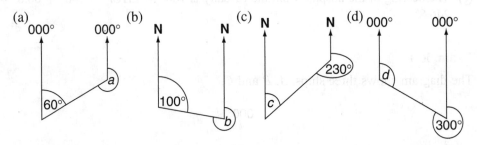

2. The bearing of A from B is 060°. Make a sketch showing A and B and their North lines and find the bearing of B from A.
3. Repeat question 2 for C and D where the bearing of C from D is 195°.
4. A, B and C are three villages.
   The bearing of B from A is 050°, and the bearing of C from B is 135°.
   The distance of B from A is 10 kilometres, and the distance of C from B is 15 kilometres.
   Make an accurate scale drawing using a scale of 1 cm represents 1 km, and find:
   (a) the distance of C from A, and
   (b) the bearing of A from C.
5. A ship sails 6 nautical miles due West from port, and then 15 nautical miles due North. (A nautical mile is a unit of distance used at sea.)
   Draw a diagram to show this journey and **calculate** the distance of the ship from port, and the bearing on which it will have to sail to return to port.
6. A ship's captain sees a marker buoy on a bearing of 073°, and a port on a bearing of 090°. The captain knows that the marker buoy is 5 kilometres due North of the port. Draw a sketch to show this information and **calculate** the distance, in kilometres, that the ship has to sail to reach port.
7. (a) Change the 3-figure bearings (i) 135° (ii) 315° to the mariner's compass.
   (b) Express (i) E, (ii) SW as 3-figure bearings.

Exercise 9.10

Mixed Exercise

1. Calculate the angles or lengths of sides marked with letters in these diagrams.

(a)

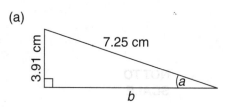

(b)

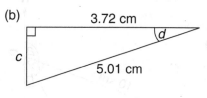

(c)

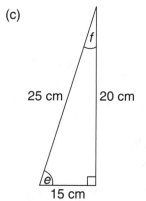

(d)

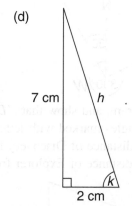

2. The diagram shows a ladder leaning against a house. The safe angle between a ladder and the ground (to minimize the risk of sliding or toppling) is 75°. The ladder is 3 metres long. How high up the wall will it reach when it is leaning at the safe angle?

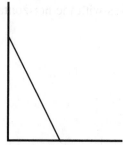

3. Calculate the height of the isosceles triangle shown below.

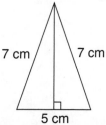

4. Two ships are carrying divers who are going to investigate an ancient wreck. The ship 'Explorer' (*E*) is on a bearing of 030° from the wreck (*W*), and the ship 'Discovery' (*D*) is on a bearing of 300° from the wreck.

   The bearing of Explorer from Discovery is 070°, and the distance between the ships is 10 kilometres.

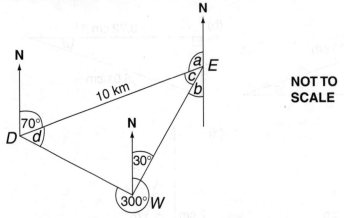

NOT TO SCALE

   (a) Copy the diagram, and show that ∠*DWE* is 90°.
   (b) Find all the angles marked with letters.
   (c) Calculate the distance of Discovery from the wreck.
   (d) Calculate the distance of Explorer from the wreck.

5.

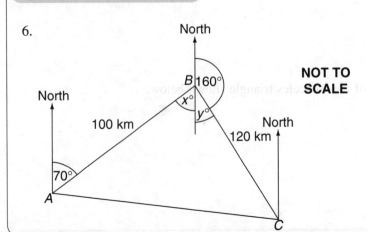

   The gradient of a hill is $\dfrac{1}{5}$. Calculate the angle (*a*) it makes with the horizontal.

Examination Questions

6.

NOT TO SCALE

A ship sails 100 kilometres from *A* on a bearing of 070° to *B*.
It then sails 120 kilometres on a bearing of 160° to *C*.

  (i) Show that $x + y = 90°$.

 (ii) Use trigonometry to calculate the size of angle *BAC*.

(iii) Find the three-figure bearing of *C* from *A*.

(iv) Find the three-figure bearing of *A* from *C*.

                                                (0580/03 May/June 2004 q 7b)

7.

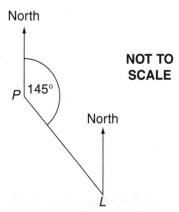

The bearing of a lighthouse, *L*, from a port, *P*, is 145°.
Find the bearing of *P* from *L*.

                                             (0580/01 May/June 2007 q 7)

8.

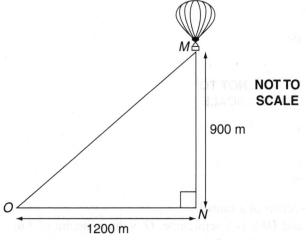

A hot air balloon, *M*, is 900 metres vertically above a point *N* on the ground.
A boy stands at a point, *O*, 1200 metres horizontally from *N*.

(a) Calculate the distance, *OM*, of the boy from the balloon.

(b) Calculate angle *MON*.

                                         (0580/01 May/June 2007 q 18)

9. Write as a 3-figure bearing the direction:
   (a) West,
   (b) North-East.

(0580/01 Oct/Nov 2004 q 6)

10. A square *ABCD*, of side 8 cm, has another square, *PQRS*, drawn inside it. *P,Q,R* and *S* are at the midpoints of each side of the square *ABCD*, as shown in the diagram.

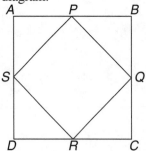

NOT TO SCALE

(a) Calculate the length of *PQ*.
(b) Calculate the area of the square *PQRS*.

(0580/02 May/June 2005 q 6)

11.

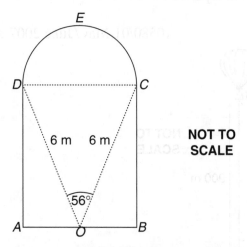

NOT TO SCALE

ABCED is the cross-section of a tunnel.
ABCD is a rectangle and *DEC* is a semicircle. *O* is the midpoint of *AB*.
OD = OC = 6m and angle *DOC* = 50°.

(a) (i) Show that angle *COB* = 62°.
   (ii) Calculate the length of *OB*.
   (iii) Write down the width of the tunnel, *AB*.
   (iv) Calculate the length of *BC*.

(b) Calculate the area of:
   (i) the rectangle *ABCD*,
   (ii) the semicircle *DEC*,
   (iii) the cross-section of the tunnel.

(c) The tunnel is 500 metres long.
   (i) Calculate the volume of the tunnel.
   (ii) A car travels through the tunnel at a constant speed of 60 kilometres per hour.
   How many seconds does it take to go through the tunnel?

(0580/03 May/June 2007 q 6)

12.

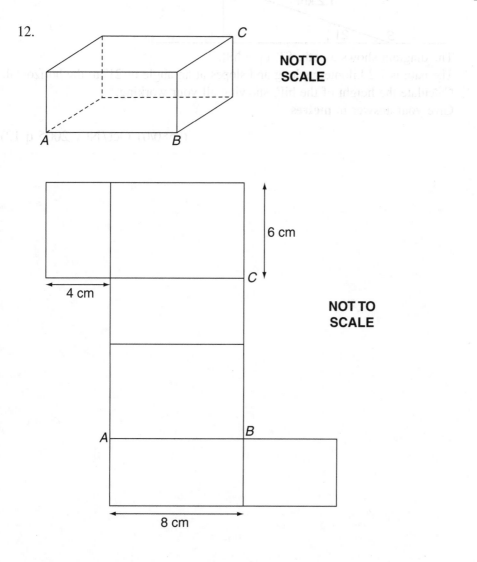

The diagram above shows a cuboid and its net.
(a) Calculate the total surface area of the cuboid.
(b) Calculate the volume of the cuboid.
(c) An ant walks directly from *A* to *C* on the surface of the cuboid.
    (i) Draw a straight line on the net to show this route.
    (ii) **Calculate** the length of the ant's journey.
    (iii) **Calculate** the size of angle *CAB* on the net.

(0580/03 May/June 2005 q 8)

13.

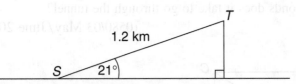

The diagram shows a path, *ST*, up a hill.
The path is 1.2 kilometres long and slopes at an angle of 21° to the horizontal.
Calculate the height of the hill, showing all your working.
Give your answer in **metres**.

(0580/01 Oct/Nov 2005 q 12)

Chapter **10**

Transformations and Vectors

Transformations are movements and changes of a shape on a plane (flat surface) according to various rules.

In this chapter you will discover how to draw the results of different transformations, and how to describe transformations that have already been drawn.

Most students find this subject quite straightforward, and provided you learn a few key words and descriptions you will find this is an easy mark earner in your examination.

This chapter should give you an interesting break after all the trigonometry you studied in Chapter 9.

For this chapter you require squared paper, a ruler, pencil, eraser and tracing paper. A small mirror would be helpful. It is advisable to draw your transformations in pencil so that any mistakes can be corrected. Neatness and reasonable accuracy in your drawings will help you avoid mistakes.

Essential Skills

1. Match the equation of each line to a graph.
   a. $y = x$ b. $x = 1$ c. $y = -3$
   d. $x = -3$ e. $y = -x$ f. $y = x + 2$

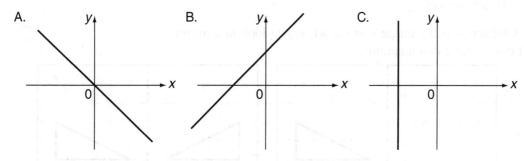

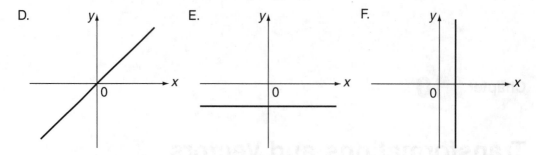

D.   E.   F.

2. Write down the equation of the vertical line which passess through 2 on the *x*-axis.
3. Write down the equation of the *y*-axis.
NOTE: What number does it pass through on the *x*-axis?

Answers

1.

Equation	a	b	c	d	e	f
Graph	D	F	E	C	A	B

2. $x = 2$
3. $x = 0$

Transformations

You need to study each of a set of four transformations:
{reflection, translation, rotation and enlargement}.
Students sometimes confuse the words transformation and translation.

NOTE: Try to remember that the word 'trans*for*mation' refers to the set of *four* members. The individual transformations are reflection, trans*lation*, rotation and enlargement.

NOTE: For maximum marks learn the four proper names, and their spelling.

Reflection

A reflection is the image you see when you look in a mirror.
Look at these two diagrams.

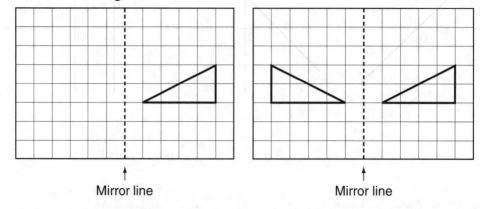

Mirror line   Mirror line

If you hold a small mirror upright so that its lower edge is on the mirror line in the first diagram, with its mirror surface pointing towards the triangle you should see an image that looks like the second diagram.

The triangles are the same, but they face in opposite directions. In fact they are **mirror images** of each other. If you put the mirror on the mirror line in the second diagram, facing either way, you will see the same picture.

A *reflection* is the first of the transformations we will study.

In the diagrams you will see two different reflections of triangle *A* drawn. In each case triangle *A* is called the **object** and the result of the transformation (*B* or *C*) is called the **image**.

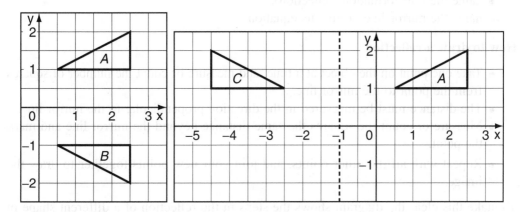

A maps to B under reflection
in the x-axis.

A maps to C under reflection in the
line x = −1.

Underneath each diagram is the correct description of the transformation.

As you will have seen we say '*A* maps to *B*' rather than '*A* is transformed to *B*'.

NOTE: When you are asked to *describe* a transformation you are expected to use the correct words, and not just describe it in your own words!

Note that examination questions at Core Level will always ask for the description of a **single** transformation, and you will get no marks for describing one transformation after another. We will discuss this again later.

Looking at these diagrams you should see that the shape and size of the object and image are exactly the same, but the object has been turned over to create the image.

Check this by tracing the triangles on tracing paper and then folding your drawing along the mirror line. In each case they should fit exactly on top of each other.

This also means that the object and image are the same distance from the mirror line, but on opposite sides.

How to recognize a reflection

A reflection is a transformation in which:

- the **object** and **image** are exactly the same shape and size,
- the object and image face in opposite directions.
- Also under **reflection** the **image** is as far behind the **mirror line** as the **object** is in front of it.

How to describe a reflection

To describe a reflection you must:

- name the transformation (reflection),
- name the mirror line or give its equation.

How to draw a reflection

- Take one point on the object at a time, and measure or count the number of squares from the point to the mirror line.
- The distance must be measured in the direction propendicular to the mirror line.
- Measure or count the squares the same distance beyond the mirror line and mark the point.
- Repeat with all the other points, and join your new points together to draw the image.

To make this clear the diagram shows the steps in the reflection of a different shape in the line LM.

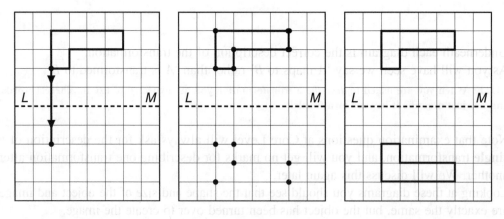

Count 2 squares to the mirror   Repeat for all corners   Draw the image
line and 2 squares beyond

For a diagonal mirror line you need to count the diagonals instead of the squares, as shown in the diagram.

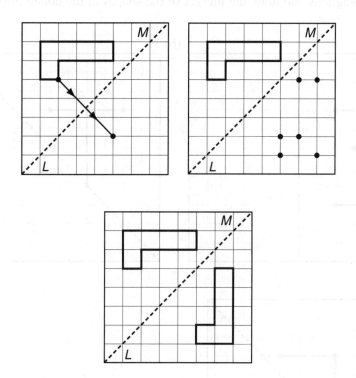

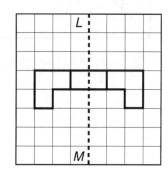

If the mirror line goes through the object the reflection produces what at first seems like a completely different picture.

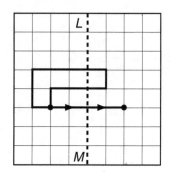

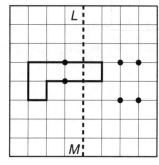

Example 1

Copy these diagrams and draw the images of the shapes in the dotted mirror lines.

(a)

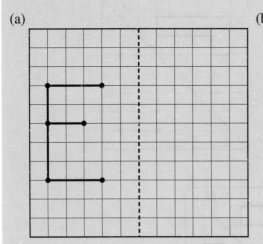

(b)

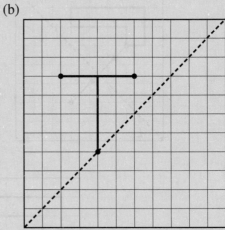

(c)

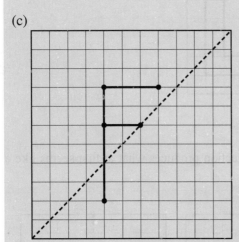

NOTE: In (b) and (c) remember to count the diagonals instead of the sides of the squares.

Answer 1

(a)

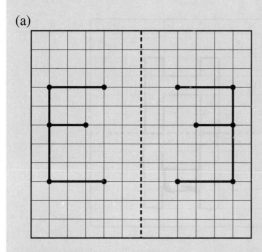

(b)

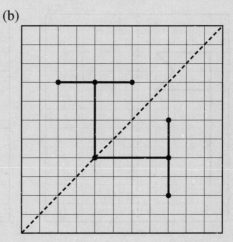

(c)

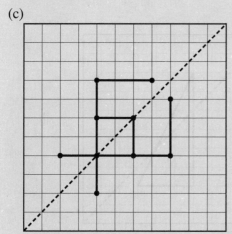

NOTE: In (b) and (c) notice that any point *on* the mirror line does not move under the reflection.

NOTE: It pays to be NEAT in your diagrams. Preferably use a ruler unless you can draw *very good* straight lines for the smaller diagrams!

Exercise 10.1

1. Copy and complete the reflections in the following diagrams.

NOTE: Remember to count the squares or diagonals so that you get the images in the right places as well as pointing in the right direction.

(a)

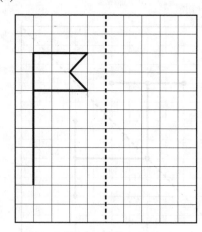

(b)

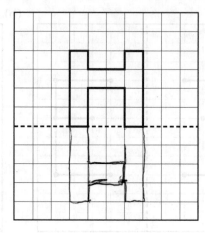

(c)

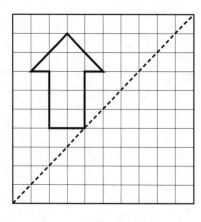

(d)

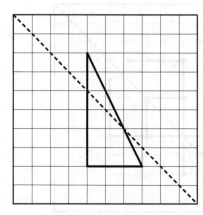

2. Draw axes x and y from -5 to $+5$, with 1cm representing each unit on each axis.
   (a) Plot the following points:
       $(1,1)$, $(3,1)$, $(3,2)$.
       Join up the points to form a triangle and label the triangle A.
   (b) Reflect A in:
       (i) the y-axis, labelling the image B,
       (ii) the line $y = -1$, labelling the image C,
       (iii) the line $y = x$, labelling the image D,
       (iv) the line $y = -x$, labelling the image E.

NOTE: It can be confusing reflecting in the axes because the scales on the axes tend to get in the way. Remember that you are reflecting in the *line*, not in the line and numbers. The diagrams below will show this. The first diagram is wrong, the second one is correct.

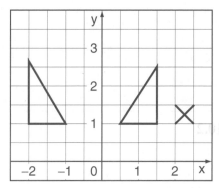

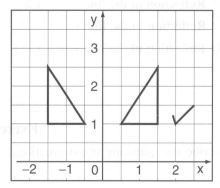

Reflection in the *y*-axis

Example 2

Describe the following transformations.

(a)

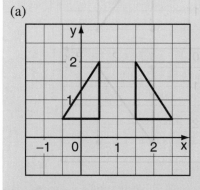

(b)

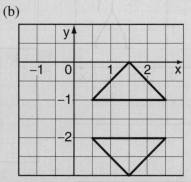

(c)

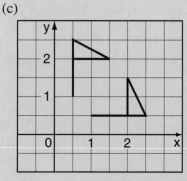

(d)

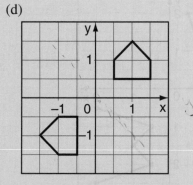

NOTE: Use tracing paper to check. If you fold your tracing along the mirror line the object and image should coincide exactly.

Answer 2

(a) Reflection in the line $x = 1$

(b) Reflection in the line $y = -1.5$

(c) Reflection in the line $y = x$

(d) Reflection in the line $y = -x$

Exercise 10.2

Describe the following transformations.

1.

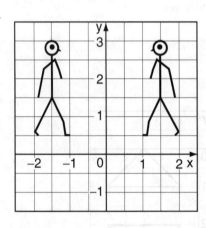

2.

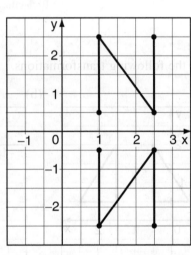

3.

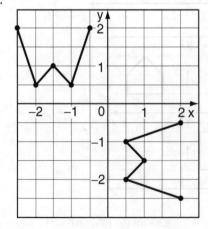

4.

5.

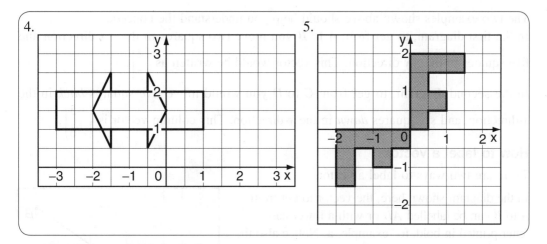

Translations and Vectors

A **translation** is a movement in which only the position of the object changes.
The image remains facing the same way as the object, and stays the same size and shape.
For example, in the diagram on the right the object A has been moved two squares to the
right ($+x$ direction) and 4 squares down ($-y$ direction) to give the image B.

NOTE: Follow one point, for example, the point marked with a dot, to count
how many squares along and how many down.

This translation is described by breaking it down into the two
movements, each parallel to each of the two axes.
These movements are written in the form of a **vector,** also called a
column vector, so in this diagram the vector would be written in
tall brackets with the x-direction at the top: $\begin{pmatrix} 2 \\ -4 \end{pmatrix}$.

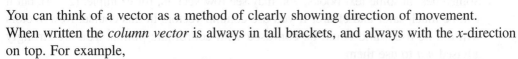

You can think of a vector as a method of clearly showing direction of movement.
When written the *column vector* is always in tall brackets, and always with the x-direction
on top. For example,

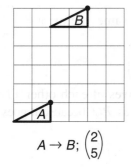

$A \rightarrow B; \begin{pmatrix} 2 \\ 5 \end{pmatrix}$

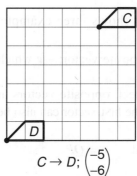

$C \rightarrow D; \begin{pmatrix} -5 \\ -6 \end{pmatrix}$

NOTE: Follow the dot!

The two examples shown above should help you understand the concept.
In the first diagram, to get from **A** to **B** you move two squares in the $+x$ direction and
five squares in the $+y$ direction. This vector would be written as $\begin{pmatrix} 2 \\ 5 \end{pmatrix}$.

In the second diagram, to get from **C** to **D** you would move five squares *back* in the
x-direction, and six squares *down* in the y-direction. This column vector is $\begin{pmatrix} -5 \\ -6 \end{pmatrix}$.

How to label a vector

There are two ways to label a vector.

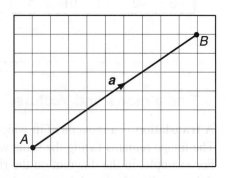

In the diagram shown here, the vector to get from
A to B can be labelled $\overrightarrow{AB}$ or with a lower case
letter printed in bold, for example, **a**. Notice also the
arrow in the diagram.
The arrow over the $\overrightarrow{AB}$ shows that the direction is
from A to B.
In hand writing the bold **a** is written with an
underline, $\underline{a}$, because it is not easy to write a
recognizable bold letter by hand.

Important points to notice when working with vectors:

- The vector **a** does not have to have just one *position*, but can be applied anywhere
  in the diagram, always in the same direction.
- The vector written as $\overrightarrow{AB}$ shows that this particular vector is the one joining A to
  B in the direction A to B.
- The vector is not a fraction, and should *not* be written with a fraction line.
- The vector should always be written with the x-direction above the y-direction.
- When you draw a vector in a diagram remember to add the arrow to show its direction.
- Sometimes, in some text books, you may see row vectors, for example, $(2\ -4)$, but it
  will be much safer and clearer if you use column vectors in this course. Row vectors
  will not be used, and could be confused with coordinate points. You are strongly
  advised *not* to use them.

The next diagram shows three examples of the same vector: $v = \begin{pmatrix} 4 \\ -3 \end{pmatrix}$, and also an

example of the opposite vector: $-v$ which is $\begin{pmatrix} -4 \\ 3 \end{pmatrix}$.

As you might expect, opposite vectors are negatives of each other. (That is, they have
opposite signs). The first vector can also be written as $\overrightarrow{PQ}$ because it goes from the point P
to the point Q.
So $v = \overrightarrow{PQ}$ and $-v = \overrightarrow{QP}$

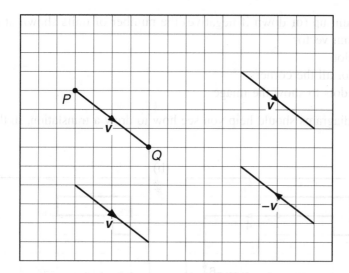

We will look at vectors in a little more detail later in the chapter.

How to recognize a translation

A translation is a transformation in which:

- the object and image have exactly the same shape and size,
- the *orientation* (direction in which the object and image face) remains the same.

How to describe a translation

A translation is described by:

- the name of the transformation (translation),
- a column vector.

How to use a column vector

- The vector shows the direction of movement broken down into a movement parallel to the x-axis and another movement parallel to the y-axis.
- Mark a point on the object and the corresponding point on the image and count the number of units moved in the x-direction and the number moved in the y-direction.
- The column vector is written in tall brackets with the x-direction on top of the y-direction.
- A column vector is not a fraction and has no fraction line.

How to draw a translation

- Mark one corner of the image with a dot.
- Count along (or back if the number is negative) the number of units shown on top in the column vector.

- Then count up (or down if negative) the number of units shown at the bottom of the column vector.
- Mark a dot.
- Repeat for all the corners.
- Join the dots to show the image.

This series of diagrams should help you see how to draw a translation, in this case $\begin{pmatrix} 4 \\ -6 \end{pmatrix}$.

(a)

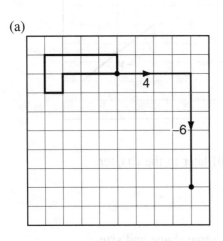

(b)

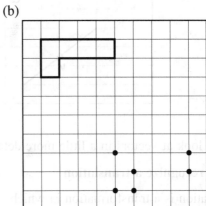

(c)

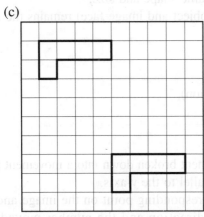

Example 3

Draw a grid on 5mm squared paper with both the *x*-and *y*-axes from −5 to 5, using one centimetre (2 squares) per unit.

(a) (i) *A*: {(−4,3), (−4,5), (−3,3)}   *B*: {(3,1), (4,1), (3,3)}
          Mark these two sets of points and join to form two triangles, *A* and *B*.
    (ii) Describe fully the single transformation which maps *A* onto *B*.
    (iii) Describe fully the single transformation which maps *B* onto *A*.

(b) (i) Draw the image of A under the translation $\begin{pmatrix} 2 \\ -3 \end{pmatrix}$.

   Label the image C.

   (ii) Draw the image of B under the translation $\begin{pmatrix} -1 \\ -2.5 \end{pmatrix}$.

   Label the image D.

(c) Plot and label the points E (-4, -1) and F (-2, -4).

   (i) Write $\overrightarrow{EF}$ as a column vector.

   (ii) Write $\overrightarrow{FE}$ as a column vector.

(d) Describe fully the single transformation which maps A onto D.

NOTE: Here the units to be counted are centimetres, not small squares. Always check the scale!

Answer 3

(a) (i)

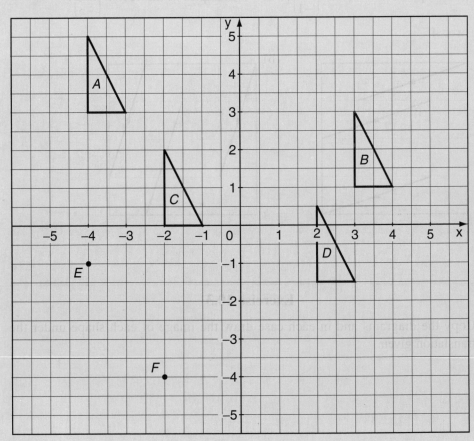

(a) (ii) Translation $\begin{pmatrix} 7 \\ -2 \end{pmatrix}$ (iii) Translation $\begin{pmatrix} -7 \\ 2 \end{pmatrix}$

(c) (i) $\overrightarrow{EF} = \begin{pmatrix} 2 \\ -3 \end{pmatrix}$ (ii) $\overrightarrow{FE} = \begin{pmatrix} -2 \\ 3 \end{pmatrix}$

(d) Translation $\begin{pmatrix} 6 \\ -4.5 \end{pmatrix}$

Example 4

On 1 centimetre squared paper draw:

(a) 3 examples of the vector $\begin{pmatrix} 3 \\ -1 \end{pmatrix}$,   (b) 3 examples of the vector $\begin{pmatrix} 1 \\ -3 \end{pmatrix}$.

Answer 4

(a) 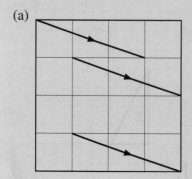   (b)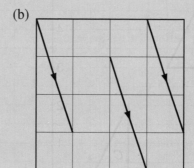

Exercise 10.3

1. Copy the diagrams and in each case draw the image of each shape under the translation given.

(a)

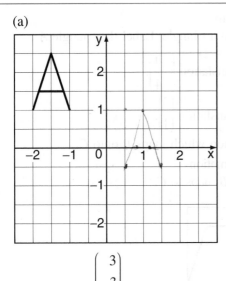

$$\begin{pmatrix} 3 \\ -3 \end{pmatrix}$$

(b)

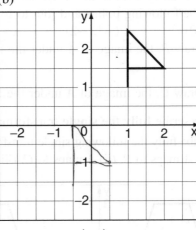

$$\begin{pmatrix} -2 \\ -3 \end{pmatrix}$$

(c)

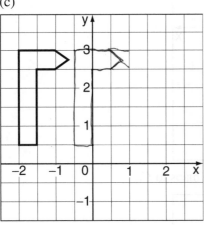

$$\begin{pmatrix} 3 \\ 0 \end{pmatrix}$$

2. Copy the diagram and draw two examples of the vector $\mathbf{c} = \overrightarrow{AB}$, and one example of $-\mathbf{c}$.

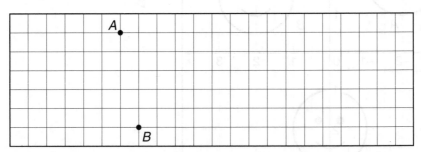

3. Draw x- and y-axes from -5 cm to $+5$ cm.
   Plot the following points and join to form a triangle. Label the triangle A.
   $(-1, 0)$ $(-1, -2)$ $(-2, -2)$
   (a) Draw B, the image of A under the translation $\begin{pmatrix} -1 \\ 4 \end{pmatrix}$.

   (b) Draw C, the image of A under the translation $\begin{pmatrix} 3 \\ 1 \end{pmatrix}$.

   (c) Describe **fully** the **single** transformation which maps B onto C.

4. Describe fully the following single transformations.
   (a) (i) A maps to B. (ii) A maps to C. (iii) B maps to C.

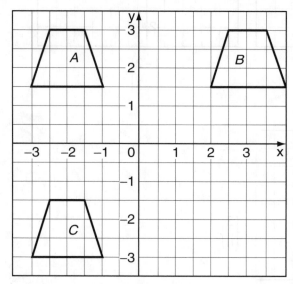

   (b) (i) A maps to B. (ii) C maps to B. (iii) B maps to C.

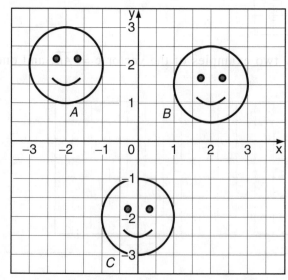

Rotation

A rotation is a turning about some central point. Think of a bicycle wheel. Each spoke turns or *rotates* about one of its ends: the one which is attached to the centre of the wheel. The valve on the rim also rotates about the centre of the wheel even though it is not directly attached to it.

A rotation of an object can be about a point on the object (as the spoke of the wheel rotates about one end) or about some other point (as the valve rotates about a point further away).

In a rotation the object and image are exactly the same size and shape, but they face different ways.

A rotation needs the most information to describe it fully.

As well as the correct word (rotation) it needs the centre of rotation, the angle of rotation and the *sense* or direction of rotation. For your examination it is best to use 'clockwise' or 'anti-clockwise' to describe the sense of the rotation.

The exception to this is a rotation of 180° because it does not matter whether the object is turned clockwise or anti-clockwise, it still arrives in the same place.

The diagram shows two examples of rotations and their correct descriptions.

The rotations are in the plane of the paper.

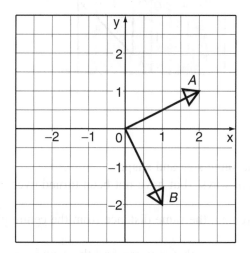

A maps to B under a rotation of 90° clockwise about the origin.

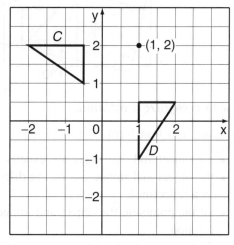

C maps to D under a rotation of 90° anti-clockwise about (1, 2).

For your Core examination, to find the centre of rotation it is best to use a piece of tracing paper. There is another geometrical method but it is not necessary at this level.

Trace the object and the axes, and mark the axes *x* and *y*.

Keeping the two pieces of paper together put your pencil point down firmly at different points on the tracing paper and carefully rotate the tracing paper until you find the point where the tracing of the object fits exactly over the image.

It can take a bit of time to find the centre, but try the obvious places first. The centre will usually be either the origin (0, 0), or a point on the object itself.

Now compare the positions of the *x*-axis before and after the rotation to see the angle and sense through which direction the object has been rotated.

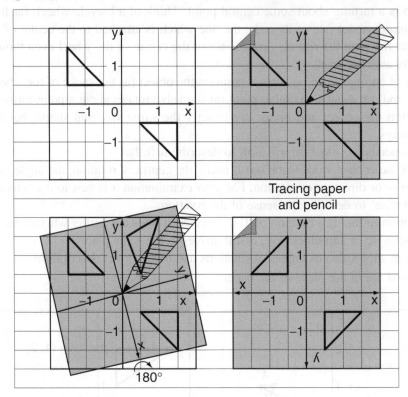

Tracing paper and pencil

180°

How to recognize a rotation

A rotation is a transformation in which:

- the object and the image are exactly the same size and shape,
- the object and image are facing different ways, (but have not been turned over as in a reflection),
- all the points on the object and image remain the same distance from the centre of rotation.
- To be sure, check with a piece of tracing paper that you can find the centre, angle and sense of the rotation.

How to fully describe a rotation

A rotation is a transformation described by:

- the name of the transformation (rotation),
- the centre of rotation,
- the angle and sense of the rotation.

How to draw a rotation, using tracing paper

- Trace the object and at least one of the axes onto your tracing paper.
- Put the point of your pencil firmly down on the centre of rotation.
- Turn the paper through the given angle and sense. The axis that you have traced will help here.
- Copy the image from the tracing onto the original diagram.

How to draw a rotation without tracing paper

- Draw vectors in the x and y directions from the centre of rotation to the various points on the object.
- Rotate the vectors to find the new positions of these points.
- The vectors must all turn through the same angle, and remain of the same length.

NOTE: It is much easier to use tracing paper. Remember that the points on the object are usually on grid lines, and so the points of the image will also be on grid lines. This should help you to draw accurate images. You will be allowed tracing paper in your examination.

Example 5

Copy the diagrams.

(a) Rotate A 180° about the origin (0,0), draw the image and label it P.

(b) Rotate B 90° anticlockwise about the point marked with a cross. Draw the image and label it Q.

(c) Describe fully the single transformation that will map C onto D.

(a)

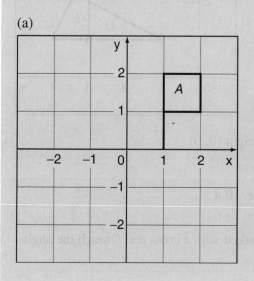

(b)

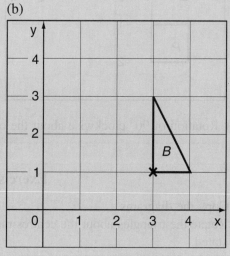

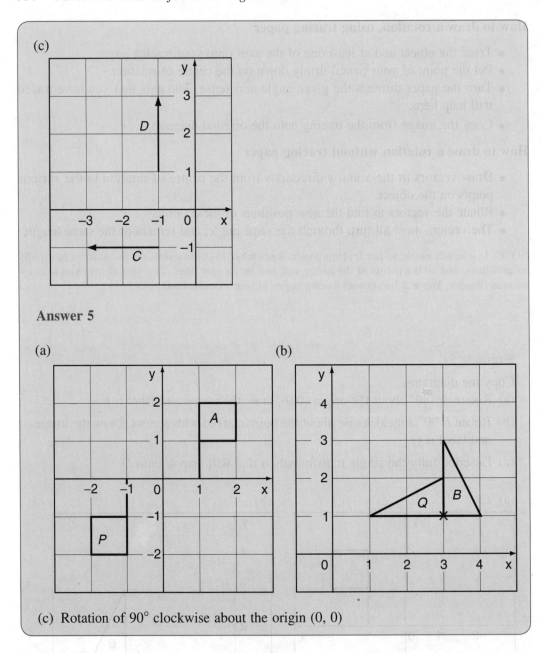

(c)

Answer 5

(a)   (b)

(c) Rotation of 90° clockwise about the origin (0, 0)

Exercise 10.4

1. Copy the diagrams.
   Rotate the triangles about the centres marked with a cross and through the angles stated.

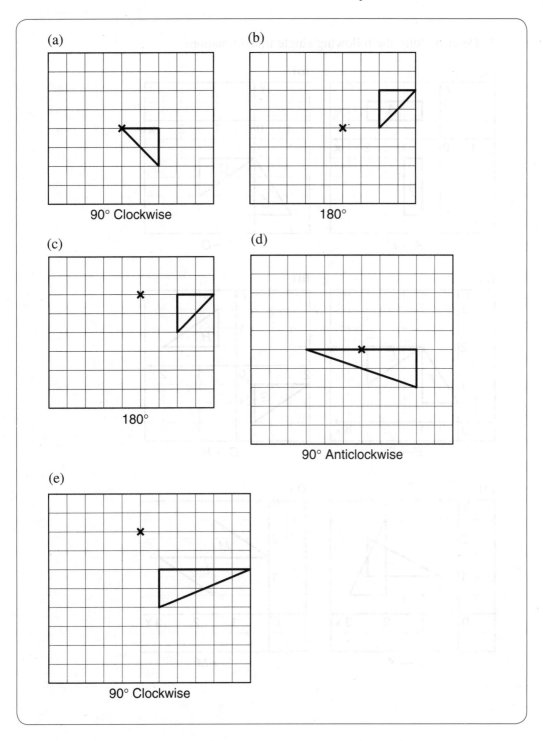

(a)

90° Clockwise

(b)

180°

(c)

180°

(d)

90° Anticlockwise

(e)

90° Clockwise

2. Describe fully the following single transformations.

(a)

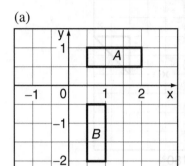

$A \rightarrow B$

(b)

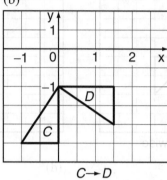

$C \rightarrow D$

(c)

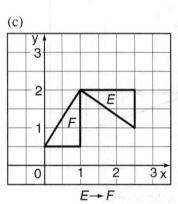

$E \rightarrow F$

(d)

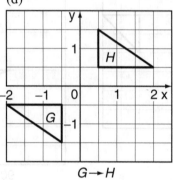

$G \rightarrow H$

(e)

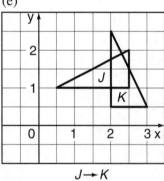

$J \rightarrow K$

(f)

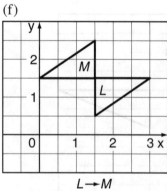

$L \rightarrow M$

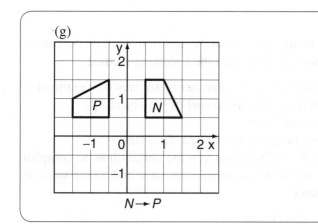

(g)

$N \rightarrow P$

Enlargement

An enlargement can, as its name suggests, make the image larger than the object. Confusingly it is also called an enlargement if it makes the image smaller than the object!

The difference is in the **scale factor** of the enlargement. A scale factor greater than 1 will make the image larger than the object, but a scale factor less than 1 will make it smaller.

You can demonstrate an enlargement by shining a torch onto a shape, such as your hand, and looking at the shadow of your hand on a wall. Your hand should be parallel to the wall.

The size of the shadow changes according to how close your torch is to your hand, and the position of the shadow changes according to the position of the torch.

The torch is called the **centre** of the enlargement, and you can see that it is important that the centre is defined when you are describing an enlargement.

How to recognize an enlargement

An enlargement is a transformation in which:

- the image is larger or smaller than the object,
- the image is exactly the same *shape* as the object so all angles remain the same and the sides of the image are all multiplied by the same scale factor,
- the image remains facing the same way as the object,
- the position of the image depends on the position of the centre of enlargement.

How to describe an enlargement

An enlargement is described by giving:

- the name of the transformation (enlargement),
- the scale factor,
- the centre of the enlargement.

How to draw an enlargement

An enlargement may be drawn by drawing 'rays' from the centre to each of the corners of the object and beyond, but this is less accurate than the following method.

- From the centre of the enlargement count the number of squares (or centimetres) along (or back) and up (or down) to get to one corner of the object.
- Multiply these distances by the scale factor.
- *Go back to the centre* and count along and up the new distances.
- Repeat for the other corners of the object, until you are certain how to complete the image, bearing in mind that all the angles remain the same and all the sides are multiplied by the same scale factor.

As usual, a set of diagrams will make this clearer.
The diagrams show the centre of the enlargement marked with a cross. The scale factor is 2.

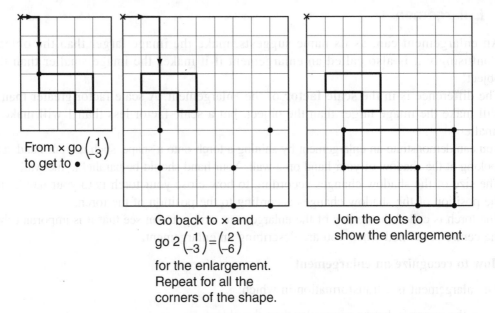

From × go $\begin{pmatrix} 1 \\ -3 \end{pmatrix}$ to get to •

Go back to × and go $2\begin{pmatrix} 1 \\ -3 \end{pmatrix} = \begin{pmatrix} 2 \\ -6 \end{pmatrix}$ for the enlargement. Repeat for all the corners of the shape.

Join the dots to show the enlargement.

The next three diagrams show how important the centre of enlargement is, and how moving the centre of enlargement completely changes the position of the image. In each case the small square is enlarged to the big square with a scale factor 3.

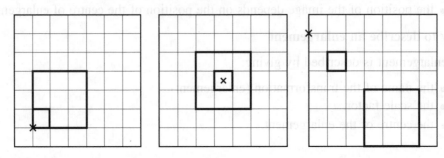

Example 6

Copy the diagrams.

(a) Draw an enlargement of *A*, centre the origin and scale factor 2. Label the image *B*.

(b) Draw an enlargement of *C*, centre the origin and scale factor $\frac{1}{2}$. Label the image *D*.

(c) Describe the single transformation which maps *E* onto *F*.

(a)

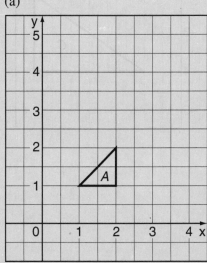

(b)

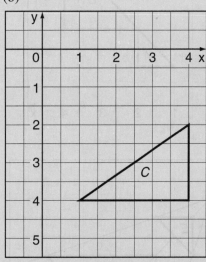

(c)

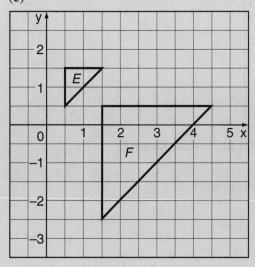

Answer 6

(a)

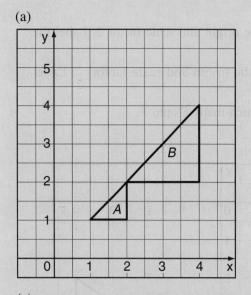

(b)

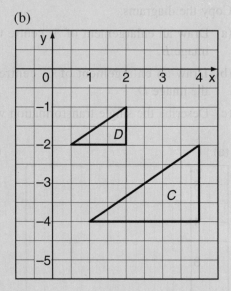

(c)

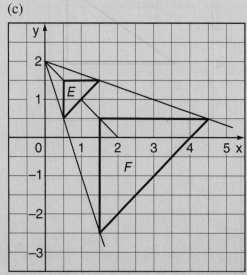

(d) *E* maps to *F* under enlargement, centre (0,2), scale factor 3

NOTE: To find the centre of the enlargement join corresponding points on object and image with straight lines. The lines meet at the centre of the enlargement.

Exercise 10.5

1. Copy the diagrams.

   Draw the enlargements with the given scale factors and the centres marked with a cross.

   (a)

   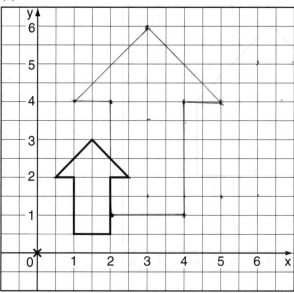

   Scale factor 2

   (b)

   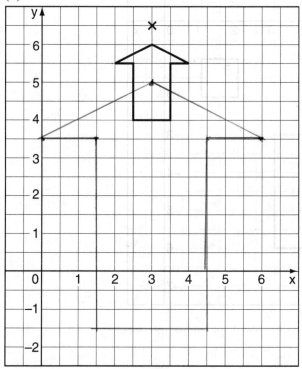

   Scale factor 3

2. For each of the diagrams below describe fully the single transformations which map *A* onto *B*.

(a)

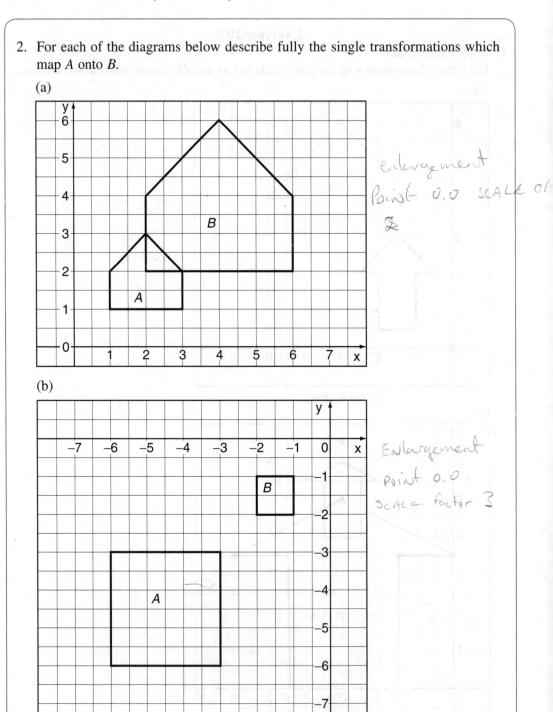

enlargement
Point . 0,0 SCALE Of
2

(b)

ENlargement
Point 0,0.
SCALE factor 3

More about Vectors

Negative vectors

As we have seen, if the vector $\boldsymbol{a} = \begin{pmatrix} 2 \\ -3 \end{pmatrix}$, then the same vector, pointing in the opposite direction

is $-\boldsymbol{a}$ or $\begin{pmatrix} -2 \\ 3 \end{pmatrix}$. Notice that both components of the vector have had their signs changed.

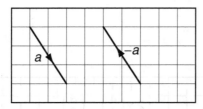

Adding vectors

Vectors may be added to produce a new vector which has the same effect as the original two applied one after another. The x components are added, and the y components are added to produce the new vector, taking account of the signs of the components.

$$\begin{pmatrix} -1 \\ 2 \end{pmatrix} + \begin{pmatrix} 3 \\ 5 \end{pmatrix} = \begin{pmatrix} -1+3 \\ 2+5 \end{pmatrix} = \begin{pmatrix} 2 \\ 7 \end{pmatrix}$$

The first diagram shows this effect.

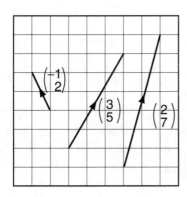
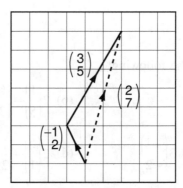

You will see in the second diagram that if we move the two vectors which are to be added until they are joined 'head to tail' (you can follow the arrows round) the result of the addition joins the 'tail' of the first to the 'head of the second'.

Vectors may also be subtracted by subtracting the x components and the y components. For example,

$$\begin{pmatrix} -2 \\ 6 \end{pmatrix} - \begin{pmatrix} 1 \\ 3 \end{pmatrix} = \begin{pmatrix} -2-1 \\ 6-3 \end{pmatrix} = \begin{pmatrix} -3 \\ 3 \end{pmatrix}$$

Multiplying a vector by a number

A vector may be multiplied by a number to change its length. Both components are multiplied by the same number.

The diagram shows the vectors $\begin{pmatrix} 4 \\ 1 \end{pmatrix}$, $2\begin{pmatrix} 4 \\ 1 \end{pmatrix}$ and $\frac{1}{2}\begin{pmatrix} 4 \\ 1 \end{pmatrix}$. The last two vectors may be

simplified to $\begin{pmatrix} 8 \\ 2 \end{pmatrix}$ and $\begin{pmatrix} 2 \\ 0.5 \end{pmatrix}$. You will see that the vectors are all parallel to each other, but

are different lengths.

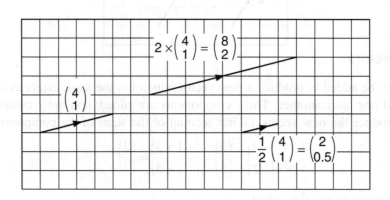

$$2 \times \begin{pmatrix} 4 \\ 1 \end{pmatrix} = \begin{pmatrix} 8 \\ 2 \end{pmatrix}$$

$$\begin{pmatrix} 4 \\ 1 \end{pmatrix}$$

$$\frac{1}{2}\begin{pmatrix} 4 \\ 1 \end{pmatrix} = \begin{pmatrix} 2 \\ 0.5 \end{pmatrix}$$

Example 7

Simplify the following:

(a) $3\begin{pmatrix} 5 \\ -2 \end{pmatrix}$

(b) $2\begin{pmatrix} 1 \\ 5 \end{pmatrix} - \begin{pmatrix} 0 \\ 3 \end{pmatrix}$

(c) $\frac{1}{2}\begin{pmatrix} 5 \\ 6 \end{pmatrix} + 3\begin{pmatrix} -1 \\ -2 \end{pmatrix}$

(d) $2\begin{pmatrix} 5 \\ 1 \end{pmatrix} - \frac{1}{2}\begin{pmatrix} -6 \\ 4 \end{pmatrix}$

NOTE: $\begin{pmatrix} 0 \\ 3 \end{pmatrix}$ is a translation of 3 units parallel to the y-axis, because it moves zero units in the x-direction.

Answer 7

(a) $\begin{pmatrix} 15 \\ -6 \end{pmatrix}$

(b) $\begin{pmatrix} 2-0 \\ 10-3 \end{pmatrix} = \begin{pmatrix} 2 \\ 7 \end{pmatrix}$

(c) $\begin{pmatrix} 2.5+(-3) \\ 3+(-6) \end{pmatrix} = \begin{pmatrix} -0.5 \\ -3 \end{pmatrix}$

(d) $\begin{pmatrix} 10-(-3) \\ 2 - 2 \end{pmatrix} = \begin{pmatrix} 13 \\ 0 \end{pmatrix}$

Example 8

(a)   From this list of vectors pick two that are parallel to a.

$$a = \begin{pmatrix} 1 \\ 2 \end{pmatrix} \qquad b = \begin{pmatrix} -1 \\ 2 \end{pmatrix} \qquad c = \begin{pmatrix} -2 \\ -4 \end{pmatrix} \qquad d = \begin{pmatrix} 3 \\ 6 \end{pmatrix} \qquad e = \begin{pmatrix} 2 \\ 1 \end{pmatrix}$$

(b)   Comment on the vectors you have chosen.

Answer 8

(a)   First take out any common factors in the vectors.

$$c = -2 \begin{pmatrix} 1 \\ 2 \end{pmatrix} \qquad\qquad d = 3 \begin{pmatrix} 1 \\ 2 \end{pmatrix}$$

So $c = -2a$ and $d = 3a$.

c and d are parallel to a

(b) c is twice as long as a and points in the opposite direction.

d is three times as long as a and points in the same direction.

Vectors can form equations which can be solved algebraically to find unknown quantities. Remember that if two vectors are equal then the x-directions must be equal and the y-directions must be equal. The next Example shows this.

Example 9

(a) $$2 \begin{pmatrix} a \\ 5 \end{pmatrix} + 3 \begin{pmatrix} a \\ b \end{pmatrix} = \begin{pmatrix} 10 \\ -8 \end{pmatrix}$$

Using the above vector equation write down an equation in a and an equation in b and solve them to find a and b.

(b) $$\begin{pmatrix} k \\ 2k \end{pmatrix} + \begin{pmatrix} 2m \\ m \end{pmatrix} = \begin{pmatrix} 4 \\ 5 \end{pmatrix}$$

Form two equations in k and m and solve simultaneously to find k and m.

Answer 9

(a) $2a + 3a = 10$ $2 \times 5 + 3 \times b = -8$

      $5a = 10$ $10 + 3b = -8$

      $a = 2$ $3b = -18$

                                                $b = -6$

(b) $k + 2m = 4$ $\xrightarrow{\times 2}$ $2k + 4m = 8$

      $2k + m = 5$ $\longrightarrow$ $\underline{2k + m = 5}$ subtract

                                          $3m = 3$

                                          $m = 1$

substituting in the first equation: $k + 2 \times 1 = 4,$ $k = 2$

Exercise 10.6

1. Simplify the following:

   (a) $3\begin{pmatrix} \frac{1}{2} \\ \frac{1}{3} \end{pmatrix}$
   (b) $-2\begin{pmatrix} 5 \\ 0 \end{pmatrix}$
   (c) $\frac{1}{2}\begin{pmatrix} -6 \\ 12 \end{pmatrix} + \begin{pmatrix} 5 \\ 3 \end{pmatrix}$
   (d) $6\begin{pmatrix} 5 \\ 4 \end{pmatrix} + \frac{1}{2}\begin{pmatrix} 60 \\ 8 \end{pmatrix}$

2. $\begin{pmatrix} a \\ 10 \end{pmatrix} + \begin{pmatrix} a \\ -5 \end{pmatrix} = \begin{pmatrix} 8 \\ 5 \end{pmatrix}$

   Find a.

3. $\begin{pmatrix} 2x \\ 15 \end{pmatrix} - \begin{pmatrix} 8 \\ y \end{pmatrix} = \begin{pmatrix} -4 \\ 12 \end{pmatrix}$

   Find x and y.

4. Draw these vectors on squared paper and list those that are parallel to each other.

   $a = \begin{pmatrix} -1 \\ 2 \end{pmatrix}$, $b = \begin{pmatrix} -2 \\ 4 \end{pmatrix}$, $c = \begin{pmatrix} 2 \\ -1 \end{pmatrix}$, $d = \begin{pmatrix} 1 \\ -2 \end{pmatrix}$, $e = \begin{pmatrix} -2 \\ -4 \end{pmatrix}$, and $f = \begin{pmatrix} 3 \\ -6 \end{pmatrix}$.

5. How can you tell that the vectors $\begin{pmatrix} 1 \\ 4 \end{pmatrix}$ and $\begin{pmatrix} 3 \\ 12 \end{pmatrix}$ are parallel and that one is three times longer than the other?

6. What can you say about the vectors $\begin{pmatrix} -1 \\ 3 \end{pmatrix}$ and $\begin{pmatrix} -2 \\ 6 \end{pmatrix}$?

NOTE: Remember that you will probably be asked to describe a transformation *fully*. To help you be sure you have given all the information imagine asking someone else to draw it using your description. Would they know exactly what to do?

NOTE: Remember that in your Core examination you will be asked to describe a *single* transformation. You will get no marks for using one transformation followed by another.

Recognising Transformations

NOTE: Make sure you can recognise transformations by working through the next example before going on to the Mixed Exercise.

Example 10

(a) Describe **fully** the **single** transformations shown in each of the diagrams. *A* maps onto *B* in each case.

(i)   (ii)

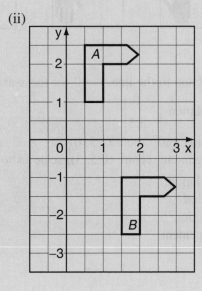

(iii)

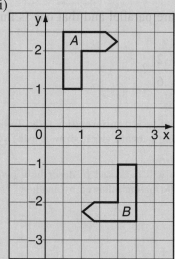

(iv)

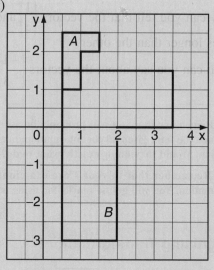

(b) Name each of the following transformations of the outline of a left hand.

(i)    (ii) 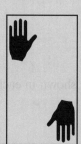   (iii)    (iv)

Answer 10

(a) (i) Reflection in the line $y = 0$ (or the x-axis)

   (ii) Translation $\begin{pmatrix} 1 \\ -3.5 \end{pmatrix}$

   (iii) Rotation 180° about (1.5, 0)

   (iv) Enlargement, centre (0.5, 3), scale factor 3

(b) (i) Translation

   (ii) Rotation

   (iii) Reflection

   (iv) Enlargement

Exercise 10.7

Mixed Exercise

1. Copy the diagrams.

    (a)

    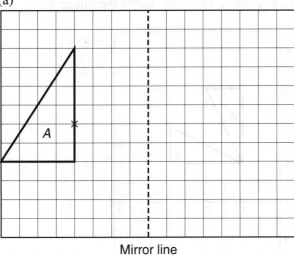

    Mirror line

    (i) Rotate A 180° about X.

    (ii) Reflect A in the dotted mirror line.

    (b)

    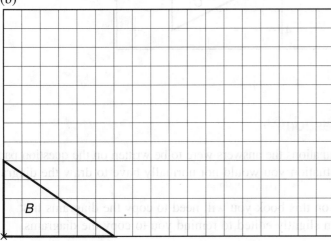

    (i) Enlarge B, centre X, scale factor 3.

    (ii) Translate B using the vector $\begin{pmatrix} 4 \\ 3 \end{pmatrix}$.

2. Copy the diagram.
   (a) Describe fully the single transformation which maps
   (i) *A* onto *B*,   (ii) *B* onto *C*.

   (b) (i) Enlarge *D*, scale factor 2, centre the orgin. Label the image *E*.
   (ii) Rotate *B*, 90° anticlockwise, centre the orgin. Label the image *F*.

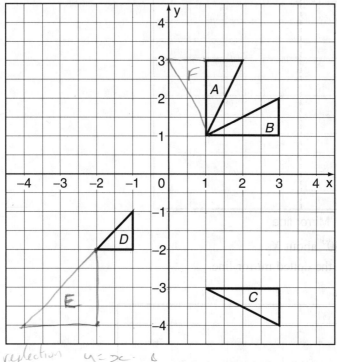

a reflection   y = x . b
b reflection   y = -1

Examination Questions

Note, in the actual examinations the answers were to be written on the question paper, so in an examination situation you would not normally have to draw the diagrams yourself.
Here, to avoid drawing on this book you will need to copy the diagrams first.
The questions have been slightly edited to remind you to copy the diagrams.

3.

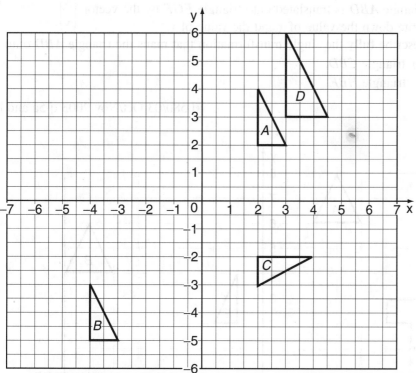

(a) Describe fully the single transformation that maps triangle *A* onto triangle *B*.

(b) Describe fully the single transformation that maps triangle *A* onto triangle *C*.

(c) Find the centre and scale factor of the enlargement that maps triangle *A* onto triangle *D*.

(d) Copy the diagram and

(i) draw the image of triangle *A* under a reflection in the line $x = -1$,

(ii) draw the image of triangle *B* under a rotation of 180° about $(-4, -3)$.

(0580/03 Oct/Nov 2004 q 2)

4.

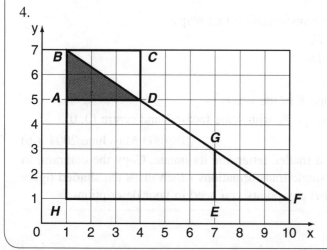

(a) Triangle **ABD** is translated onto triangle **EGF** by the vector $\begin{pmatrix} x \\ y \end{pmatrix}$.
   Write down the value of *x* and the value of *y*.

(b) Describe **fully** the single transformation that maps the triangle **ABD** onto
   (i) triangle *CBD*,
   (ii) triangle *HBF*.

(0580/03 Oct/Nov 2003 q 5a and b)

5.

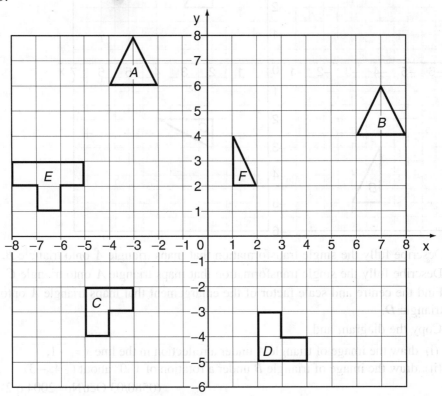

(a) Describe fully the single transformation that maps:
   (i) shape *A* onto shape *B*,
   (ii) shape *C* onto shape *D*.

(b) Copy the grid and draw:
   (i) the reflection of shape *E* in the *y*-axis,
   (ii) the enlargement of shape *F*, with scale factor 2 and centre (0, 0).

(0580/03 May/June 2004 q 6)

6. Each diagram below shows a shaded letter and its image. Copy the diagram. In
   each case describe fully the single transformations which maps the **shaded** figure
   onto its image. Mark and label any points you need in your descriptions.

(a)

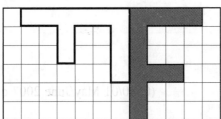

(b)

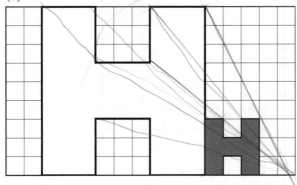

(0580/03 Oct/Nov 2005 q 1 (part only))

7. $\overrightarrow{AB} = \begin{pmatrix} -1 \\ 4 \end{pmatrix}$ and $\overrightarrow{CD} = 3\overrightarrow{AB}$.

(a) Write $\overrightarrow{CD}$ as a column vector.

(b) Make two statements about the relationship between the lines AB and CD.

(0580/01 May/June 2006 q 15)

8. The points A and B are marked on the diagram.

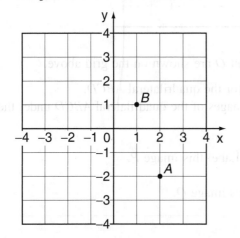

(a) Write $\overrightarrow{AB}$ as a column vector.

(b) $\overrightarrow{BC} = \begin{pmatrix} -3 \\ -2 \end{pmatrix}$.

Write down the coordinates of *C*.

(0580/01 May/June 2007 q 8)

9.

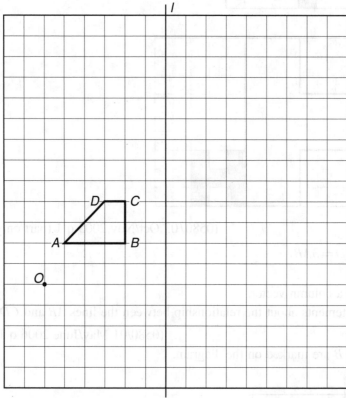

A quadrilateral *ABCD*, a line *l* and a point *O* are shown on the grid above.

(a) Write down the mathematical name for the quadrilateral *ABCD*.

(b) Copy the grid above, and draw the images of the quadrilateral *ABCD* under the following transformations.

  (i) Translation by the vector $\begin{pmatrix} 9 \\ -3 \end{pmatrix}$. Label this image *P*.

  (ii) Reflection in the line *l*. Label this image *Q*.

    (iii) Rotation, centre *A*, through 90° anti-clockwise. Label this image *R*.

    (iv) Enlargement, centre *O* and scale factor 3. Label this image *S*.

(0580/03 May/June 2007 q 7)

10.

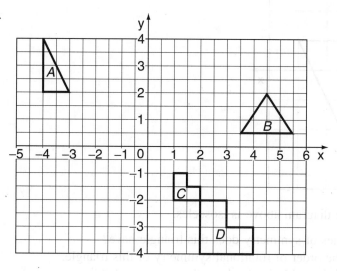

(a) A translation is given by $\begin{pmatrix} 6 \\ 3 \end{pmatrix} + \begin{pmatrix} -3 \\ -4 \end{pmatrix}$.

   (i) Write this translation as a single column vector.

   (ii) On a copy of the grid, draw the translation of triangle *A* using this vector.

(b) Another translation is given by $-2\begin{pmatrix} 1 \\ -1 \end{pmatrix}$.

   (i) Write this translation as a single column vector.

   (ii) On your copy of the grid, draw the translation of triangle *B* using this vector.

(c) Describe fully the single transformation that maps shape *C* onto shape *D*.

(d)

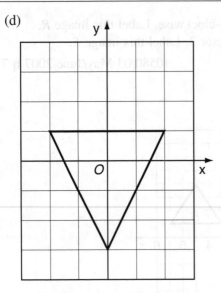

The triangle in the diagram above is isosceles.

(i) How many lines of symmetry does this triangle have?

(ii) Write down the order of rotational symmetry of this triangle.

(iii) On a copy of the grid above, draw the rotation of this triangle about *O* through 180°.

(iv) Describe fully another single transformation that maps this triangle onto your answer for part **(d) (iii)**.

(0580/03 May/June 2005 q 4)

11. $\mathbf{p} = \begin{pmatrix} 2 \\ -3 \end{pmatrix}$ and $\mathbf{q} = \begin{pmatrix} 3 \\ 1 \end{pmatrix}$

(a) Write **p** + **q** as a column vector.

(b) The point *O* is marked on the grid below.

Draw the vector $\overrightarrow{OP}$ where $\overrightarrow{OP} = \mathbf{p}$.

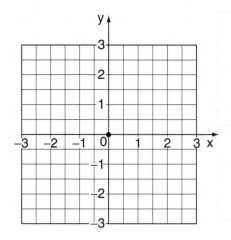

(0580/01 Oct/Nov 2005 q 11)

12.

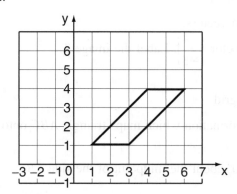

On the grid, draw the reflection of the parallelogram in the line $x = 3$.

(0580/11 May/June 2008 q 15)

13.

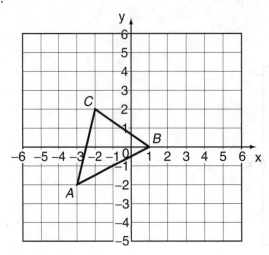

Triangle *ABC* is drawn on the grid.

(a) (i) Write down the coordinates of *A*.

(ii) Write $\overrightarrow{AB}$ and $\overrightarrow{BC}$ as column vectors.

(b) Translate triangle *ABC* by the vector $\begin{pmatrix} 4 \\ -3 \end{pmatrix}$. Label the image *T*.

(c) $\overrightarrow{AP} = 2\overrightarrow{AB}$ and $\overrightarrow{AQ} = 2\overrightarrow{AC}$.

(i) Plot the points *P* and *Q* on the grid.

(ii) Describe fully the single transformation which maps triangle *ABC* onto triangle *APQ*.

(d) Rotate triangle *ABC* through 180° about the midpoint of the side *AB*. Label the image *R*.

(0580/03 Oct/Nov 2008 q 8)

Chapter 11

Statistics

Statistics are all around us in the modern world, so it is necessary to have some knowledge of how statistics are used and the strengths and failings of statistics.

Statistics are used by governments to determine such diverse things as how many schools and hospitals need to be built in the next decade, how the proportion of elderly people in the population will change, and consequently how much support for the elderly will be needed, or whether the undesirable side effects of a particular drug outweigh its benefits. Businesses may need to predict their stock requirements for the coming years, based on population statistics and current trends. Universities will need to predict the proportion of students who will want to study mathematics next year. You may need to know the likelihood of rain tomorrow. Insurance companies need to know things such as life expectancy of individuals of different ages.

All these things will use statistics. Proper use of statistics involves many steps such as:

- defining the problem that needs to be answered,
- collecting data and organising it in a way that can be used,
- illustrating the results with pictures, graphs or charts in a way that can be easily understood,
- interpreting the data so that useful conclusions can be drawn,
- and finally, testing the conclusions to see how valid they may be.

Statistics is different from the rest of the mathematics you are studying because it can only assess the probability of certain conclusions from given data. However, it is an enormously important branch of mathematics for today's world.

It is very important that you should understand that interpreting the statistics can sometimes lead to wrong conclusions. People may want to prove a favourite theory and in some way distort the statistics to support their theory. This does not necessarily mean fraud, but it can mean that other people are misinformed. In particular the public can be misinformed by the distortion of statistical diagrams as you will see later in the chapter.

Good statistical arguments depend on the collection of sufficient data in an unbiased manner.

For example, you could not predict the number of people in the general population who would be interested in going to the Olympic Games by only asking the members of a Sports Club. Also you could not make a sensible prediction based on asking such a small number of people.

Essential Skills

Spend a little time (with your calculator where necessary) working on these questions. It will help you answer some of the questions in this chapter. Check each answer with your calculator as you go along.

1. $\dfrac{360}{18} = 20$ $\qquad\qquad \dfrac{18}{360} = 0.05$

   Using the above information copy and fill in the blanks:

   (a) $20 \times 18 = $ ......
   (b) $360 \times 0.05 = $ ......
   (c) $360 \div $ ...... $= 18$
   (d) $18 \div $ ...... $= 360$
   (e) $20 \times $ ...... $= 1$

2. Copy and fill in the blanks:

   (a) $360 \times $ ...... $= 30$
   (b) $360 \div $ ...... $= 30$
   (c) $360 \times $ ...... $= 12$
   (d) $360 \div $ ...... $= 12$
   (e) $10 \times $ ...... $= 360$
   (f) $360 \times $ ...... $= 10$
   (g) $360 \div $ ...... $= 10$
   (h) $360 \times $ ...... $= 720$
   (i) $360 \div $ ...... $= 540$
   (j) $1440 \times $ ...... $= 360$

3. Copy and fill in the blanks:

   (a) $100 \times $ ...... $= 360$
   (b) $72 \times $ ...... $= 54$
   (c) $690 \div $ ...... $= 23$
   (d) $690 \times $ ...... $= 23$
   (e) $72 \div $ ...... $= 24$
   (f) $42 \div $ ...... $= 28$
   (g) $55 \times $ ...... $= 99$
   (h) $81 \div $ ...... $= 36$
   (i) $81 \times $ ...... $= 36$
   (j) $90 \times $ ...... $= 72$
   (k) $90 \div $ ...... $= 72$

Answers

1. (a) $20 \times 18 = 360$
   (b) $360 \times 0.05 = 18$
   (c) $360 \div 20 = 18$
   (d) $18 \div 0.05 = 360$
   (e) $20 \times 0.05 = 1$

2. (a) $360 \times 0.08\dot{3} = 30$
   (b) $360 \div 12 = 30$
   (c) $360 \times 0.0\dot{3} = 12$
   (d) $360 \div 30 = 12$
   (e) $10 \times 36 = 360$
   (f) $360 \times 0.02\dot{7} = 10$
   (g) $360 \div 36 = 10$
   (h) $360 \times 2 = 720$
   (i) $360 \div 0.\dot{6} = 540$
   (j) $1440 \times 0.25 = 360$

3. (a) $100 \times 3.6 = 360$
   (b) $72 \times 0.75 = 54$
   (c) $690 \div 30 = 23$
   (d) $690 \times 0.0\dot{3} = 23$
   (e) $72 \div 3 = 24$
   (f) $42 \div 1.5 = 28$

(g) $55 \times 1.8 = 99$ (h) $81 \div 2.25 = 36$
(i) $81 \times 0.\dot{4} = 36$ (j) $90 \times 0.8 = 72$
(k) $90 \div 1.25 = 72$

Collecting and Organising Data

A data set consists of individual pieces of information that you may collect in order to solve a problem.

A typical collection of data would be made by a **survey**.

A survey is an organised collection of sufficient and relevant data that can be used to help solve a problem. It is collected from a **sample** of the whole possible set of data (which is the **population**). It can then be used to make predictions for the whole population.

Examples of Surveys

SURVEY 1 Traffic congestion

Traffic congestion through a village is causing so much hardship that the authorities are considering widening the road, or even building a new road to bypass the village.

The first step towards deciding whether to spend the money necessary to build the bypass is to find the number of cars passing a particular point (point A) in the village at different times of the day.

The authorities commission a survey. The cars are counted throughout the day, over many days to collect enough data. The method used is to count every car that passes point A during each hour during the day. The term 'car' in this survey includes all motorised vehicles, even lorries, motorbikes and tractors.

An easy way to count cars in this situation is to use a **tally chart**. Using a tally chart could be as simple as making a mark on paper every time a car passes. In order to make it easier to add up the marks it is usual to put a strike through the previous four marks to represent the fifth mark for each set of five marks. The diagram shows how such a chart might look.

You will notice that the 'Time of day' column shows entries such as 0700 to before 0800. This means that a car passing at exactly 0800 goes into the next row labelled 0800 to before 0900.

A more technical method would be to have some sort of hand-held data logging device which could be clicked every time a car passes and would record all the data and the time of day automatically.

A survey would normally have much more data than this, but this will be sufficient for our purposes.

SURVEY 1

Cars passing point *A* at different times of the day

DATE		
PLACE		
Time of day	Tally	Total
0700 to before 0800	ЖЖ ЖЖ ЖЖ ЖЖ ЖЖ	25
0800 to before 0900	ЖЖ ЖЖ ЖЖ ЖЖ ЖЖ ЖЖ ЖЖ ЖЖ ЖЖ I	46
0900 to before 1000	ЖЖ ЖЖ ЖЖ ЖЖ ЖЖ ЖЖ ЖЖ ЖЖ ЖЖ ЖЖ ЖЖ ЖЖ III	63
1000 to before 1100	ЖЖ ЖЖ ЖЖ ЖЖ ЖЖ ЖЖ ЖЖ ЖЖ I	41
and so on		

The table above is an example of a **frequency distribution**. The frequency of an event is another way of expressing how often it occurs. The column labelled Total is the same as the frequency, so a **frequency table** of the data above might look like the next table.

Cars passing point *A* at different times of the day

Time of day	Frequency
0700 to before 0800	25
0800 to before 0900	46
0900 to before 1000	63
1000 to before 1100	41
and so on	

This is actually a **grouped frequency table** because the individual times at which the cars passed are not given, but they have been grouped into hourly intervals or classes.

SURVEY 2 Wild flower meadow

A botanist wants to estimate the number of wild flowers in a particular meadow. To save having to survey the whole meadow **sample** squares are taken at different places across the meadow. Each section is 1 metre square. For this survey a tally chart is not necessary as the flowers can just be counted. In the survey the sections were labelled *A, B, C, D* and *E* and the results were as shown in the frequency table below.

SURVEY 2

Wildflowers per square metre

Section of meadow	Frequency
A	15
B	14
C	21
D	8
E	20

Again, a proper survey would need much more data, but this will be sufficient for our work. Further statistical work could make a prediction of the number of flowers in the whole meadow.

SURVEY 3 Reduction of pollution

Our third survey has been commissioned by a city council because they are trying to find ways of cutting down on pollution in a city. They would like people taking children to school in the mornings to try to share transport.

They commission a survey to find how many passengers there are in each car, in the morning rush hour, entering the city past a certain point on one of the main roads into the city.

The results are shown in the frequency table below.

SURVEY 3

Numbers of passengers per car

Number of passengers	0	1	2	3	4	5
Number of cars	64	38	41	27	10	2

The first thing you may notice about this frequency table is that, unlike the previous two surveys, it is presented horizontally instead of vertically. You must be able to recognise which row is the frequency, and which is the data you are counting. In this case the number of cars is the frequency; the data you are counting is the number of passengers per car.

Types of Data

In the first survey we are counting the numbers of cars which pass a particular point at different times of the day. You have seen that we have had to make a decision about where a car passing at 0800 belongs. The time of day is **continuous** data. Continuous data is measured data. Examples would include heights of buildings, volumes of liquids and so on.

In the second survey we are counting the numbers of flowers in different sections of a meadow. There is no doubt about to which section any particular flower belongs. The sections are separate, or **categorical**. Categorical data could include colours of sweets, countries of the world and so on.

In the third survey we are counting the numbers of cars which carry passengers. Again there is no doubt as to where each car belongs. It must carry either 0, 1, 2, 3, 4 or 5 passengers. This is **discrete** data. Discrete data can be counted. Examples could be scores at a games match, examination marks and so on.

Illustrating the Data

One possible method to illustrate the data would be a **pictogram.**

A pictogram uses a small picture or symbol, to represent a given number of the pieces of data. For example, using the frequency distribution from Survey 1 (Traffic congestion), we might use a small car to represent a frequency of, say, ten cars. Less than ten cars would have to be represented by a part of a car, so this is not a very accurate method, but it does present the data in a pictorial way, which gives everyone an immediate idea of the scale of the problem.

Cars passing point *A* at different times of the day

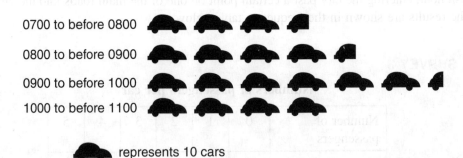

0700 to before 0800

0800 to before 0900

0900 to before 1000

1000 to before 1100

represents 10 cars

Points to note about pictograms:

- the pictogram needs a key to show how many objects are represented by each whole symbol,
- parts of a symbol are used to represent smaller numbers, but these are not very accurate,
- the pictogram needs a title.

Example 1

Draw a pictogram to illustrate the data in Survey 2, the Wildflower survey.

Answer 1

Numbers of wild flowers per square metre of meadow

A ✿ ✿ ✿

B ✿ ✿ ✾

C ✿ ✿ ✿ ✿ ˒

D ✿ ˺

E ✿ ✿ ✿ ✿

✿ represents 5 flowers

Pictograms are not very accurate representations of the data. Precise illustrations are provided by **bar graphs** (or **bar charts**) or simple **histograms.**

Bar charts illustrate data which can be divided into completely *separate* categories.

The wildflower survey can be illustrated by a bar chart because the data is divided into separate sections *A, B, C, D* and *E* of the meadow.

In a bar chart the vertical axis represents frequency, and the horizontal axis distinguishes each bar with a label. The bars can be separated because there is no actual connection between them.

Numbers of wild flowers per square metre of meadow

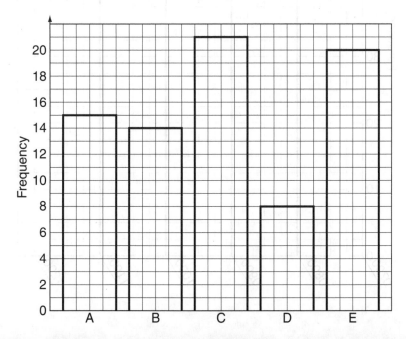

You will also see bar graphs with the bars drawn horizontally and the frequency on the horizontal axis.

Points to note about bar graphs:

- the bars can be separated by blank areas,
- the bars are labelled, not the grid lines,
- the height of each bar represents the frequency,
- the bars are all the same width,
- the bar chart needs a title.

The Traffic congestion survey (Survey 1), would be illustrated by a simple histogram. This is very similar to a bar chart, but is used to represent data that is **continuous.** As we have seen, continuous data is not divided into separate parts but can take any value.

. The data is still represented by bars, but the bars cannot be separated by blank spaces because there is no gap in the possible measurements.

Cars passing point *A* at different times of the day

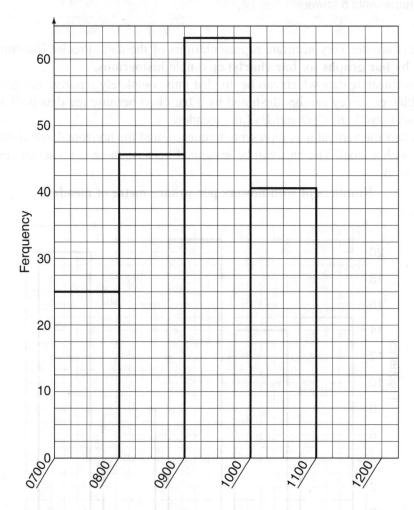

It is important to note that a *simple* histogram can only be used if the data is divided into groups of *equal width* (as in the above example where the groups are each of one hour). For groups of varying widths the histograms are slightly more complicated, but these histograms are not required for CORE.

Points to notice about simple histograms:

- the bars touch each other and are not separated by blank spaces,
- the grid lines are labelled, not the bars,
- **the widths of the bars are all the same**,
- the height of each bar represents the frequency,
- the histogram needs a title.

Example 2

Draw a bar chart to illustrate the data in Survey 3, the Reduction of pollution survey.

Answer 2

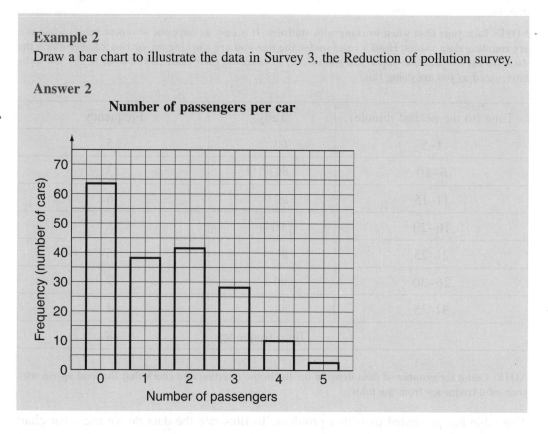

Some data has so many values that it is better to group it into **classes** before drawing any conclusion, as the next survey shows.

SURVEY 4 Waiting times in a surgery

The data set below records the time patients had to wait in a surgery before they were able to see a doctor. The times are given to the nearest minute. The data is an example of **raw data** which has not yet been sorted in any way.

SURVEY 4

Waiting time for patients in a doctor's surgery

10	3	15	7	8	4	21	33	11	31	27	25
7	5	9	11	27	17	28	16	14	8	23	35
19	10	28	21	16	15	19	19	19	27	6	9
26	7	2	18	23	14	10	10	9	32	29	3

This data is shown in the frequency table below with the waiting times grouped into classes.

NOTE: Take your time when working with statistics. It is easy to leave out or repeat values when you are counting data values. Hold a ruler under the line you are working on, or run a finger along the data. On the examination paper you will be able to cross out each value as you enter it. Try not to be interrupted as you are doing this.

Time (to the nearest minute)	Tally	Frequency
1–5	IIII I	5
6–10	IIII IIII III	13
11–15	IIII I	6
16–20	IIII III	8
21–25	IIII I	5
26–30	IIII II	7
31–35	IIII	4
	Total frequency	48

NOTE: Count the number of data items in the list in the question and check that the total agrees with your total frequency from the table.

This table has presented us with a problem. To illustrate the data do we use a bar chart or histogram?

Time is continuous, but these classes appear to be for discrete data because the times have been rounded to the nearest minute. However, this means that a time of 5.4 minutes would be rounded to 5 minutes, and a time of 5.5 minutes would be rounded to 6 minutes. Just as measurements are rounded according to our known rules, so the classes into which continuous items of data will be entered have **upper** and **lower class boundaries**, and the data is effectively rounded into these boundaries. The class boundaries in the histogram are calculated as $\dfrac{5+6}{2}, \dfrac{10+11}{2}, \dfrac{15+16}{2}$ and so on.

If the data had not previously been rounded it could have been entered into the table below, with the classes shown. The result is still the same, but the table now makes it clear that we should draw a histogram.

Time (t minutes)	Frequency
$0.5 \leqslant t < 5.5$	5
$5.5 \leqslant t < 10.5$	13
$10.5 \leqslant t < 15.5$	6
$15.5 \leqslant t < 20.5$	8
$20.5 \leqslant t < 25.5$	5
$25.5 \leqslant t < 30.5$	7
$30.5 \leqslant t < 35.5$	4
Total frequency	48

The waiting times are illustrated in the histogram below, and as you can see *the class boundaries must be shown on the horizontal axis.*

Waiting time for patients in a doctor's surgery

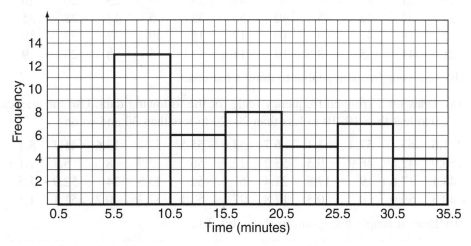

When do we draw a bar chart and when do we draw a simple histogram?

- **Bar charts** are drawn to illustrate **discrete** or **categorical** data which are separate or *counted* data.

- Examples could be numbers of students, colours, votes for political parties, types of farm animals.
- The data can still be grouped into classes.
- The bars can be separate.

- **Histograms** are drawn to illustrate **continuous** data which is *measured* data.
- Examples could be journey times, long jump records, weights of bags of flour.
- The data can be grouped into classes, but decisions must be made as to where values fit in the boundaries between classes.
- There is no space between the bars (unless one class has a frequency of zero).

Exercise 11.1

1. (a) Construct a frequency table for the following scores on the spinning of a spinner numbered from 1 to 6.

2	5	6	3	3	3	4	1	5	1	5
5	5	6	3	4	4	2	2	3	4	5
1	6	1	3	2	3	3	2	5	2	5
2	6	4	1	3	5	5	2	1	4	3

   (b) Draw a bar chart to illustrate the frequency of each score.

2. (a) Draw up a grouped frequency table for the following *discrete* data.
   Use the classes 1–10, 11–20, 21–30, 31–40, 41–50.

1	45	50	37	3	3	36	5	48	38	17
19	21	12	36	46	49	29	29	29	20	40
34	31	42	45	40	50	32	5	11	28	35
2	8	41	9	10	17	18	49	37	33	30
21	41	5	8	10	11	13	5	8	23	9

   (b) Use your frequency table to draw a bar chart.

3. (a) Draw up a grouped frequency table for the following data. Use the classes:
   0 to less than 10, 10 to less than 20, 20 to less than 30 and so on.

1.6	63.1	57.3	3.1	6.7	54.6	55.5	52.7	13.6	41.7	8.1
56.9	42.8	46.9	9.5	53.2	12.7	56	3.9	8	5.7	1.1
44.1	17.6	9	17.8	27.4	57.3	52	33.8	34	52.9	7.5
49.3	59.9	0	0.5	17.8	27	53.1	37.3	0.7	51.1	1.1

   (b) Draw a simple histogram to show this data.

4. The heights of 25 students were measured, and the results are shown below.

SURVEY 5

Heights of students

160	155.5	128.5	161	152	152.5	153	154.7	141	163.4	129.2
164.2	151.9	150	150.3	145	138	136.6	129	149	148.2	135.1
132.8	143	141.5								

(a) Draw up a frequency table with the data grouped into the following classes:
120 to < 130, 130 to < 140, 140 to < 150, 150 to < 160 and 160 to < 170.

NOTE: This means that 129.9 would go in the first class and 130 would go in the second class and so on.

(b) Draw a simple histogram with the horizontal axis labelled from 120 to 170.

5. The bar chart shown below shows the numbers of a group of students taking examinations in Maths, Physics, Chemistry, English and Economics in one week in an examination period. 26 students are taking the Chemistry examination, and 15 are taking the Economics examination.

(a) Copy the bar chart and complete the frequency scale on the vertical axis. (The scale runs from zero.)

(b) Draw in the missing bar for Economics.

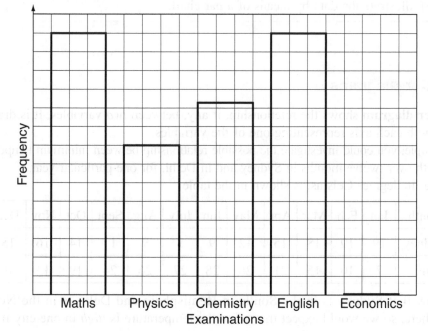

(c) How many examinations will be taken all together?

6. Liam counts the numbers of different colours in a bag of sweets.
   He finds 12 red sweets, 9 green sweets, 4 yellow sweets and 7 purple sweets.
   Draw a bar chart to illustrate the numbers of sweets of each colour.

7. A hotel manager leaves a questionnaire in each room for visitors to rate their
   stay. Feedback from the questionnaire should give the hotel managers valuable
   insight into how they can improve their customer satisfaction.
   The questionnaire asks the visitors to tick boxes which best describe their
   satisfaction with different aspects of the hotel.
   The boxes are numbered   0: very dissatisfied

       1: fairly dissatisfied

       2: neither satisfied nor dissatisfied

       3: fairly satisfied

       4: very satisfied

       5: don't know.

   After a week the questionnaires were studied.
   The answers to the question: "How satisfied were you with the hotel
   breakfast?" are shown below.

5	4	1	0	1	1	0	4	4	3	2
5	2	5	2	3	4	1	0	0	1	2
2	3	3	4	3	2	5	1	0	3	3
4	0	1	3	3	4	2	2	3	3	4

   (a) Use the data above to draw up a frequency table.
   (b) Illustrate the data by means of a bar chart.

Scatter Diagrams

A **scatter diagram** shows the relationship, if any, between *two* variables. It is drawn on
a grid, with each axis representing one of the variables.

For example, you could investigate the possible relationship between minimum temperatures
per month over twelve months in Sydney and in Delhi, for one particular year.
The data, in degrees Celsius, is shown in the table.

Month	Jan	Feb	Mar	Apr	May	Jun	July	Aug	Sept	Oct	Nov	Dec
Sydney	19	19	18	15	12	9	8	9	11	14	16	18
Delhi	7	10	15	21	26	28	27	26	24	19	13	8

We know that Sydney is in the Southern Hemisphere and Delhi is in the Northern
Hemisphere, so we would expect that when the temperature is *high* in one city it would

be *low* in the other. This is called **negative correlation**. To see if this is the case, and to investigate how strongly the data conforms to this expectation we draw a scatter diagram.

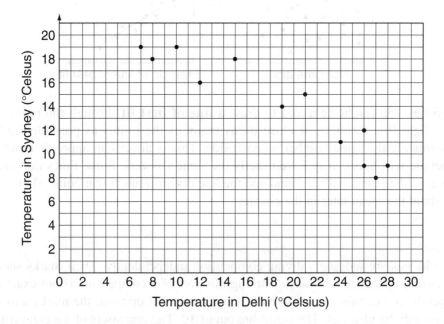

Each dot on the diagram represents one month.

The diagram shows that there is indeed a negative correlation, and it is a strong correlation because the results cluster close to a line. For our purposes we could say that they are close to a straight line, although, in this case it looks as though it could possibly be a curve and would need further investigation to establish whether this is so. We are not concerned with that amount of detail here.

The following diagrams show the types of scatter diagrams you might see, and how the correlation would be described in each case.

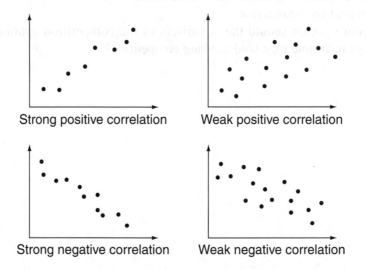

Strong positive correlation   Weak positive correlation

Strong negative correlation   Weak negative correlation

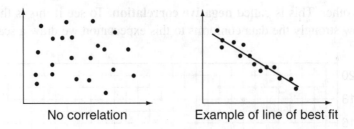

No correlation   Example of line of best fit

Sometimes you might be asked to draw a **line of best fit** through the points. This is a straight line through, or close to, as many as the points as possible, and with approximately the same number on each side. This is always an approximation so no two people will necessarily have exactly the same answers. Some times the mean, or average, of the points will be found and the line drawn through that point. There will be more about the mean later in the chapter.

Example 3

Two judges are judging a diving competition independently. Their marks should be close together, but because marking this sort of a competition is not exact but depends to a certain extent on each judge's personal opinion, the marks will not necessarily be identical. The marks are out of 10. The organisers of the competition want to make sure that the judges are reasonably consistent.

The table shows the results.

Competitor number	1	2	3	4	5	6	7	8	9	10
Judge A	5	4	3	7	9	9	10	3	1	4
Judge B	6	4	5	6	8	9	9	2	2	4

(a) Draw a scatter diagram to show these results.
(b) Comment on the diagram.
(c) In your opinion should the organisers of the competition appoint these two judges next time they hold a diving competition?

Answer 3

(a)

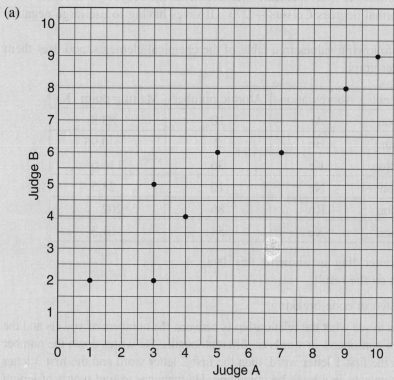

Marks in diving competition

(b) The diagram shows a reasonably strong consistency between the two judges.

(c) The organisers could safely ask the judges back next time.

Exercise 11.2

1. The table shows the average maximum and minimum temperatures in degrees Celsius for each month in a city in the Northern Hemisphere.

Anita assumes that there will be a positive correlation between the maximum and minimum temperatures.

	Jan	Feb	Mar	Apr	May	Jun	July	Aug	Sept	Oct	Nov	Dec
Av Max	21	22	29	37	39	37	34	33	34	33	28	23
Av Min	7	9	15	22	26	28	27	26	25	19	13	8

Draw a scatter diagram and comment on the result.

2. As part of her Chemistry project Zaida is investigating the properties of the noble gases. She suspects that there is an association between the melting point in K and the atomic number (K is temperature measurement with respect to absolute zero, so it is the normal degrees Celsius + 273°. It saves having to use large negative numbers).

   She finds the following values in a table of the chemical elements, and uses them to plot a scatter graph.

Element	Abbreviation	Atomic number	Melting point (K)
Argon	Ar	18	83.8
Helium	He	2	0.95
Krypton	Kr	36	116.6
Neon	Ne	10	24.5
Radon	Rn	86	202
Xenon	Xe	54	161.3

   (a) Draw a scatter diagram showing this data.

   (b) Comment on the result.

3. Joel is interested in code-breaking.

   (a) He decides to see what the relationship is between the numbers of vowels and the numbers of consonants in words of different lengths. He writes down the number of each in the first 1 letter word, then the first 2 letter word and the first 3 letter word he comes to in a book he is reading. He manages to find words of length up to 14 letters, and one of 18 letters.

   His results are shown in the table.

Word length	1	2	3	4	5	6	7	8	9	10	11	12	13	14	15	16	17	18
Vowels	1	1	1	1	2	2	4	3	3	5	4	4	7	6				8
Consonants	0	1	2	3	3	4	3	5	6	5	7	8	6	8				10

   (i) Draw a scatter diagram to show these results.

   (ii) Comment on the diagram.

   (b) Joel decides that this is not going to help very much.

   He thinks that the most commonly used letters in the English language could be **a e d r s t**.

   Joel makes a tally chart by looking at the number of times these letters occur in the first paragraph of his book.

From the tally chart he draws up the following frequency table.

Letter	a	e	d	r	s	t
Frequency	35	59	10	30	36	36

Draw a bar chart to show this data.
(c) How do you think he could improve his investigation?

4. Pierre and Mignon have to be in school by 0900.
Their father drives them to school but the rush hour traffic causes delays.
They keep a note of the time they leave home each day and the length of time it takes to get to school (to the nearest minute) for 12 days.
The results are shown in the table.

Time of day	0800	0712	0805	0815	0825	0845	0724	0746	0738	0700	0835	0750
Time taken	18	10	22	25	23	27	13	25	20	9	22	22

(a) Draw a scatter diagram to show these times.
(b) Draw the line of best fit.
(c) Use your line of best fit to estimate how long the journey will take if they leave home at 0730.
(d) Estimate how many minutes late will they might be if they leave home at 0840.

Practical Investigations

1. Investigate the possible relationship between people's hand span and the length of their feet. The hand span is the furthest distance between the outstretched thumb and little finger that a person can spread their fingers.
2. Investigate whether people find it easier to estimate lengths than areas.
3. Investigate a possible relationship between the length of a person's middle finger and their musical ability.
4. Investigate a possible relationship between a person's ability in mathematics and their ability in music.

Pie Charts

A pie chart is so called because it looks like a pie divided into slices. Who gets the biggest slice? It is used to represent how the whole of a group is divided up into categories. The sizes of the 'slices' are proportional to the numbers in each category. The sizes are dependent on the angle at the centre of each, so for example, if the total number in the group is 36 and the number in one of the categories is 5, then the angle at the centre of the slice is a fraction $\left(\dfrac{5}{36}\right)$ of the complete turn.

For example, suppose you want to illustrate the composition of a local orchestra which has the following members:

32	string players,
8	woodwind players,
5	brass players,
3	others (percussionist, conductor and pianist).

The first step is to make a table.

PLAYERS	NUMBER	CALCULATION	ANGLE
Strings	32		
Woodwind	8		
Brass	5		
Percussionist, conductor, and pianist	3		
TOTAL	48	$\dfrac{360}{48} = 7.5$	360°

The total number of players is 48, which is represented by the whole circle (360°). This is a question on proportion, so find the multiplier to get from 48 to 360.

This is $\dfrac{360}{48} = 7.5$, so $48 \times 7.5 = 360°$. Now we can calculate the angles: $32 \times 7.5 = 240°$, and so on.

The table becomes:

PLAYERS	NUMBER	CALCULATION	ANGLE
Strings	32	$32 \times 7.5 =$	240°
Woodwind	8	$8 \times 7.5 =$	60°
Brass	5	$5 \times 7.5 =$	37.5°
Percussionist, conductor, and pianist	3	$3 \times 7.5 =$	22.5°
TOTAL	48	$48 \times 7.5 =$	360°

The pie chart can now be drawn.

Pie chart showing composition of an orchestra

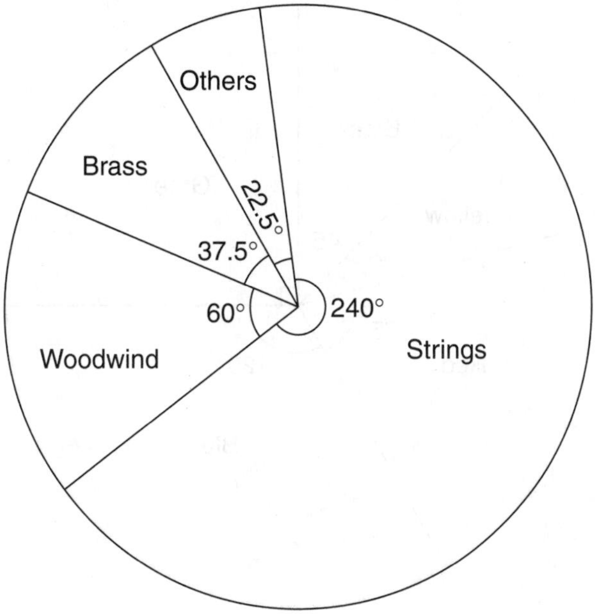

Points to remember when drawing a pie chart:

- Be as accurate as possible.
- Mark the centre as soon as you have drawn the circle (it is easy to lose the centre when you remove your compass point!).
- It does not matter where you start, but the first thing to do is to draw in one radius and start measuring angles from there.
- Draw each slice as soon as you have measured it. (Do not try to measure all the angles at once unless you have a 360° protractor, and even then it is easy to make a mistake.)
- A pie chart measurer is not necessarily the best thing to use as it will probably be marked in percentages instead of angles.
- Each slice must be labelled.
- The pie chart needs a title.
- If you end up with some pie left over you have miscalculated. Check your original calculation and that you have found the correct multiplier or divisor.

Example 4
The pie chart shows the proportion of pencils of various colours in Ethan's pencil case. Ethan has 5 red pencils.

Colours of pencils in pencil case

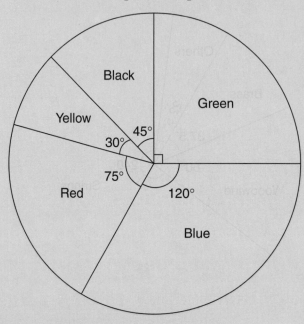

The angles are shown in the diagram.
(a) Draw up a table to show the information given.
(b) Calculate the numbers of pencils of each colour and complete the table.

Answer 4

(a)

COLOUR	ANGLE	NUMBERS OF PENCILS
Green	90°	
Blue	120°	
Red	75°	5
Yellow	30°	
Black	45°	
TOTAL	360°	

(b) **NOTE: We are not given the total number of pencils, but we are told that 75° represents 5 pencils.**

First we must find the multiplier or divisor to get from 75 to 5.

$\dfrac{75}{5} = 15$, so $\dfrac{75}{15} = 5$, and *dividing* by 15 will convert each angle to the number of pencils it represents.

NOTE: Check that dividing by 15 gives a reasonable answer: $90 \div 15 = 6$ which is reasonable, but $90 \times 15 = 1350$, which is *not* reasonable!

COLOUR	ANGLE	CALCULATION	NUMBERS OF PENCILS
Green	90°	90 ÷ 15	6
Blue	120°	120 ÷ 15	8
Red	75°	75 ÷ 15	5
Yellow	30°	30 ÷ 15	2
Black	45°	45 ÷ 15	3
TOTAL	360°	360 ÷ 15	24

NOTE: Check your answers by adding the number of pencils (24) and by calculating $360 \div 15$ ($= 24$).

Exercise 11.3

1. The owners of a village store need to decide how much floor space to devote to different classes of goods.

   These are fresh produce, groceries, household products, magazines and stationery, and frozen goods.

   As an initial study the contents of the shopping baskets of customers are analysed over a week and the results entered into a computer spreadsheet.

   The total numbers of items in each category are printed out and the results are shown below.

Fresh produce	Groceries	Household products	Magazines and stationery	Frozen goods	TOTAL
351	183	66	315	165	

   (a) Copy and complete the table.

   (b) Draw a pie chart to show this information.

   (c) If the store has 100 square metres of floor space available calculate the area which should be dedicated to each category, showing your results in a table. Give your answers to 1 decimal place.

2. The electricity consumption on a farm is recorded in Units of Energy/day for four consecutive quarters of a year.
   The results, in the order they were recorded, are: 23   11   21   65
   (a) Draw up a table to show these results and calculate the angles required to show them on a pie chart. The pie chart will show how the electricity consumption for one year is used in the different seasons.
   (b) Draw and label the pie chart.
3. A newsagent stocks the following groups of items.
   Newspapers   35% of total stock.
   Magazines   50% of total stock.
   Snacks   15% of total stock.
   (a) Copy and complete the table.

	Percentage of total stock	Angle on pie chart
Newspapers		
Magazines		
Snacks		
TOTAL		

   (b) Draw and label a pie chart to show this information.
4. (a) Copy and complete this table showing the angles on a pie chart which is being drawn up to represent the numbers of students taking Psychology, Sociology, Economics and History out of a group of students.

	Number of students	Angle on pie chart
Psychology	14	
Sociology	20	
Economics		110
History		80
TOTAL	72	

   (b) Draw the pie chart.

Mean, Median, Mode and Range

The mean, median and mode are used to find the average of a set of data. They represent the *central tendency* of the data. For example, you might want to find the average of your set of examination results to compare with those of your friend (or rival!).

Suppose this is your set of results (all in percentages):

$$35 \quad 56 \quad 81 \quad 19 \quad 73 \quad 49 \quad 5 \quad 76 \quad 56 \quad 82 \quad 90 \quad 50$$

The three averages will usually give different results, and are used in different circumstances.

We will start with the **mean.**

The mean is calculated by adding up each value in the set of data (examination results), and dividing by the number of values.

So to calculate the mean of your results calculate

$$\frac{35+56+81+19+73+49+5+76+56+82+90+50}{12} = \frac{672}{12} = 56$$

The mean of your examination results is 56%.

Now the **median**.

The median is found by putting the values in order of size, and then finding the middle value. The median should divide the ordered set of data into two equal groups.

If there is an even number of values in the set, say 10, then the mean of the middle two values (5th and 6th values) is the median. If there is an odd number, say 59, take one out to represent the middle value and divide the remainder (58) into two equal groups (29 in each), then the median is the 30th value.

Rearranging the results:

$$5 \quad 19 \quad 35 \quad 49 \quad 50 \quad 56 \quad 56 \quad 73 \quad 76 \quad 81 \quad 82 \quad 90$$

There are 12 values, so there is no particular middle value. The values divide equally into two sets of 6. The median in this case is by taking the sixth and seventh values and finding their mean (or the value half way between them).

5 19 35 49 50 56 ... 56 73 76 81 82 90

In this case the required numbers are the same (56 and 56), so their mean is 56.

The median value is 56%.

Lastly the **mode**.

The mode is the value which appears the most frequently. It is easiest to see in the ordered values used for the median. Once again it is 56.

The mode is 56%.

It was only by chance that these examination results had the same mean, median and mode.

We will now look at your friend's results.

$$37 \quad 45 \quad 32 \quad 76 \quad 65 \quad 48 \quad 79 \quad 24 \quad 79 \quad 35 \quad 76 \quad 85$$

It would be difficult to compare them with yours without finding an average.

$$\text{The mean} = \frac{37+45+32+76+65+48+79+24+79+35+76+85}{12}$$

$$= \frac{681}{12} = 56.75\%$$

The median:

24 32 35 37 45 48 65 76 76 79 79 85

Again there are 12 values so we need to find the mean of the sixth and seventh values.

24 32 35 37 45 48 ... 65 76 76 79 79 85

$$\frac{48+65}{2} = \frac{113}{2} = 56.5\%$$

This time there are two modes! This is not unusual.

There is one more useful statistic which can be quoted, and that is the **range**.

The range gives an idea of the *spread* of the values, and is simply the difference between the largest and the smallest values.

The range of your marks is $90 - 5 = 85$.

The range of your friend's marks is $85 - 37 = 48$.

Let us look again at your marks. There is one very low mark of only 5%. Perhaps you were feeling rather ill when you took that examination, or perhaps you really do not like that particular subject. How would it change your averages if we left it out?

$$\text{First the mean} = \frac{35+56+81+19+73+49+76+56+82+90+50}{11}$$

$$= \frac{667}{11} = 60.6\%$$

This is looking much better!

The median:

19 35 49 50 56 56 73 76 81 82 90

Now that there are 11 values we can pick out a middle value and divide the remaining 10 into two groups of 5, making the median the 6th value:

19 35 49 50 56 ... 56 ... 73 76 81 82 90

The median is again 56%, so that has made no difference.

Also the mode is still 56%, so no difference there either.

Your new averages:

Mean   60.6%
Median 56%
Mode   56%

Your new range is $90 - 19 = 71$.

This shows that the mean can be so affected by unusual or extreme values (at either end of the set of values) that it is not always the best average to use. The median and the mode are not so affected, sometimes not at all as you can see.

Another student has the following set of results:

23 45 78 56 23 79 34 98 80 82 57 89

The mean is 62%.
The median

23 23 34 45 56 57 ... 78 79 80 82 89 98

$$\frac{57+78}{2} = 67.5\%$$

But the mode is 23%! Perhaps this student had two bad days!
The mode is clearly not a suitable average in this case, as it says nothing about the overall ability of the student.

Now consider the following.
A shopkeeper keeps a record of the sizes of shoes sold in one day. They were:

5 6 3 8 5 4 5 5 5 8 4 5 5 6 5 4 5 7 8 5 5

Which is the most popular size?
The mode is clearly 5, and this would be the size that this shopkeeper would need to keep in stock in larger numbers than the other sizes. In this case the mode is the most suitable average to consider.

To summarise:

- The **mean** is found by adding up all the values and dividing by the number of values.
- The **median** is found by arranging the values in size order and finding the middle value if there are an odd number of values, or the mean of the two middle values if there are an even number of values.
- The **mode** is the most frequently occurring value.
- The mean is probably the most used average.
- The median can be a better average if there are non-typical values at either end of the set of values, which could distort the mean.
- The mode is best for deciding which of a set of values is most popular.
- The mean may not be a whole number even if, for example, it represents people. (One old example of this is the strange report that the average family in Britain has 2.4 children.) The mean can be rounded, but not necessarily to a whole number. The usual rounding rule should be applied.
- The mean can be used to make further predictions about the data as you will see later.
- The **range** is the difference between the largest and smallest values, and shows the spread of the data.

You should appreciate by now that the popular phrase "on average" is imprecise, and may be misleading.

NOTE: If you have difficulty remembering which average is which try the following:
- the <u>medi</u>an is the <u>mid</u>dle value (when arranged in order)
- the <u>mode</u> is the <u>mo</u>st popular
- which leaves the mean, which is the other one!

Example 5

Find

(a) the mean,   (b) the median,   (c) the mode,   (d) the range,

of the following data.

7   1   9   6   2   8   1   4   5   1   6

9   4   6   2   7   6

Answer 5

(a) The mean $= \dfrac{84}{17} = 4.94$

(b) 1   1   1   2   2   4   4   5   6   6   6

6   7   7   8   9   9

NOTE: Check that you have not left any values out when you rearrange them by counting the total in the rearranged list.

There are 17 values so the values divide into 2 groups of 8 values with one value in the middle, like this:

first group of 8   1   last group of 8

The median is the 9th value, which is 6.

The median = 6

(c) The mode = 6

(d) The range = 9 − 1 = 8

Exercise 11.4

1. Find:
   (i) the mean   (ii) the median   (iii) the mode   (iv) the range for each of these sets of data.
   (a) 5   6   9   4   10   3   1   9
   (b) 1   1   1   2   3   3   5
       7   7   7   7   8   8   9
   (c) 2.5   2.6   2.6   3.1   4.2   4.8   5.1   5.3   5.6

2. The mean height of a group of 5 students is 160 cm.
   (a) What is their total height?
   2 new students join them. Their mean height is 156 cm.
   (b) What is the new mean height of the group?

NOTE: Remember that the mean height will be the total height of the enlarged group divided by the new total number of students.

3. The total number of students in a school is 856.
   (a) There are 30 classrooms available. Calculate the mean number of students per classroom, giving your answer to 3 significant figures.
   (b) It has been decided that 25 should be the maximum number of students per classroom. Calculate how many new classrooms should be built.
   (c) What is the new mean number of students per classroom?
4. The mean number of potatoes in each of 25 bags of potatoes is 60.
   (a) How many potatoes are there altogether?
   (b) The potatoes are going to be put into smaller bags, and the new mean is 15. Calculate the number of bags used.

The Mean, Median and Mode from a Frequency Table

How can these averages be calculated from data which is already drawn up into a frequency table?
We will start by examining the following set of data.

2	7	5	3	9	1	7	2	5	5	3	4
7	1	9	1	5	8	2	6	3	8	6	1
2	4	6	8	2	0	1	7	8	7	7	0

To calculate the **mean** we need to add up all the numbers and divide by 36:

$$\frac{2+7+5+3+9+1+7+2+5+5+...}{36}$$

This is going to take some time and probably be prone to errors.
It would be quicker to look along the rows of data and find the numbers of each item, for example, there are 5 ones, 5 twos, 3 threes and so on.

$$\frac{5\times1+5\times2+3\times3+2\times4+4\times5+3\times6+6\times7+4\times8+2\times9}{36}$$

$$=\frac{5+10+9+8+20+18+42+32+18}{36}$$

$$=\frac{162}{36}$$

$$=4.5$$

But the frequency table sets it up for us, and all we have to do is add another column where the number can be multiplied by the number of times it appears (the frequency).

Number	Frequency	Number × Frequency
0	2	0
1	5	5
2	5	10
3	3	9
4	2	8
5	4	20
6	3	18
7	6	42
8	4	32
9	2	18
Totals	36	162

Now all that has to be done is to divide the sum of all the values by the total frequency:

$$\text{mean} = \frac{162}{36} = 4.5$$

The **median** is also easy from the frequency table, because the table has automatically ordered the data for us.

There are 36 items of data, which will divide into two groups without a middle value:

first group has 18 items .... second group has 18 items.

The median is the mean of the 18th and 19th items.

Using the frequency column we can count down to the 18th and 19th entries, by making a running total.

There are two zeroes, 5 ones, 5 twos, 3 threes, and 2 fours, which accounts for the first 17 items $(2 + 5 + 5 + 3 + 2 = 17)$, so the first 17 data items are made up of the numbers 0 to 4.

The 18th and 19th items are both 5, so the median is 5.

The **mode** is simply the most frequent, that is the number with the highest frequency.

Looking down the frequency column we see that the highest frequency is 6, and it corresponds to the data item 7 (there are 6 sevens).

The mode is 7.

NOTE: Remember that the mode is the item from the data set which has the highest frequency, and it is not the frequency itself. The highest frequency is 6, but this corresponds with the data item which is 7, so the mode is 7.

Exercise 11.5

Copy each of these frequency tables, and use them to:
(a) calculate the mean,
(b) work out the median,
(c) find the mode
for each data set.

1.

Data Value	Frequency	
100	7	
110	10	
120	15	
130	2	
140	6	
150	3	
160	7	

2.

Data Value	25	26	27	28	29	30	31
Frequency	51	70	69	32	15	43	15

3.

Data Value	Frequency	
12.4	3	
12.5	5	
12.6	2	
12.7	1	
12.8	0	
12.9	5	
13.0	0	
13.1	2	

Statistics in the Media

Statistics should be treated with caution, and you should try to think behind those statistics that are published in the newspapers and other media, and which can cause alarm or panic in the general population. The following headlines are imaginary, but you may have read similar examples.

People with a BMI of greater than 30 are officially obese!

What does this mean?

BMI stands for Body Mass Index and is calculated by dividing your mass in kilograms by the square of your height in metres.

This actually makes some rugby players or other athletes obese!

Examinations are getting easier! The number of students getting A grades is the highest for 3 years!

What could be the possible reasons for this result?

1. Examinations are getting easier.
2. Teaching is improving.
3. Students are working harder.
4. Discipline in schools is improving.
5. Homework is being checked more rigorously.
6. The results are within the normal statistical variation.

Can you think of any other possible reasons?

Numbers of unemployed have fallen under the Hot Air Party!

(This would be just before the next election is due.)

1. How do these figures get counted?
2. Has the school leaving age been raised so that fewer young people are looking for jobs?
3. Has the Government made more University places available?
4. Have some other group of people been excluded from these figures, for example, the over 60's?

Can you think of any other reasons?

Look at these two pie charts, both showing the same information. What do you think about the way the second one is presented?

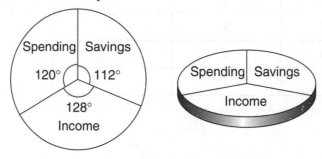

What about these two line graphs?

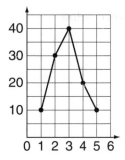
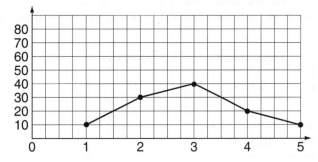

And these two bar charts, one of which appeared in a newspaper.

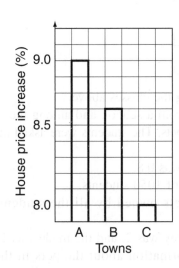
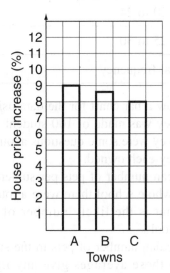

All this shows that one needs to be a bit sceptical, particularly when the statistics seem to prove something sensational. However, we still need to use statistics in every walk of life, so it is an essential subject and a powerful tool if used correctly.

Exercise 11.6

Mixed Exercise

1. The numbers of days each of a group of people were absent from work due to sickness in one year are shown below.

6	8	21	3	24	4	10	11	8	3	15
12	18	0	2	25	3	10	14	8	9	4
7	0	1	11	5	8	9	15	2	13	5

(a) Find the mode and range of these values.

(b) Copy and complete the grouped frequency table shown below.

Days absent	Frequency
0 to 4	10
5 to 9	
10 to 14	
15 to 19	
20 to 24	
25 to 30	
Total frequency	

(c) Draw a bar chart for the data using the classes above.

2. A survey was made of 100 students from a school, and among the topics on the survey were some questions about pets. The students were asked how many pets they each owned.

The mean number of pets per student was 0.8.

In the whole school there were a total of 1050 students.

(a) What is the likely number of pets owned by all the students in the school?

The median number of pets in the survey was 2, and the mode was 1.

(b) Do these averages give any information about the pets in the whole school?

3. You are given the following information about a set of data.

There are 5 items in the set. They are: 2 a b c c,

arranged in order of increasing size.

The mean = 5

The mode = 7

The median = 6

The range = 5

Use the information given to work out the values of a, b and c.

4. The time (to the nearest minute) some students took to complete a test was recorded, and the results are shown in the table below.

Time to the nearest minute	Number of students
1–10	2
11–20	10
21–30	6
31–40	2
Total	

Using class boundaries (0.5 to less than 10.5, and so on), draw a simple histogram to show these results.

5. Calculate the mean and, median from the frequency table you drew up in Exercise 11.1, question 1. What can you say about the mode?

6. 5   4   7   8   3   1   7   4   4   6

Calculate the mean and find the median and mode of the data given above.

7. Calculate the mean number of passengers per car in Survey 3 (page 355).

8. The table shows the distance travelled per litre of fuel for the Gofaster car at different speeds. We will call this the 'fuel economy'.

Speed (Kilometers per hour)	Fuel economy (Kilometers per litre)
80	12
96	11
112	10
128	9
135	8
144	7
160	6

(a) Draw a scatter diagram to show this information.

(b) Calculate   (i) the mean speed,   (ii) the mean fuel economy.

(c) Plot the point showing the mean speed and mean fuel economy on your scatter diagram, and draw a line of best fit through this point.

(d) Comment on the correlation shown between fuel economy and speed.

(e) Use your line of best fit to estimate the fuel economy when the car is travelling at 115 kilometres per hour.

(f) Estimate the possible speed when the fuel economy is 7.5 kilometres per litre.

Examination Questions

9. Ahmed selected a sample of 10 students from his school and measured their hand spans and heights. The results are shown in the table below.

Hand span (cm)	15	18.5	22.5	26	19	23	17.5	25	20.5	22
Height (cm)	154	156	164	178	162	170	154	168	168	160

He calculated the mean hand span to be 20.9 cm and the range of hand spans to be 11 cm.

(a) Calculate:

    (i) the mean **height**,

    (ii) the range of **heights**.

(b) In order to compare the two measures, he used a scatter diagram.
    The first three points are plotted on the grid.

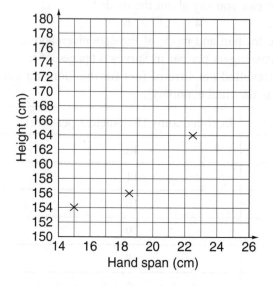

    (i) Copy and complete the scatter diagram by plotting the remaining 7 points.

    (ii) Draw the line of best fit on the grid.

    (iii) Use the line of best fit to estimate the height of a student with hand span 21 cm.

    (iv) Which of the following words describes the correlation?

        Positive   Negative   Zero

    (v) What does this indicate about the relationship between hand span and height?

(0580/3 May/June 2006 q 6)

10. A country has three political parties, the Reds, the Blues and the Greens.
    The pie chart shows the proportion of the total vote that each party received in an election.

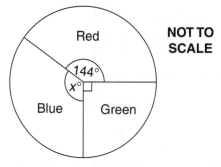

Red NOT TO SCALE

144°

x°

Blue   Green

(a) Find the value of *x*.
(b) What percentage of the votes did the Red party receive?

(0580/1 Oct/Nov 2003 q 4)

11. Which word describes the correlation in the scatter graph below?
    positive   negative   none

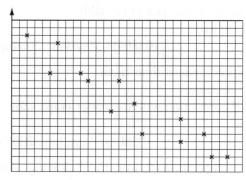

(0580/01 Oct/Nov 2006 q 3)

12. In a school, the number of students taking part in various sport is shown in the table below.

Sport	Number of students
Basketball	40
Soccer	55
Tennis	35
Volleyball	70

Draw a bar chart below to show this data.
Show your scale on the vertical axis and label the bars.

(0580/01 May/June 2004 q 17)

13. Grades were awarded for an examination.
    The table below shows the number of students in the whole school geting each grade.

Grade	Number of students	Angle on a pie chart
A	5	
B	15	
C	40	
D	20	
E	10	
Totals	90	

   (a) Complete the table above by calculating the angles required to draw a pie chart.
   (b) Draw an accurate pie chart to show the data in the table.
       Lable the sectors A, B, C, D and E.

                                                (0580/03 May/June 2004 q 1b)

14. Daniel plots a scatter diagram of speed against time taken.
    As the time taken increases, speed decreases.
    Which one of the following type of correlation will his scatter graph show?
    Positive   Negative   Zero

                                                (0580/01 May/June 2007 q 5)

15. The table gives the average surface temperature (°C) on the following planets.

Planet	Earth	Mercury	Neptune	Pluto	Saturn	Uranus
Average temperature	15	350	−220	−240	−180	−200

   (a) Calculate the range of these temperatures.
   (b) Which planet has a temperature 20°C lower than that of Uranus?

                                                (0580/02 May/June 2006 q 3)

16. (a) The list shows the rainfall in millimetres in Prestbury for the 12 months of 2002.

       61   146   22   54   67   94   141   22   167   87   170

(i) Write down the mode.
(ii) Find the median.
(iii) Calculate the mean.

(0580/03 Oct/Nov 2004 q 8a)

17. 15 students estimated the area of the rectangle shown below

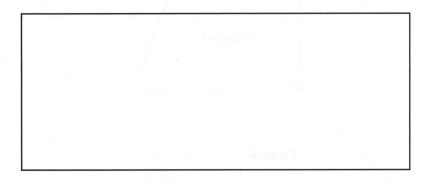

Their estimates, in square centimetres, were:

45 44 50 50 48
24 50 46 43 50
48 20 45 49 47

(a) Work out:
   (i) the mode,
   (ii) the mean,
   (iii) the median.

(b) Explain why the mean is not a suitable average to represent this data.

(0580/01 May/June 2007 q 20)

18. Yousef asked 24 students to choose their favourite sport.
He recorded the information in the table below so that he could draw a pie chart
(a) Complete the table.

Sport	Volleyball	Football	Hockey	Cricket
Number of students	6	9	7	2
Angle on pie chart	90°	135°		

(b) Complete the pie chart accurately to show this data.

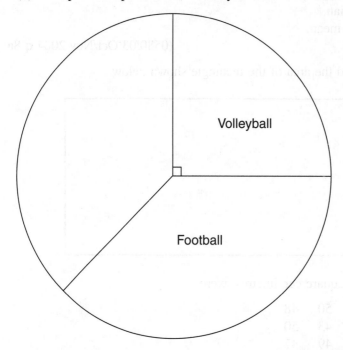

(c) Which is the modal sport?

(0580/01 May/June 2006 q 16)

19. Jane records the number of telephone calls she receives each day for two weeks.

   5   6   10   0   15   6   12   2   13   16   0   16   6   10

   (a) Calculate the mean.
   (b) Find the median.
   (c) Write down the mode.
   (d) Complete the frequency table below.

Number of calls	0–4	5–9	10–14	15–19
Frequency				

(0580/03 Oct/Nov 2005 q 4a, b, c and d)

20. A dentist recorded the number of fillings that each of a group of 30 children had in their teeth. The results were:

    2 4 0 5 1 1 3 2 6 0

    2 2 3 2 1 4 3 0 1 6

    1 4 1 6 5 1 0 3 4 2

    (a) Copy and complete this frequency table.

Number of fillings	Frequency
0	
1	
2	
3	
4	
5	
6	

    (b) What is the modal number of fillings?
    (c) Find the median number of fillings.
    (d) Work out the mean number of fillings.
    (e) These 30 children had been chosen from a larger group of 300 children. Estimate how many in the larger group have no fillings in their teeth.

    (0580/03 Nov 2003 q 4 (part))

21. Fifty students take part in a quiz.
    The table shows the results.

Number of correct answers	5	6	7	8	9	10	11	12
Number of students	4	7	8	7	10	6	5	3

(a) How many students had 6 correct answers?

(b) How many students had less than 11 correct answers?

(c) Find

(i) the modal number of correct answers,

(ii) the median number of correct answers,

(iii) the mean number of correct answers.

(d) A bar chart is drawn to show the results.
The height of the bar for the number of students who had 5 correct answers is 2 cm.
What is the height of the bar for the number of students who had correct answers?

(e) A pie chart is drawn to show the results.
What is the angle for the number of students who had 11 correct answers?

(f) The students who had the most correct answers shared a top prize of $22.50.
How much did each of these students receive?

(g) Work out the percentage of students who had **less than** 7 correct answers.

(0580/3 May/June 2003 q 1)

22. Marie counts the number of people in each of 60 cars one morning.

(a) She records the first 40 results as shown below.

Number of people in a car	Tally	Number of cars
1	ⵏⵏⵏⵏ	
2	ⵏⵏⵏⵏ ⵏⵏⵏⵏ	
3	ⵏⵏⵏⵏⵏ	
4	ⵏⵏⵏⵏⵏ	
5	ⵏⵏⵏⵏⵏⵏ	
6	ⵏⵏⵏⵏⵏ	

The remaining 20 results are:
2, 2, 5, 2, 2,   4, 2, 6, 5, 3,   4, 5, 4, 6, 2,   5, 3, 2, 1, 6.

(i) Use these results to complete the frequency table above.

(ii) On the grid below, draw a bar chart to show the information for the 60 cars.

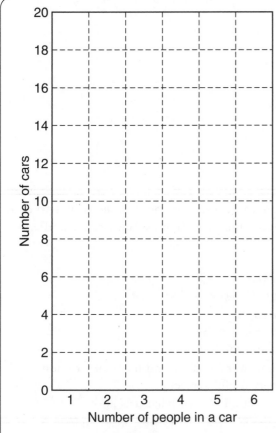

(iii) Write down the mode.
(iv) Find the median.
 (v) Work out the mean.
(b) Manuel uses Marie's results to draw a pie chart.
   Work out the sector angle for the number of cars with 5 people.

(0580/03 May/June 2008 q3)

Chapter 12

Probability

At the end of this chapter you will know the words commonly used in the study of probability and will be able to work out simple probabilities, both calculated and by experiment.

> *How likely is it to rain tomorrow?*
> *Which side might bat first in the next Test Match?*
> *What are the chances of my reaching my 60th birthday?*
> *What is the most likely score if I throw two dice?*
> *How sure is the airline that their plane will arrive on time?*

We often hear questions like these in everyday life.

Mathematicians attempt to find numerical answers to these and many other questions by studying Probability. It is a complex but very useful subject.

We will look at the basics so that you can get an idea of what it is about.

Essential Skills

Take a little time to check that you can do the following questions correctly before you start to study Chapter 12.

1. Work out:

    (a) $\dfrac{2}{3} + \dfrac{1}{6}$ (b) $\dfrac{5}{8} - \dfrac{2}{5}$ (c) $\dfrac{3}{10} + \dfrac{4}{5} - \dfrac{3}{20}$ (d) $1 - \dfrac{3}{5}$ (e) $1 - \dfrac{17}{40}$

2. Simplify:

    (a) $\dfrac{4}{6}$ (b) $\dfrac{9}{27}$ (c) $\dfrac{44}{110}$

3. Change to decimals:

    (a) $\dfrac{3}{5}$ (b) $\dfrac{3}{4}$ (c) $\dfrac{17}{100}$ (d) $\dfrac{5}{8}$ (e) $\dfrac{6}{25}$

4. Change to percentages:

(a) $\dfrac{1}{5}$ (b) $\dfrac{3}{8}$ (c) 0.39   (d) 0.165

5. Work out:

(a) $\dfrac{1}{2} \times \dfrac{2}{3}$ (b) $\dfrac{2}{5} \times \dfrac{3}{4}$ (c) $\dfrac{3}{10} \times \dfrac{5}{6}$ (d) $\dfrac{4}{7} \times \dfrac{3}{5}$ (e) $\dfrac{2}{3} \times \dfrac{4}{5} \times \dfrac{6}{7}$

Answers

1. (a) $\dfrac{5}{6}$ (b) $\dfrac{9}{40}$ (c) $\dfrac{19}{20}$ (d) $\dfrac{2}{5}$ (e) $\dfrac{23}{40}$

2. (a) $\dfrac{2}{3}$ (b) $\dfrac{1}{3}$ (c) $\dfrac{2}{5}$

3. (a) 0.6   (b) 0.75   (c) 0.17   (d) 0.625   (e) 0.24

4. (a) 20%   (b) 37.5%   (c) 39%   (d) 16.5%

5. (a) $\dfrac{1}{3}$ (b) $\dfrac{3}{10}$ (c) $\dfrac{1}{4}$ (d) $\dfrac{12}{35}$ (e) $\dfrac{16}{35}$

Some Terms Used in the Study of Probability

Events and outcomes

If you toss a coin it can land on either of its two faces. The two faces are called 'heads' and 'tails'.

Experiment 1

Take a coin, decide which side is 'heads' and which is 'tails'. Now toss the coin 20 times, each time recording in a table whether it lands with 'heads' or 'tails' on top. Keep your table for later.

How many times did you expect to get a 'head'?

How many times did you actually get a 'head'?

One possible *outcome* or *event* is that the coin lands 'heads' up, the other possible outcome is that it lands 'tails' up.

The word *event* usually means the actual tossing of the coin, and the *outcome* is whether the coin lands heads up or tails up. However, the word *event* is sometimes used to refer to the outcome. You may meet this in textbooks.

Probability scale

We measure probability on a scale of zero to one. A probability of zero is for an impossible outcome, while a probability of one is for a certain outcome. The probabilities of other outcomes lie somewhere in between. The numbers in between can be fractions or decimals, but are *never* given as a ratio.

Probability is sometimes measured as a percentage. Then the scale is zero to one hundred. We could safely say that:

(a) the probability that you will fly to the moon tomorrow is zero (no chance),

(b) the probability that the world will still be turning tomorrow is one (certain).

(c) Also the probability that the next baby born in the world will be a boy is just over 0.5 (there are slightly more boys born than girls).

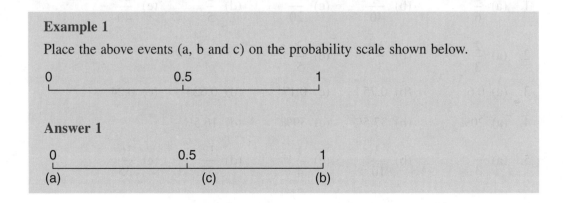

Example 1

Place the above events (a, b and c) on the probability scale shown below.

```
0                  0.5              1
|_____|_____|
```

Answer 1

```
0                  0.5              1
|_____|_____|
(a)                (c)             (b)
```

Try to think of some more examples, and show them on the probability scale.

Bias

Bias describes how fair an event is. For example, an unbiased coin is equally likely to land heads up or tails up. But if there is something in the coin which makes it more or less likely to land heads up it is a biased coin.

Do you think your coin in experiment 1 was biased or fair?

If we toss an unbiased coin we expect either heads or tails to be equally likely. The probability of it landing heads up is $\frac{1}{2}$, or 0.5 or 50%.

You might hear someone say "There is a fifty-fifty chance of the coin landing heads up".

The toss of a coin is often used to choose between two options. For example, the side to bat first in a cricket match is decided by tossing a coin.

Random

Random means completely without order. If you make a list of random numbers then every number is equally likely to be anywhere in the list. One way of getting a list of random numbers is to use a specially written calculator or computer program. These programs are called *random number generators*. Find out if your calculator can produce random numbers.

You cannot make a proper list of random numbers just by writing down the next number to come into your head. Why do you think this is?

If you choose a card from a pack of cards that have been shuffled and laid face down on the table, and if there is absolutely nothing to make you choose one more than any other you will be making a random choice.

Probability may be calculated or found by experiment.

Theoretical (or Calculated) Probability

We use the following definition:

$$\text{Theoretical probability of an outcome} = \frac{\text{number of ways we can get this outcome}}{\text{total number of possible outcomes}}$$

In your experiment with the coin what is the *theoretical* probability of getting a head? How many heads might you have expected to get in 20 tosses?

When we calculate a theoretical probability it is important that we know that each outcome is equally likely.

Example 2

Tariq's pencil case contains 10 coloured pencils. There are 3 blue pencils, 2 red pencils, and the rest are green. Tariq chooses one pencil without looking in the case.

The pencils are identical apart from colour so he is making a random choice, and each choice is equally likely.

Work out the probability that he chooses a blue pencil.

Answer 2

There are 3 blue pencils and 10 pencils altogether, so P(blue pencil) $= \dfrac{3}{10}$

As you can see from this answer it is useful to refer to the probability of choosing a blue pencil as P(blue pencil) or even just P(blue).

Try to work through the next example before you look at the answers given below.

Example 3

(a) In example 2 above how many green pencils are there in Tariq's pencil case?

(b) What is the probability of Tariq choosing a green pencil?

(c) What is the probability of Tariq choosing a yellow pencil?

(d) Find P(red pencil).

(e) Calculate P(blue pencil) + P(green pencil) + P(red pencil).

(f) Find P(*not* a red pencil)   **NOTE: How many pencils are not red?**

(g) Tariq needs either a blue pencil or a green pencil, it does not matter which.

     Find P(blue *or* green).

Answer 3

(a) There are 5 green pencils in the pencil case

(b) P(green pencil) $= \dfrac{5}{10}$

(c) P(yellow pencil) $= 0$ (there are no yellow pencils in the pencil case)

(d) P(red pencil) $= \dfrac{2}{10}$

(e) P(blue) + P(green) + P(red) $= \dfrac{3}{10} + \dfrac{5}{10} + \dfrac{2}{10} = 1$

(f) There are 8 pencils which are not red, so

    P(*not* red) $= \dfrac{8}{10}$

(g) There are 3 blue and 5 green pencils so P(blue *or* green) $= (3 + 5) \div 10 = \dfrac{8}{10}$

You can see from part (e) in the example above that if all the possible outcomes have been accounted for then the probabilities must add up to 1. This is an important result.

Also, in part (f) we calculated the probability of *not* red to be 8/10.

So because the pencils can either be red or not red and there is no other possibility, then:

$$P(\text{\textit{not} red}) + P(\text{red})$$
$$= \frac{8}{10} + \frac{2}{10}$$
$$= 1$$

Of course, if we are working in percentages,

$$P(\text{\textit{not} red}) + P(\text{red})$$
$$= 80\% + 20\%$$
$$= 100\%$$

You can see that:

probability of an event happening = 1 − probability of the event *not* happening.

We can also see in part (g) that we can find the probability of blue *or* green either by counting the pencils, or use the fact that P(blue *or* green) = P(blue) + P(green).

This is because picking a green pencil and picking a blue pencil are *mutually exclusive* events. They cannot happen together because the pencils are *either* blue *or* green but not both.

Two mutually exclusive events cannot happen at the same time.

If A and B are mutually exclusive events then P(A or B)= P(A) + P(B).

Example 4

A game consists of a circular board divided into six equal sectors, with each sector numbered 1, 2 or 3, as shown on the diagram. A dart is thrown and lands on the board in one of the sectors. The score is the number written in that sector.

Write down the probabilities of scoring 1, 2 or 3.

Answer 4

The probability of the dart landing on any particular number depends on the total area for that number, in this case, how many sectors there are for that number.

$$P(1) = \frac{2}{6} \qquad P(2) = \frac{1}{6} \qquad P(3) = \frac{3}{6}$$

Exercise 12.1

1.

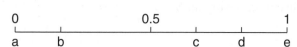

From this list of words choose those that could be used on the above number line in the positions a, b, c, d, and e.

  (i) unlikely   (ii) certain   (iii) impossible

  (iv) quite likely   (v) very likely

2. Choose suitable words from the list below to complete the sentences.

  mutually exclusive   outcome   random   biased   bias

  (a) When Iravan threw a 6 ten times in twenty throws of a die he decided that the die was probably ....

  (b) Scoring 6 and scoring 3 with a single roll of a die are .... events.

(c) Choosing a tall girl and choosing a girl with glasses to partner you in a game of tennis doubles are not .... events.

(d) A .... choice is one in which every .... is equally likely.

3. Which of the following pairs of events are always mutually exclusive?
   (a) Throwing a five or a six with a six-sided die.
   (b) There are some clouds in the sky and the sun is shining.
   (c) Paula won the high jump competition and Paul broke the record for the high jump in the same competition.
   (d) Tomorrow is Wednesday and tomorrow is Thursday.
   (e) Tomorrow is Wednesday and yesterday was Monday.

4. The weather forecasters have said that there is a 40% chance of rain tomorrow. What is the probability that it will be dry?

5. A bag contains 12 coloured discs. There are 3 yellow, 5 red and the rest are blue. Paris takes one disc out without looking.
   (a) Write down the probability that it is a red disc.
   (b) Find P(blue disc).
   (c) What is the probability of a green disc?
   (d) Find P(not blue).

6. A 'lucky dip' has two sorts of prizes hidden in a tub. One sort is a model car and the other a model aeroplane. They are in identical boxes so it is not possible to feel which is which. The probability of picking an aeroplane is 3/5. What is the probability of picking a car?

7. 'Drawing the short straw' is a way of picking a person to do an unpopular job. To decide who does the washing up Theresa holds 5 straws concealed in her hand with only the ends showing. One of the straws is shorter than the others. I pick a straw. What is the probability that I will be doing the washing up?

8. The letters of the word STATISTICS are written on cards which are then shuffled and laid face down on a table. One card is chosen.
   (a) Write down the probability of choosing an S.
   (b) Find P(S or T).
   (c) Find P(a vowel)
   (d) What is the probability of choosing a B?

9. (a) A six-sided spinner is spun once.

 NOTE: The sections on the spinner are numbered 1 to 6.

Write down the probability of the spinner landing on 3.
(b) A five-sided spinner is spun once.
Write down the probability of this spinner landing on 3.

10. A word game consists of small tiles, each with a letter of the alphabet on it. The tiles are put in a bag. The bag contains three letter 'A', five letter 'B', and ten letter 'C'. There are no other tiles in the bag.

Erin chooses one letter at random without looking in the bag.

What is the probability that Erin chooses a 'B'?

11. Another game consists of throwing a counter to land on the rectangle shown below.

4	3
2	1

The score is the number shown on the area on which the counter lands.

Give a reason why the probability of scoring 3 is not $\dfrac{1}{4}$.

12. The pie chart shows the distribution of the colours of the cars already sold and awaiting collection in a manufacturer's car park. The cars are either red, blue or black. There are no other colours.

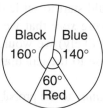

(a) Why is the probability that the next car collected will be blue not $\dfrac{1}{3}$?

(b) What is the probability that the next car collected will be red?

(c) What is the probability that the next car collected will be white?

(d) What is the probability that the next car collected will not be black?

Experimental Probability or Relative Frequency

I have just done an experiment. I had 10 counters in a bag, coloured red, blue or green. I did not know how many of each. I picked out a counter at random, noted its colour on a tally chart, and replaced it. I repeated the experiment many times, with the following results.

Number of trials	Red	Blue	Green
30	13	15	2
100	48	43	9
200	83	97	20

The relative frequency of an outcome or event is the number of times it happens divided by the total number of trials.

$$\text{Relative frequency of an outcome} = \frac{\textbf{number of times that outcome occurs}}{\textbf{total number of trials}}$$

So the relative frequency of red $= \dfrac{\text{number of times red was chosen}}{\text{the number of trials}}$

The table below shows the results of calculating the relative frequencies (to 2 decimal places) in the experiment above.

For example, the relative frequency of red after 30 trials $= \dfrac{13}{30} = 0.43.$

Number of trials	Red	Blue	Green
30	0.43	0.50	0.07
100	0.48	0.43	0.09
200	0.42	0.49	0.10

If the number of trials were increased the relative frequencies would start to settle down. If we were able to do the experiment enough times and if we were *sure* that we were picking the counters at random (so that each outcome is equally likely) we would find that the relative frequencies would eventually come out very close to the calculated or theoretical probabilities.

So the finding the relative frequencies of events is a way of *approximating* to the probabilities of those events,

After 200 trials I looked in the bag and discovered that I had 4 red counters, 5 blue counters and 1 green counter. I calculated the theoretical probabilities and compared them with the relative frequencies from the experiment.

	Theoretical Probability	Relative frequency after 200 trials
Red	$\dfrac{4}{10} = 0.40$	0.42
Blue	$\dfrac{5}{10} = 0.50$	0.49
Green	$\dfrac{1}{10} = 0.10$	0.10

Have another look at your table from experiment 1.

What was the relative frequency of heads after 20 trials?

Calculating the probabilities as decimals makes it easier to compare the values.

Experiment 2

Repeat your experiment (Experiment 1 on page 399), noting your results after 20, 50, 100 and 200 trials to see how close you get to the theoretical probabilities.

Experiment 3

Use a bag with 20 coloured counters in it to repeat my experiment. Ask someone else to put in the counters so that you do not know what the colours are, or how many there are of each colour.

Take out one counter at a time, note its colour and then replace it. Repeat 100 times, and see if you can work out how many counters of each colour there are in the bag.

You should see that:

the expected number of any coloured counter = relative frequency of that colour × the total number of counters in the bag.

Example 5

Akash and Namita are tossing an unbiased coin. They have repeated the experiment 30 times and the results are 11 heads and 19 tails. Akash says that the next coin will be more likely to land heads up. Namita says she is wrong. Who is correct and why?

Answer 5

Namita is correct because each toss of the coin is *independent* of the one before and so each time the coin is tossed it has exactly the same probability of landing heads up.

This is another important aspect of probability. If the trials are independent of each other the probability of a certain outcome remains the same for each trial. The *trend* is for the relative frequency to come close to the expected probability if enough trials are carried out.

Independent events are not affected by other events.

Exercise 12.2

Mixed Exercise

1. The manufacturers of a certain type of computer part know from experience that the relative frequency of faults in the component is 0.05.
   (a) What is the probability that the next component picked will not be faulty?
   (b) (i) How many of these components would you expect to be faulty in a consignment of 1000?
   (ii) Would you expect there to be *exactly* this number faulty?
2. On average 51% of babies born are boys. What is the probability of the next baby to be born being a girl?
3. An examination paper is made up of 25 multiple-choice questions.
   Each question has five different possible answers given for it, only one of which is correct.

Ryan guesses the answer to each question at random.

(a) What is the probability that he gets question 1 correct?

(b) How many questions could you expect him to get correct on the whole paper?

(c) What would his mark be as a percentage?

4. A manufacturing company runs a check on one of its components by taking 100 of them from the assembly line at random, and then checking each of these to see how many are faulty.

The company finds that the relative frequency of faulty components is $\frac{3}{100}$ or 3%.

How many perfect components would the company expect to get in a production run of 4000 components?

5. Nesip picks a sweet at random from a jar containing 15 red sweets, 20 orange sweets and 12 yellow sweets.

(a) What is the probability that he picks his favourite orange sweet?

(b) What is the probability that he picks a red or an orange sweet?

6. Anaya rolls a seven-sided spinner, with each of the numbers 1 to 7 on it.

(a) What is the probability that she gets an even number?

(b) What is the probability that she does not get a seven?

7. The average number of wet days in Mumbai in June is 14.

What is the probability that June 13th next year will be wet?

Examination Questions

8. Aminata has a bag containing 35 beads. The beads are either blue, yellow or red. One bead is chosen at random.

The probability of choosing a blue bead is $\frac{2}{7}$ and the probability of choosing a yellow bead is $\frac{3}{5}$.

Calculate:

(a) the number of blue beads in the bag,

(b) the probability of choosing a red bead.

(0580/01 Oct/Nov 2004 q 20)

9.

S	U	S	A	N

Susan writes the letters of her name on five cards.

One of the five cards is chosen at random.

Find the probability that the letter on the card is:

(a) S   (b) E

(0580/01 Oct/Nov 2003 q 3)

10. Grades were awarded for an examination.
The table below shows the number of students in the whole school getting each grade.

Grade	Number of students	Angle on pie chart
A	5	
B	15	
C	40	
D	20	
E	10	
Total	90	

(a) Copy and complete the table above by calculating the angles required to draw a pie chart.

(b) Draw an accurate pie chart to show the data in the table.
Label the sectors A,B,C,D and E.

(c) What is the probability that a student chosen at random from the group taking the examination was awarded
  (i) grade C,
  (ii) grade D or E?

(0580/03 May/June 2004 q 1b)

11. A dentist recorded the number of fillings that each of a group of 30 children had in their teeth. The results were:

2	4	0	5	1	1	3	2	6	0
2	2	3	2	1	4	3	0	1	6
1	4	1	6	5	1	0	3	4	2

(a) One of these children is chosen at random.
Find the probability that this child has:
  (i) exactly one filling,
  (ii) more than three fillings.

(b) These 30 children had been chosen from a larger group of 300 children.
Estimate how many in the larger group have no fillings in their teeth.

(0580/03 Oct/Nov 2003 q 4e and f)

12. (a) 85% of the seeds in a packet will produce red flowers.
One seed is chosen at random. What is the probability that it will **not** produce a red flower?

(b) A box of 15 pencils contains 5 red, 4 yellow and 6 blue pencils. One pencil is chosen at random from the box. Find the probability that it is
  (i) yellow   (ii) yellow or blue   (iii) green.

(0580/01 Oct/Nov 2008 q 20)

13. A bag contains 24 discs.
    10 discs are red, 9 discs are green and 5 discs are yellow.
    (a) A disc is chosen at random.
        Find, **as a fraction**, the probability of each of the following events.
        (i) Event A: the disc is red.
        (ii) Event B: the disc is red or yellow.
        (iii) Event C: the disc is **not** yellow.
    (b)

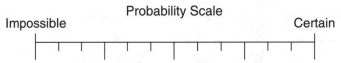

Probability Scale

Impossible   Certain

The diagram shows a horizontal probability scale. Copy the diagram and show
on the diagram the probability of
(i) an impossible event
(ii) a certain event.
    (c) Mark the positions of A, B and C, your answers to part (a), on your probability
        scale.

(0580/03 May/June 2007 q 5b, c and d)

14. (a) There are 11 boys and 12 girls in a choir.
        The teacher chooses one choir member at random.
        What is the probability that it is a girl? Write your answer as a fraction.

    (b) The probability that Carla arrives at school before 08 00 is $\frac{9}{20}$.

        What is the probability that Carla does not arrive before 08 00?
        Write your answer as a fraction.

(0580/1 May/June 2008 q 12)

15. The diagram shows a six-sided spinner.

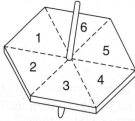

    (a) Amy spins a biased spinner and the probability she gets a two is $\frac{5}{36}$.
        Find the probability she
        (i) does not get a two,
        (ii) gets a seven,
        (iii) gets a number on the spinner less than 7.

(b) Joel spins his blue spinner 99 times and gets a two 17 times.
Write down the relative frequency of getting a two with joel's spinner.

(c) The relative frequency of getting a two with Piero's spinner is $\dfrac{21}{102}$.

Which of the three spinners, Amy's Joel's or Piero's, is most likely to give a two?

(0580/ 01 Oct/Nov 2006 q 23)

Revision and Examination Technique

Revision

Give yourself time to do some systematic revision before the examination. The points outlined here should give you some ideas for a routine you could follow.

- Have a special notebook for revision, some coloured pencils and a highlighter or two. Make sure that you have some tracing paper, a protractor, compasses and a ruler ready before you start.
- Make sure that you have a suitable calculator for the examination, which has a fully charged battery and you can use it for all your revision. You must know exactly how it works.
- Read each chapter through carefully, noting down in your revision notebook anything you think you might not remember. In particular, write down any formulae that you will need to learn. It is helpful to underline things in different colours, or use different coloured pens to write them down. Make your notebook as visually attractive as you can.
- Try the examination questions at the end of each chapter, checking each answer as you go along. It does not matter if you have already done these questions because you are unlikely to remember them all! If you need to look anything up in your revision notebook highlight it, so that you can learn it later.
- Get someone to test you on the formulae that you need to know. You *must* know the formulae or you will not be able to answer the questions! Do not just hope that you will remember them, LEARN them, and keep checking that you know them. For example, learn the trigonometry ratios, the mean, median and mode in statistics, the formulae for areas and volumes.
- Finally, try to get hold of some past examination papers *and their answers*, so that you can get the feel of the real examination pattern. Work through these without your notebook, keeping an eye on the time. If you do need to look anything up, make a special effort to learn it straight away.

Examination Technique

You have learned and practised as much as possible, now you are ready for the examination. How can you maximise your marks? Every mark is valuable and you do not want to waste them unnecessarily. Most marks are awarded for either method or accuracy.

Method marks

- Look at the marks available for each part of the question: if there is more than one mark available you must *earn* the extra marks. This means that you *must show your working*, or you are in danger of getting no marks for that part of the question.
- In some questions you will be asked to show that some statement or result is correct. In these questions you will have to show your working. This means that you must write down the steps in the working in a way that the examiner can understand. Pretend that the answer is not given, and work it out. The reason that some questions are worded in this way is that the result is needed in the rest of the question. This gives you a chance to finish the question even if you get the first step wrong, but do use the result given the question even if you have not been able to prove it!

Accuracy marks

- Some accuracy marks are awarded for getting exactly the right answer, and some for giving the answer rounded to a required degree of accuracy. In the general instructions on the front of the paper you are told to give answers to three significant figures if they are not exact, or unless the question specifies otherwise, so read each question carefully to see if a required degree of accuracy is specified. Angles should be given correct to one decimal place. Answers should be rounded, not truncated (for example, 12.36 should be given as 12.4, not 12.3), and you should not round in a stepwise way, (for example, 14.345 should be given as 14.3, not rounded to 14.35 and then 14.4.)
- However, you must not round in the middle of your working or you will lose the accuracy in the final answer. When you use your calculator write down most of the figures on your calculator display for each step of your working, and try to keep the answer in your calculator ready for the next step (for example, in a trigonometry question, $\sin x = 0.64571 \ldots$ should not be rounded before entering *shift* sin 0.64571 ..., to obtain the answer $x = 40.2189 \ldots$, which is then given as $x = 40.2°$). In this way you should not lose accuracy as you work through the question.

General points

- Write your answers clearly in ink. If you make a mistake cross through the work which is wrong and replace it. Do not use correcting fluid. If the work is crossed out, is still readable, and has not been replaced, it will be marked.
- Graphs and accurate drawings should be done in pencil. They must be recognisable, which means that the pencil marks must not be too faint.
- Remember that your examiner wants to award you marks, but cannot do so if your writing is not readable.

- Your working must be done on the examination paper, in the space provided, and not on rough paper.
- Remember to check that your answer seems reasonable. Do you expect the hypotenuse in a right-angled triangle to be larger or smaller than the other sides? If money is earning interest in the bank should you end up with more or less than when you started? Is it reasonable for your journey to school to be 200 km? Could the mean age of the students in your class possibly be 3 years 5 months?
- Remember that there are 60 minutes in one hour, not 100. So 6.5 hours is 6 hours and 30 minutes, not 6 hours and 50 minutes.
- Leave in your construction lines when you are drawing accurately. Remember that these are awarded method marks.

You will already have been told many times to *read the question carefully*. This means that you should make sure that you are answering the question that has been set, and also giving the answer in the form required.

You now have done all that you can to ensure that you get the best possible result, so go into your examination knowing that you are well prepared and are going to do your best.

Answer Key

Chapter 1

Exercise 1.1

1. (a) $5, -100, -3.67, \pi, 0, 1507, \dfrac{99}{7}, \dfrac{6}{1}$

   (b) $5, -100, -3.67, 0, 1507, \dfrac{99}{7}, \dfrac{6}{1}$

   (c) $5, -100, 0, 1507, \dfrac{6}{1}$ (d) $5, 1507, \dfrac{6}{1}$ (e) π

2. (a) 1, 2, 3, 5, 6, 10, 15, 30   (b) 2, 3, 5

   (c) $30 = 2 \times 3 \times 5$

   (d) for example, 60, 90, 120

3. (a) 30, 45, 15, 1500   (b) 1, 5, 15, 3

4. (a) 2, 3, 5

   (b) $2 \times 2 \times 2 \times 2 \times 3 \times 5$ or $2^4 \times 3 \times 5$

5. 23, 29, 31, 37

6. 37, 53, 101

7. (a) 83, 89   (b) 80, 85, 90   (c) 87

Exercise 1.2

1. (a) $\{1, 2, 4, 8\}$ $\{1, 2, 3, 4, 6, 12\}$ (b) 4

2. 21

3. (a) (i) 1, 3, 5, 15   (ii) 1, 5, 7, 35

      (iii) 1, 2, 4, 5, 10, 20

   (b) 5

4. (a) 12, 24, 36, 48, 60, 72

      8, 16, 24, 32, 40, 48

   (b) 24

5. 60

6. 21603:

   does not divide by 2 (not even)

   does divide by 3 (digital root is 3)

   does not divide by 5 (does not end in 5 or 0)

   does not divide by 9 (digital root is not 9)

7. 515196:

   does divide by 2 (even number)

   does divide by 3 (digital root is 9)

   does not divide by 5 (does not end in 5 or 0)

   does divide by 6 (even and divides by 3)

   does divide by 9 (digital root is 9)

Exercise 1.3

1. (a) divide   (b) add

   (c) square root   (d) cube

2. (a) 36   (b) 3   (c) 8

   (d) 5   (e) 100   (f) 1000

3. (a) 27.04   (b) 9.1

   (c) 10   (d) 10

4. (a) $\sqrt{256}, \sqrt{841}, \sqrt{449.44}$ (b) $\sqrt{6.1}, \sqrt{7}$

5. 1, 4, 9, 16, 25, 36, 49   6. 27, 125

7. 1   8. for example, 64

9.

Natural numbers	1 2 3 4 5 6 7 8 9 10 11
Prime numbers	2 3 5 7 11
Even numbers	2 4 6 8 10
Multiples of 3	3 6 9
Square numbers	1 4 9
Cube numbers	1 8
Factors of 20	1 2 4 5 10

Exercise 1.4

1. (a) -5 (b) -5 (c) -3

   (d) -7 (e) 5

2. (a) 2.5m   (b) -3.6m   (c) 5.6m

3. 85m

4. (a) $287   (b) yes   (c) $3

Exercise 1.5

1. (a) π (b) $\sqrt{}$ (c) $\sqrt[3]{}$

   (d) $\neq$ (e) $<$ (f) $\geqslant$

2. (a) $2 < 4$ (b) $-2 > -5$

   (c) $-10 < 4$ (d) $-1 < 0$

3. $-100, -89, -76, -62, -1, 0, 61, 75, 100, 101$

Exercise 1.6

1. (a) 1.2×10^4 (b) 3.65×10^2

   (c) 5.9103×10^4 (d) 6×10^3

   (e) 7.0104×10^6

2. (a) 3.5×10^{-3} (b) 1.56×10^{-1} (c) 5×10^{-4}

   (d) 4.3×10^{-6} (e) 1.02×10^{-2}

3. (a) 3.45×10^{-3} **(b)** 5.2016×10^5
(c) 1.12×10^2 **(d)** 1×10^{-3}
(e) 1.001×10^{-1} **(f)** 2×10^6

4. (a) 5600 **(b)** 0.00027 **(c)** 0.0116
(d) 600000 **(e)** 0.002

5. (a) 18 **(b)** 14 **(c)** 0 **(d)** 3

6. (a) 5 **(b)** 41 **(c)** 19

7. (a) $(5 - 3) \times 4 = 8$
(b) $9 + (50 - 24) \div 2 = 22$
(c) $(31 - 15) \div (10 - 2) = 2$

Exercise 1.7

1. (a) (i) true **(ii)** true
(iii) false **(iv)** true
(b) (i) $-4 < 3$ **(ii)** $0 > -2$
(iii) $5 > -5$ **(iv)** $3 > -2$

(c) (i) $\{2, 3, 5, 7\}$
(ii) $\{1, 3, 5, 9, 15, 45\}$
(iii) $\{3, 6, 9, 12, 15, 18\}$

2. (a) 60 **(b)** 90

3. (a) 4 **(b)** 4

4. (a) 4.41 **(b)** 27 **(c)** 5.3
(d) 9 **(e)** 5

5. $2^3 \times 3 \times 5^2$

6. 1, 2, 4, 5, 8, 10, 16, 20, 32, 40, 80, 160

7. $-14\ °C$ **8.** -2 **9.** 39

10. 842 m

11. (a) 49 **(b)** 31

12. 17

13. $(10 - 5) \times (9 + 3) = 60$

14. (a) 1, 2, 3, 5, 6, 10, 15, 30
(b) 2, 3, 5

15. 28

16. 3.62×10^{-3}

Chapter 2

Exercise 2.1

1. (a) $3\frac{4}{5}$ **(b)** $20\frac{1}{10}$ **(c)** $16\frac{1}{2}$

2. (a) $\frac{31}{8}$ **(b)** $\frac{201}{2}$ **(c)** $\frac{47}{12}$

3. $\frac{5}{15} = \frac{10}{30} = \frac{1}{3} = \frac{7}{21} = \frac{21}{63}$

4. (a) $\frac{70}{100}$ **(b)** $\frac{16}{100}$ **(c)** $\frac{95}{100}$

(d) $\frac{26}{100}$ **(e)** $\frac{9}{100}$

5. (a) $\frac{5}{7}$ **(b)** $\frac{1}{5}$ **(c)** $\frac{5}{12}$

(d) $\frac{35}{36}$ **(e)** $3\frac{19}{20}$ **(f)** $1\frac{9}{10}$

(g) $\frac{1}{7}$ **(h)** $\frac{7}{12}$

6. (a) $\frac{2}{7}$ **(b)** $\frac{5}{6}$ **(c)** $\frac{3}{4}$

(d) $\frac{1}{8}$

7. 21
8. 15

Exercise 2.2

1. (a) $\frac{3}{5}$ **(b)** $1\frac{1}{5}$ **(c)** $7\frac{1}{2}$

2. (a) $\frac{1}{18}$ **(b)** $\frac{3}{28}$ **(c)** $\frac{15}{32}$

3. (a) $\frac{1}{14}$ **(b)** $\frac{7}{20}$ **(c)** $\frac{1}{6}$

4. (a) $8\frac{1}{8}$ **(b)** $11\frac{9}{10}$ **(c)** $3\frac{1}{9}$

5. (a) $7\frac{1}{2}$ **(b)** $\frac{4}{9}$ **(c)** $\frac{1}{14}$

6. (a) $1\frac{1}{6}$ **(b)** $\frac{6}{7}$ **(c)** $\frac{35}{54}$

7. (a) $\dfrac{1}{3}$ **(b)** 3   **(c)** $1\dfrac{2}{3}$

8. (a) $1\dfrac{5}{6}$ **(b)** $7\dfrac{1}{3}$ **(c)** $2\dfrac{4}{25}$

9. (a) 6   **(b)** $\dfrac{18}{25}$ **(c)** $5\dfrac{4}{9}$

Exercise 2.3

1. 13.86   **2.** 502.97   **3.** 16.55
4. 4.109   **5.** 13410   **6.** 16.9
7. 0.06017   **8.** 31.62   **9.** 15.8

Exercise 2.4

	Fraction	Decimal	Percentage
1	$\dfrac{1}{2}$	0.5	50%
2	$\dfrac{1}{4}$	0.25	25%
3	$\dfrac{3}{4}$	0.75	75%
4	$\dfrac{1}{10}$	0.1	10%
5	$\dfrac{3}{10}$	0.3	30%
6	$\dfrac{1}{5}$	0.2	20%
7	$\dfrac{1}{8}$	0.125	12.5%

Exercise 2.5

Example methods

1. $75\% \text{ of } 64 = \dfrac{3}{4} \text{ of } 64 = 3 \times \dfrac{1}{4} \times 64 = 3 \times 16 = 48$

2. $30\% \text{ of } 1550 = 0.3 \times 1550 = 465$

3. $9\% \text{ of } 3400 = 9 \times \dfrac{1}{100} \times 3400 = 9 \times 34 = 306$

4. $55.5\% \text{ of } 680 = 50\% \text{ of } 680 + 5\% \text{ of } 680$
$\qquad\qquad\qquad + 0.5\% \text{ of } 680$
$\qquad\qquad = 340 + 34 + 3.4 = 377.4$

5. $3\% \text{ of } 73 = 3 \times \dfrac{1}{100} \times 73 = 3 \times 0.73 = 2.19$

Exercise 2.6

1. 25%   **2.** 12%
3. 46%   **4.** 32%
5. 6.8%   **6.** 50%
7. 200%   **8.** 2.9%

Exercise 2.7

1. $\dfrac{7}{10}$

2. 4.098, 4.105, 4.51, 4.579

3. $\dfrac{2}{3}, \dfrac{3}{4}, \dfrac{4}{5}, \dfrac{17}{20}$

4. $\dfrac{3}{50}, \dfrac{3}{25}, \dfrac{33}{100}, 33\dfrac{1}{3}\%, \dfrac{67}{200}$

Exercise 2.8

1. (a) $\dfrac{3}{4}$ **(b)** $\dfrac{7}{100}$

2. $4\dfrac{1}{2}$

3. $\dfrac{3}{5} \div \dfrac{7}{10} = \dfrac{3}{5} \times \dfrac{10}{7} = \dfrac{6}{7}$

4. (a) $23 < 32$ **(b)** $9\% = 0.09$

5. (a) (i) 0.28 **(ii)** 0.275 **(iii)** 0.2857

   (b) $\dfrac{275}{1000}$, 28%, $\dfrac{2}{7}$

6. (a) $\dfrac{33}{50}$, 67%, 0.68

   (b) $\dfrac{17}{25}$

7. Example working:

   (a) $\dfrac{1}{2} + \dfrac{2}{3} = \dfrac{3}{6} + \dfrac{4}{6} = \dfrac{7}{6} = 1\dfrac{1}{6}$

   (b) $1\dfrac{1}{5} \times 1\dfrac{3}{4} = \dfrac{6}{5} \times \dfrac{7}{4} = \dfrac{21}{10} = 2\dfrac{1}{10}$

8. 0.58 $\dfrac{3}{5}$ 62%

9. 1278

Chapter 3

Exercise 3.1

(a) (i) $3x$ **(ii)** 90
(b) (i) y **(ii)** 154
(c) (i) z^3 **(ii)** 27
(d) (i) $3x - y$ **(ii)** 8
(e) (i) $2x + 2y$ **(ii)** 22
(f) (i) 0   **(ii)** 0
(g) (i) $x^2 + y^2$ **(ii)** 25
(h) (i) $8x - 2y$ **(ii)** 392

Exercise 3.2

1. $2m + n$
2. $6.5 - m$
3. (a) $T = 10 + t$ **(b)** $T = 25$
4. (a) $x = 21$ **(b)** $x = 17$ **(c)** $x = 51$
5. (a) $L = 2a + 3$ **(b)** $L = 23$
     (c) (i) and **(v)** would not make triangles
6. (a) $C = \dfrac{e}{2} + b + \dfrac{t}{2}$ **(b)** $C = 75$
7. (a) 6   **(b)** 1   **(c)** 5
   (d) 33

Exercise 3.3

1. $13x$ **2.** $11x$ **3.** $7x - 4y$
4. $4a + 6b$ **5.** $3x + y$ **6.** $3x + 3$
7. $4z + 3w$ **8.** $6c - 3$ **9.** $3 + 2a$
10. x **11.** $x^2 + 2y^2$
12. $3x^2 + 3x$ **13.** $5x^2 + 3xy$
14. $3x^2 + y^2 - xy$ **15.** $3x^2 + xy - 4y^2$
16. $3x^2 - 5x^3 + 3x^2y$ **17.** $4x^2y^2 - 2x^2y$

Exercise 3.4

1. $15ab$ **2.** $24yz$ **3.** $6x^2$
4. $60x^2$ **5.** $6xyz$ **6.** a
7. $60abd^2$ **8.** $\dfrac{4}{x}$ **9.** $2d$
10. $2x$ **11.** $\dfrac{1}{2d}$ **12.** $\dfrac{3}{2}$
13. $20abcd$ **14.** 12

Exercise 3.5

1. -7 **2.** 7   **3.** 0
4. 0   **5.** 0   **6.** 0
7. -6 **8.** 7   **9.** 1
10. 5

Exercise 3.6

1. (a) 5   **(b)** 1   **(c)** -1
   (d) -5 **(e)** 1   **(f)** 1
   (g) -5 **(h)** -5
2. (a) 6   **(b)** -6 **(c)** 6
   (d) -6 **(e)** -6
3. (a) 2   **(b)** -2 **(c)** 2   **(d)** -2
4. (a) $\dfrac{1}{2}$ **(b)** $-\dfrac{1}{2}$ **(c)** $\dfrac{1}{2}$ **(d)** $-\dfrac{1}{2}$
5. 10   **6.** -2 **7.** 10
8. 10   **9.** -9
10. (a) 1   **(b)** -2 **(c)** 1
    (d) -1 **(e)** 1
11. 14

Exercise 3.7

1. $-2xy$ **2.** $2xy$ **3.** $-x - 2y$
4. $-x + 2y$ **5.** $\dfrac{a^2}{2}$ **6.** $-\dfrac{a}{2b}$
7. $8xy$ **8.** $\dfrac{x}{2y}$ **9.** $-3x^2y$
10. $xy - 3x$ **11.** $-\dfrac{y}{3}$ **12.** $-6x^2y$
13. $-6z + 3x^2$ **14.** $2a + 3b$

Exercise 3.8

1. (a) 1024   **(b)** 25   **(c)** 32
   (d) 4   **(e)** 512
2. (a) x^9 **(b)** $x^4 + x^5$ **(c)** $2x^3$
   (d) x^6 **(e)** x^4 **(f)** $6x^{11}$
   (g) $5x^5$ **(h)** $4x^2$ **(i)** $4x^2$
   (j) $4x^6$ **(k)** $4x^6$ **(l)** x^{45}
3. (a) $6x^{10}$ **(b)** $2x^{13}y^3$ **(c)** $9x^6$
   (d) $24x^5y$

Exercise 3.9

1. (a) 16   **(b)** $\dfrac{1}{16}$ **(c)** 4
   (d) $\dfrac{1}{16}$ **(e)** $\dfrac{1}{25}$ **(f)** $\dfrac{2}{5}$

(g) $\dfrac{1}{10}$ **(h)** $\dfrac{4}{25}$ **(i)** 8

(j) $\dfrac{4}{25}$ **(k)** $\dfrac{49}{4}$ or $12\dfrac{1}{4}$

(i) $\dfrac{64}{343}$

2. (a) $\dfrac{y}{x^2}$ **(b)** $\dfrac{y^4}{x^8}$ **(c)** 1

(d) 1   **(e)** $x^2y^3z^4$ **(f)** x^7y^8

3. (a) $n = 3$ **(b)** $n = -4$ **(c)** $n = -1$

(d) $n = 2$ **(e)** $n = 3$ **(f)** $n = -3$

(g) $n = -3$ **(h)** $n = 1$ **(i)** $n = 0$

(j) $n = -3$ **(k)** $n = -2$ **(l)** $n = 1$

(m) $n = 2$ **(n)** $n = 3$ **(o)** $n = 1$

(p) $n = 4$ **(q)** $n = k + 1$ **(r)** $n = k - 2$

4. (a) $3x^2 + 2x^3$ **(b)** $\dfrac{6}{x^5}$ **(c)** $\dfrac{1}{3}x^2$

(d) $9x^2$ **(e)** 18   **(f)** 24

Exercise 3.10

1. $2a + 2b$ **2.** $18 + 6x$

3. $3x - 3y$ **4.** $30 - 5b$

5. $12x - 8$ **6.** $7 - 21c$

7. $30x + 25y$ **8.** $8x - 8y + 32z$

9. $5x^2 + 20$ **10.** $14x^2 - 21y^2$

11. $12xy + 20z$ **12.** $2x - 3xy$

13. $a^2 + 2a$ **14.** $x^2 - xy$

15. $2c^2 + 2cd$ **16.** $6m^2 - 3mn$

17. $8x^2y - 36xy^2$ **18.** $21x^2 - 14x^2y + 28x^2z$

Exercise 3.11

1. $6 + 8x$ **2.** $-6 - 8x$ **3.** $3x^2 + 4xy$

4. $-3x^2 - 4xy$ **5.** $-3x^2 + 4xy$

6. $-14x^2 + 12x$ **7.** $-x - y$

8. $-2 + z$ **9.** $6pq + 18pr - 6ps$

10. $-6pq - 18pr + 6ps$

11. $x^2y^2 - 5x^2y$ **12.** $-6x^2y + 9x^2$

13. $-8a - 12a^2$ **14.** $8a + 12a^2$

15. $2y$ **16.** $5a + 12b$

17. $-22x - 23y$ **18.** $7x - y$

19. $-3x + 7y$ **20.** $xy - y$

Exercise 3.12

1. $4(2x + y)$ **2.** $5(3a - 5b)$ **3.** $4(x - 5)$

4. $x(y + 2)$ **5.** $x(x - 2)$ **6.** $x(x - 1)$

7. $3x(y + 3)$ **8.** $3x^2(x - 3y)$

9. $3a(a - 2b)$ **10.** $yz(x + 4)$

11. $10y(1 + 10y)$ **12.** $fg(5 + 6h)$

13. $3x(b - 2y)$ **14.** $3bx(b - 2)$

15. $2b(2b - 1)$ **16.** $b(4b - 1)$

17. $xy(xy - 1)$ **18.** $7cd^2(c - 3)$

Exercise 3.13

1. $-24x^2z$ **2.** $-3x + 3y$

3. (a) 35   **(b)** -7 **(c)** 33

(d) -1 **(e)** -10 **(f)** 0

(g) -80 **(h)** -12

4. $2x^2 - y - 5$

5. (a) 1   **(b)** 1   **(c)** 1

(d) 0   **(e)** x **(f)** 0

(g) $4x$ **(h)** 1   **(i)** 1

(j) x^4 **(k)** 1   **(l)** x^2

(m) $-x^3$ **(n)** $-x^2$

6. for example,

(a) $x = 1, y = 3$ **(b)** $x = 2, y = -3$

(c) $x = 2, y = 1$ **(d)** $x = 5, y = 7$

(e) $x = 3, y = 2$ **(f)** $x = 3, y = -1$

(g) $x = 4, y = 3$ **(h)** $x = 9, y = 4$

(i) $x = 12, y = 13$

7. (a) $\dfrac{16}{9}$ **(b)** $\dfrac{9}{19}$ **(c)** 1

(d) $\dfrac{1}{4}$ **(e)** $\dfrac{121}{25}$ or $4\dfrac{21}{25}$

8. (a) $6x + 4y$ **(b)** $-3x^5$ **(c)** $6xy$

(d) $\dfrac{1}{x}$ **(e)** x^4 **(f)** x^5

9. (a) $n = 4$ **(b)** $n = -2$ **(c)** $n = 1$

(d) $n = 2$ **(e)** $n = -1$

10. (a) $a^2b - abc$ **(b)** $5x + 8y + 2xy$

(c) $6ab + 8ac$ **(d)** $-3x^2 - 6x$

11. (a) $ab(b - a)$ **(b)** $2x(x - 3y + 2)$

(c) $2xyz(z + 2xy)$ **(d)** $2abc(1 - 2abc)$

12. (a) x^2 **(b)** ab **(c)** $2x^5$

(d) ab^3c^5 **(e)** a^6 **(f)** x^2

13. (a) $f = 10 - \dfrac{b}{2}$ **(b)** $f = 7$ **(c)** $b = 12$

14. (a) 2   **(b)** -9 **(c)** $-x + 2$

(d) $2y$ **(e)** $\dfrac{1}{3}$ **(f)** $\dfrac{2}{x}$

15. $2x(2y - 3z)$

16. $y = 13$

17. (a) 100   **(b)** 400

18. (a) a^7 **(b)** b

19. (a) $10x^2 - 15xy$ **(b)** $6x(x + 2)$

20. (a) $n = 3$ **(b)** $n = -4$

(c) $n = 0$ **(d)** $n = -2$

21. $\dfrac{4}{9}$

22. (a) $y = -30$ **(b)** $v(4u - 3)$

23. (a) (i) 16 **(ii)** $3x + 8$
    (b) $-9a + 5b$ **(c)** $3a(2 - 3a)$

24. $\dfrac{1}{64}$

25. -9

26. (a) $4x + 17$ **(b)** $x(5x - 7)$

27. (a) 61 **(b)** 63 **(c)** 64
28. (a) $3r - 3s$ **(b)** q **(c)** p^4
29. (a) p^5 **(b)** q^7 **(c)** r^6
30. (a) $7a(c + 2)$ **(b)** $6ax(2x + 3a^2)$
31. (a) 10 **(b)** 3 **(c)** -2

Chapter 4

Exercise 4.1

1. (a) 6 km² **(b)** 45000 m²
    (c) 48 cm² **(d)** 20 m²
2. (a) 343 m³ **(b)** 1000 cm³
    (c) 12 m³ **(d)** 200 cm³
3. (a) 32 cm **(b)** 24 m

Exercise 4.2

1. 33 cm **2.** 63 g
3. 706 kg **4.** 611 m
5. 500 km **6.** 91 cm
7. 90 kg **8.** 61 m
9. 60 m **10.** 800
11. 10 **12.** 100

Exercise 4.3

1. 240 **2.** 520
3. 7400 **4.** 600
5. 3990 **6.** 8000
7. 1000 **8.** 56.1
9. 56.14 **10.** 56.136
11. 3.1 **12.** 3.10

Exercise 4.4

1. 216 **2.** 220
3. 350 **4.** 400
5. 6010 **6.** 6000
7. 81.0 **8.** 0.199
9. 0.20 **10.** 1.00
11. 0.000395 **12.** 0.0004
13. 10.1 **14.** 657000
15. 700000

Exercise 4.5

1. 23720 **2.** 8.18 to 3 sf
3. 5.6° to 1 dp **4.** 56.23

Exercise 4.6

1. 156.5 cm
2. (a) $9.5 \leqslant w < 10.5$
    (b) $18.5 \leqslant h < 19.5$
3. (a) 14.5 cm $\leqslant$ 15 cm $<$ 15.5 cm
    (b) 23.55 cm $\leqslant$ 23.6 cm $<$ 23.65 cm
    (c) $3055 \leqslant 3060 < 3065$
    (d) $99.65 \leqslant 99.7 < 99.75$
    (e) $678.85 \leqslant 678.9 < 678.95$
    (f) $55000 \leqslant 60000 < 65000$
    (g) $250 \leqslant 300 < 350$
    (h) $99.85 \leqslant 99.9 < 99.95$
4. (a) 8.5 g **(b)** 9.5 g

Exercise 4.7

1. (a) 350 cm **(b)** 58.1 cm
    (c) 0.04096 km **(d)** 570000 mm
    (e) 812 g **(f)** 300 mm²
    (g) 0.050681 km² **(h)** 6700 cm³
    (i) 0.21 litres
2. (a) 195 cm²
    (b) 1977 mm² or 19.77 cm²
3. (a) 3.038 m³
    (b) 10.44 cm³

Exercise 4.8

1. (a) $2000 + 1000 + 2000 + 2000 = 7000$ km
    (b) 35 days
2. (a) $200000 + 500 \times 1000 = 700000$
    (b) $(4+1)^2 \div (7-2) = 5$
    (c) $\dfrac{4000}{200} + \dfrac{4000}{50} = 100$
3. (a) 8.2×10^8 **(b)** 2.05×10^2
    (c) 4.272×10^7 **(d)** 9.0109×10^5

Exercise 4.9

1. Estimate: $30 + 400 \times 0.03 = 42$
   Calculator: 45.2569
2. Estimate: $7 \times 3 - 20 + 20 \times 30 \approx 600$
   Calculator: 480 to 3 sf
3. Estimate: $\dfrac{30 + 3}{10} + 30 \approx 33$

   Calculator: 31.2 to 3 sf
4. Estimate: $\sqrt{\dfrac{20 + 20}{20 - 6}} \approx 2$

   Calculator: 1.94 to 3 sf
5. Estimate: $1 \times 10^4 - 6 \times 10^3 = 4 \times 10^3$
   Calculator: 6.52×10^3
6. Estimate: $4 \times 10^5 \times 8 \times 10^3 \approx 3 \times 10^9$
   Calculator: 3.03×10^9

Exercise 4.10

1. (a) 1 : 16   (b) 2 : 3 : 5
   (c) 1 : 8   (d) 5 : 1
   (e) 5 : 1   (f) 1 : 200000
   (g) 6 : 7   (h) 55 : 28
2. 1 : 30
3. 1 : 4
4. (a) 0.9 : 1   (b) 200000 : 1
5. (a) 1 : 0.25   (b) 1 : 12
6. 1 : 150
7. (a) 1 : 38.0 to 3 sf   (b) 0.0263 : 1 to 3 sf

Exercise 4.11

1. 84 Hydrogen, 42 Oxygen
2. (a) Rs. 484 : Rs. 121 (b) 138.6 g : 693 g
   (c) 3.75 m : 0.75 m (d) $30 : $75 : $195
   (e) 14.4 : 57.6 : 144 (f) 0.735 : 0.245
3. 10 cm
4. 1 : 500
5. 315
6. (a) 6   (b) 33
7. $25
8. 1 tonne cement, 4 tonnes sand
9. 10 km
10. (a) 120 m³
    (b) (i) 96 m³   (ii) 24 m³

Exercise 4.12

1. (a) direct   (b) neither
   (c) inverse   (d) direct
2. 10 tins   3. $170

4. 31 minutes   5. 63 minutes
6. (a) 2 days   (b) 9 painters

Exercise 4.13

1. $2
2. (a) 5 hours 20 minutes   (b) 97.5°
3. 1200000 kg   4. 1 hour 45 minutes
5. 82.4 km/h   6. 0615
7. 35000 kg/m³   8. 1.25 g/cm³

Exercise 4.14

1. (a) (i) loss   (ii) $5   (iii) 10%
   (b) (i) profit (ii) $5   (iii) 10%
   (c) (i) profit (ii) Rs. 91.25 (iii) 25%
   (d) (i) profit (ii) Rs. 6.4 on (iii) 8.14%
                      each one
   (e) (i) profit (ii) £0.05 on   (iii) 10%
                      each one
2. (a) $1050   (b) $1102.50
3. (a) 5500   (b) 11406.25   (c) 4350
4. (a) 12.5 miles (b) 56 km
5. (a) 1.53 m/s   (b) 96 km/h
6. (a) 0.628 kg   (b) 0.572 kg The iron one is
                       heavier.
7. 3 km
8. 41%
9. Large: £1.80 per litre, small: £1.76 per litre. The
   small one is the better buy.
10. Penti: 9 km/litre, Quadri: 12 km/litre. The Quadri
    is more economical.
11. Maths: 60%, Science: 69%. Science is his better
    subject.
12. $15309

Exercise 4.15

1. (a) 72   (b) 144
2. (a) 120 m³   (b) 0.036 m³
3. (a) 687.5 g and 625 g   (b) 2
4. (a) 18 hours 5 minutes (b) 444 miles per hour
   (c) 21 40
5. (a) 163 minutes (b) 7 hours 9 minutes
   (c) 3.6 hours   (d) 0.45 hours
6. 7.37×10^{22} kg
7. (a) 3.844×10^5 (b) 3.844×10^8
8. 4.5 m
9. 141.5, 142.5
10. $190.48
11. (a) (i) 200   40   (ii) 5
    (b) 5.6

12. For example,
  A costs $\$0.001625$ per ml
  B costs $\$0.001533$ per ml
  B is better value for money
13. **(a)** $\$4.50$ **(b)** 56.3%
14. **(a)** 00 15   **(b)** **(i)** 7.5 hours or
                              7 hours 30 minutes
                        (ii) 749 km/h
15. 1.5 kg
16. **(a)** 1018   **(b)** $\$89.38$
17. 400 g
18. $\$2.70$
19. 1250, 1350
20. **(a)** **(i)** 2.409677419   **(b)** 19.3
         (ii) 2.41
21. $\$42$
22. 366 riyals
23. **(a)** 88   **(b)** 85.5, 86.5

24. **(a)** 20 05   **(b)** **(i)** 0.4 km/min
                          (ii) 24 km/h
25. **(a)** 19.55249345   **(b)** 19.55
26. **(a)** 12.5%   **(b)** 120 minutes
27. 40%
28. 35 : 8
29. $\$900$
30. **(a)** **(i)** $\dfrac{9-3\times2}{3}$ **(ii)** 1
        (b) 1.01
31. **(a)** $\$5.60$ **(b)** $\$2.40$
32. **(a)** 0.5   **(b)** **(i)** $10-6\times0.5=7$
                         (ii) 7.0908
33. **(a)** $\$315$ **(b)** $\$135$
34. **(a)** $\$25$ **(b)** $\$551.25$
35. **(a)** 14 23 or 2.23 pm
        (b) 94 km/h
36. 11.5 km

Chapter 5

NOTE: Unless the question states otherwise, answers to algebraic questions which are fractions greater than 1, may be given as improper (top heavy) fractions, mixed numbers or, if they are exact, decimals. They should not be given as rounded decimals. The fractions must be simplified.

Exercise 5.1

1. $x=2$ **2.** $x=4$
3. $x=8$ **4.** $x=3$
5. $x=-\dfrac{1}{2}$ **6.** $x=4$
7. $x=2$ **8.** $x=\dfrac{1}{4}$
9. $x=5$ **10.** $x=-6$
11. $x=-9$ **12.** $x=7$
13. $x=-3$ **14.** $x=-22$
15. $x=7\dfrac{1}{2}$ **16.** $x=-\dfrac{5}{3}$

Exercise 5.2

1. $x=\dfrac{4}{5}$ **2.** $x=-4$

3. $x=-\dfrac{2}{7}$ **4.** $x=\dfrac{17}{5}$
5. $x=\dfrac{1}{4}$ **6.** $x=-3$
7. $x=-\dfrac{50}{27}$ **8.** $x=\dfrac{1}{2}$
9. $x=0$ **10.** $a=-\dfrac{23}{6}$
11. $x=\dfrac{5}{4}$ **12.** $a=\dfrac{8}{17}$
13. $y=8$ **14.** $b=\dfrac{7}{5}$

Exercise 5.3

1. $x=-\dfrac{31}{9}$ **2.** $x=\dfrac{11}{3}$
3. $x=\dfrac{9}{10}$ **4.** $x=-\dfrac{14}{5}$
5. $x=-3.5$ **6.** $x=\dfrac{27}{14}$
7. $x=-\dfrac{14}{9}$ **8.** $x=-\dfrac{22}{15}$
9. $x=\dfrac{5}{3}$ **10.** $a=-\dfrac{25}{34}$

Exercise 5.4

1. $x = -\dfrac{10}{3}$　　　　**2.** $x = 3$

3. $x = \dfrac{7}{2}$　　　　**4.** $x = -\dfrac{3}{4}$

5. $x = \dfrac{9}{4}$　　　　**6.** $a = -44$

7. $y = -\dfrac{2}{7}$　　　　**8.** $b = \dfrac{1}{5}$

9. $x = 1$　　　　**10.** $x = -\dfrac{2}{3}$

11. $c = -19$　　　　**12.** $x = 2$

13. $x = 0$　　　　**14.** $x = -1$

15. $x = \dfrac{1}{3}$

16. (a) $3x + 3x + (2x + 5) = 33$
　　(b) $x = 3.5$　　(c) 10.5, 10.5 and 12 cm

17. (a) blue $= \dfrac{x}{2}$, green $= x - 2$

　　(b) $x + \dfrac{x}{2} + (x - 2) = 23$　　(c) $x = 10$
　　(d) 10 red, 5 blue and 8 green

18. (a) algebra $= x + 4$, shape $= 2x$
　　(b) $x + x + 4 + 2x = 20$
　　(c) $x = 4$　　(d) 8

19. $x + (x + 1) + (x + 2) = 114$　　$x = 37$
　　　　　　　　　　　　37, 38 and 39

20. $x + (x + 2) + (x + 4) = 135$　　43, 45 and 47

NOTE: Remember that the fraction line acts like a bracket, so that brackets in the denominator or numerator may be optional. (See question 3. (b)). Either form may be given here.

Exercise 5.5

1. (a) $x = 3$　　(b) $x = d - b$
2. (a) $x = 18$　　(b) $x = yz$

3. (a) $x = 6$　　(b) $x = \dfrac{(c+b)}{a}$ or $\dfrac{c+b}{a}$

4. (a) $x = \dfrac{3}{2}$　　(b) $x = \dfrac{c}{(a+b)}$

5. (a) $x = \dfrac{2}{3}$　　(b) $x = \dfrac{2}{(a-b)}$

6. (a) $x = 1$　　(b) $x = \dfrac{3}{(a-1)}$ or $\dfrac{-3}{(1-a)}$

7. (a) $x = -\dfrac{3}{2}$　　(b) $x = \dfrac{a}{(c-b)}$ or $\dfrac{-a}{(b-c)}$

8. (a) $x = 6$　　(b) $x = ab$

9. (a) $x = \dfrac{15}{7}$　　(b) $x = \dfrac{3}{y}$

10. (a) $x = 20$　　(b) $x = ab$

Exercise 5.6

1. $M = \dfrac{Fd^2}{Gm}$　　**2.** $d = \sqrt{\dfrac{GmM}{F}}$

3. $d = \dfrac{u - a}{n - 1}$　　**4.** $n = \dfrac{u - a}{d} + 1$

5. $a = ch$　　**6.** $h = \dfrac{a}{c}$

7. $d = st$　　**8.** $t = \dfrac{d}{s}$

9. $A = \dfrac{F}{P}$　　**10.** $C = \dfrac{5}{9}(F - 32)$

11. $a = \dfrac{2(s - ut)}{t^2}$

12. (a) $a = \dfrac{bc}{d}$　　(b) $c = \dfrac{ad}{b}$

　　(c) $b = \dfrac{ad}{c}$　　(d) $d = \dfrac{bc}{a}$

13. (a) $y = \dfrac{5 - 3x}{2}$　　(b) $y = -\dfrac{13}{2}$

14. (a) $r = \sqrt{\dfrac{A}{\pi}}$　　(b) $r = 2.52$

15. (a) $I = \dfrac{5}{3}$　　(b) $c = \dfrac{b}{I}$　　$c = 13\dfrac{1}{3}$

16. (a) $h = \dfrac{2A}{a + b}$　　(b) $h = \dfrac{10}{7}$

Exercise 5.7

1. (a) 53　　(b) nth term $= n + 3$
2. (a) 56　　(b) nth term $= n + 6$
3. (a) 12, 15　(b) nth term $= 3n - 9$　(c) 300
4. (a) 64, 81　(b) nth term $= (n + 3)^2$　(c) 3364
5. (a) 51, 66　(b) nth term $= n^2 + 2$　(c) 363
6. (a) 8, 14, 20　　　　(b) 52
　　(c) nth term $= n + 2$　　(d) 1274

Exercise 5.8

1. $x = -4$ $y = 2$
2. $x = 2$ $y = 2$
3. $x = 0$ $y = 1$
4. $x = 2$ $y = -\dfrac{6}{5}$

Exercise 5.9

1. $x = 7$ $y = -8$
2. $x = 2$ $y = 2$
3. $x = -\dfrac{1}{2}$ $y = -1$
4. $x = 2$ $y = -\dfrac{1}{2}$
5. $x = 2$ $y = 1$
6. $p = 2$ $q = 0$
7. $x = -3$ $y = -1$
8. $x = 2$ $y = 2$
9. $x = 3$ $y = -1$
10. $x = 1$ $y = 6$

Exercise 5.10

1. (a) $x = 1$ (b) $x = -\dfrac{3}{2}$
   (c) $x = -\dfrac{9}{10}$ (d) $x = 2$
   (e) $x = \dfrac{23}{9}$ (f) $x = \dfrac{3}{4}$
   (g) $x = -6$ (h) $x = 0$
   (i) $a = -6$ (j) $b = 2$

2. (a) $r = \sqrt{\dfrac{A}{\pi}}$ (b) $h = \dfrac{2V}{bl}$ (c) $t = \dfrac{d}{V}$

   (d) $t = \dfrac{2A}{a+b}$ (e) $V = \dfrac{M}{D}$

   (f) $b = \sqrt{a^2 - c^2}$ (g) $S = \dfrac{q+r}{p}$

   (h) $x = \dfrac{A - Bc}{B}$ or $x = \dfrac{A}{B} - c$

   (i) $b = \dfrac{x}{a} - c$ (j) $x = \sqrt{b^2 - 3a^2}$

   (k) $h = \dfrac{V}{a}$ (l) $b = ad - c$

   (m) $a = \dfrac{2A}{l} - b$ or $\dfrac{2A - bl}{l}$

3. (b) nth term $= 2n + 1$ (c) 199
4. (a) nth term $= 4n - 2$
   (b) nth term $= 3n - 6$
   (c) nth term $= -2n + 6$
5. (a) $x = -1$ $y = 2$
   (b) $x = 0$ $y = 3$
   (c) $x = \dfrac{1}{2}$ $y = 3\dfrac{1}{2}$
6. (a) $x + y = 57$ $x - y = 15$
   (b) $x = 36$ $y = 21$
7. $x = 3$ $y = 9$
8. (a) (i) 27 28 29 30 31 32 33 34 35 36
       (ii) (a) square
            (b) 100
            (c) n^2
       (iii) (a) 43
             (b) 871
9. (a) (i) 13   (ii) 9
   (b) $x = \dfrac{75 - 2y}{7}$ or $\dfrac{2y - 75}{-7}$
   (c) $x = 11$ $y = -1$
10. (a) (i) 27   (ii) 6   (iii) $x = \dfrac{P - 3}{6}$
    (b) (i) $4x + 3$ (ii) 10, 16, 23
11. $s = \dfrac{p+q}{t}$
12. $x = 3$
13. $a = \dfrac{P - 2b}{2}$
14. (b) 13   16   19   (c) 298
    (d) $3n + 1$ (e) 28
15. (a) (i) $x = 6$
        (ii) $y = 9$
        (iii) $z = 1.5$
    (b) (i) $p + q = 12$
        (ii) $25p + 40q = 375$
        (iii) $p = 7$ $q = 5$
16. (a) (i) 360
        (ii) 7.5
        (iii) $m = \dfrac{2E}{v^2}$
    (b) $xy(y - x)$ (c) $x = 2$
17. (b) 22   29   36
    (c) (i) 71   (ii) $7n + 1$
    (d) $7n + 1 = 113$ $n = 16$
18. $x = 5$ $y = -3$

Chapter 6

NOTE: The answers to questions involving measuring line and angles are given as a range of values (for example, 5.5 to 5.6 cm). If your work is sufficiently accurate you should get an answers in this range. If you do not, try again.

Exercise 6.1

1. 32 to 33°   **2.** 90°   **3.** 116 to 117°
4. 180°   **5.** 337 to 338°   **6.** 360°

Exercise 6.2

1. (a) $x = \angle BAC$, $y = BC$
  (b) $x = \angle EDH$, $y = \angle GFH$, $z = GH$
  (c) $x = \angle JMK$, $y = \angle KML$
  (d) $x = \angle QPR$, $y = \angle NPQ$, $z = NQ$

2. (a) 24 to 26°
  (b) (i) 54 to 56°   (ii) 124 to 126°
  (c) (i) 7.6 to 7.8   (ii) 39 to 41°
    (iii) 82 to 84°
  (d) (i) 89 to 91°   (ii) 167 to 169°

3. (a) acute   (b) reflex
  (c) acute   (d) obtuse
  (e) reflex   (f) reflex
  (g) obtuse   (h) right angle
  (i) acute   (j) reflex
  (k) obtuse

4. (a) 50°   (b) 35°, 105°, 140°
  (c) 36°, 108°, 36°   (d) 45, 90°, 135°
  (e) 20°, 70°   (f) 45°, 90°, 45°

Exercise 6.3

1. parallel   **2.** vertical
3. vertical   **4.** perpendicular, horizontal, vertical
5. intersecting   **6.** intersecting
7. vertical   **8.** parallel, horizontal

Exercise 6.4

1. (a) (i) 7.6 to 7.8 cm   (ii) asa
  (b) This is not a triangle: the third angle
    would be 0°

  (c) This is not a triangle: $2 + 7 < 10$
  (d) (i) 8.6 to 8.8 cm   (ii) sas
  (e) (i) 6.3 to 6.5 cm   (ii) asa

2. (a) $\angle A = 52$ to 54°, $\angle B = 36$ to 38°,
    $\angle C = 89$ to 91°
  (b) $\angle D = 66$ to 68°, $\angle E = 89$ to 91°,
    $\angle F = 22$ to 24°
  (c) $GH = 3.9$ to 4.1 cm, $\angle G = 89$ to 91°,
    $\angle H = 59$ to 61°
  (d) $\angle J = 79$ to 81°, $JK = 5.8$ to 6.0 cm,
    $JL = 7.8$ to 8.0 cm
  (e) $\angle M = 59$ to 61°, $\angle N = 59$ to 61°,
    $\angle P = 59$ to 61°

3. (a) 36°, 56°   (b) 65°, 65°
  (c) 36°, 72°   (d) 45°
  (e) 60°   (f) 60°, 55°, 65°

Exercise 6.5

1. $a = 145°$ $b = 35°$ $c = 35°$
  $d = 145°$ $e = 145°$
2. $a = 70°$ $b = 30°$ $c = 70°$
  $d = 110°$ $e = 150°$
3. $a = 70°$ $b = 35°$
  $c = 145°$ $d = 145°$
4. $a = 110°$ $b = 30°$
  $c = 40°$ $d = 110°$

Exercise 6.6

1. 5 lines, order 5
2. 1 line, no rotational (or order 1)
3. 2 lines, order 2
4. 2 lines, order 2
5. 1 line, no rotational (or order 1)
6. 8 lines, order 8
7. 4 lines, order 4
8. no lines, order 2
9. 4 lines, order 4

Exercise 6.7

(a) all 90°
(b) opposite angles equal
(c) no angles equal

(d) equal lengths, bisect each other but not at right angles

(e)

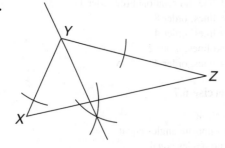

(f) different lengths, bisect each other at right angles

(g) 2   **(h)** 2   **(i)** 0   **(j)** 1

(k) 4   **(l)** 2   **(m)** 2

Exercise 6.8

1. $a = 110°$ $b = 70°$
   $c = 110°$ $d = 70°$
2. $a = 20°$ $b = 70°$ $c = 40°$
3. $a = 130°$ $b = 120°$
4. $a = 70°$ $b = 20°$ $c = 20°$
   $d = 70°$ $e = 40°$
5. $30°, 60°, y = 60°$
6. $x = 80°$ $y = 160°$ $z = 20°$
7. $∠TSR = 60°$ (equilateral triangle)
   $∠SPQ = 60°$ (angles on a straight line)
   PQ is parallel to RS (corresponding angles)

Exercise 6.9

1. 2340°   **2.** 30°   **3.** 30°
4. **(a)** 72°   **(b)** 108°   **(c)** 54°
   (d) 54°   **(e)** 144°   **(f)** 18°
   (g) 72°
   $∠CAE = ∠AEF = 72°$
   AC and DE are parallel (alternate angles)

5. n = 18
6. **(a)** 16   **(b)** 157.5°

Exercise 6.10

1. 135°
2. $b = 70°, c = 40°$
3. $d = 60°, e = 30°, f = 30°$
4. $g = 40°, h = 50°$
5. $j = 20°$
6. $k = 50°, l = 50°, m = 40°, n = 50°$

Exercise 6.11

(a) 12   **(b)** 8   **(c)** 5
(d) 9   **(e)** 6   **(f)** 4
(g) 6   **(h)** 4   **(i)** 5
(j) 8   **(k)** 5   **(l)** 1
(m) 1

Exercise 6.12

2. cone

Exercise 6.13

1. a, h; b, f; c, g; d, i
2. b, d; c, e

Exercise 6.14

1. **(b)** 11.9 to 12.1 cm   **(c)** 59.5 to 60.5 km
2. **(b)** 5.3 to 5.5 km

Exercise 6.15

NOTE: For reasons of space some of the answers to this exercise have been reduced in size. Yours should be full size.

1.

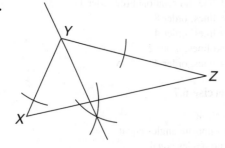

2.

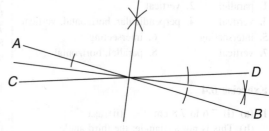

3. **(a)** to **(c)**

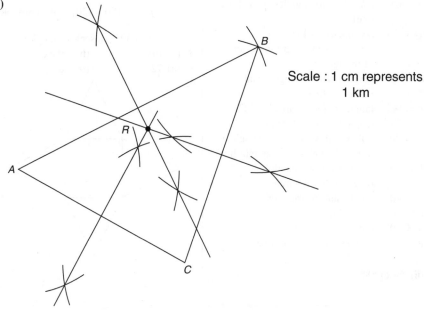

Scale : 1 cm represents
1 km

(d) 3.5 to 3.7 km

4. 3.2 to 3.5 km

Exercise 6.16

NOTE: For reasons of space some of the answers to this exercise have been reduced in size. Yours should be full size.

1. **(a)** to **(e)**

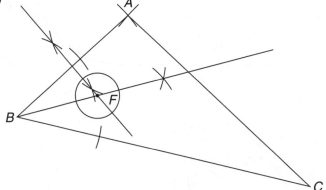

(f) **(i)** 6.4 to 6.8 m
    (ii) 22.4 to 22.8 m
2. **(a)** 1 line, no rotational (order 1)
    (b) 1 line, no rotational (order 1)
    (c) 4 lines, order 4
    (d) 4 lines, order 4
    (e) 2 lines, order 2
    (f) 1 line, no rotational (order 1)

3. **(a)** $a = b = 15$ cm, $c = 60°$ (equilateral triangle)
    (b) $a = 60°$ $\angle PST = 30°$ (angles on a straight line)
        $\angle PQT = 30°$ (symmetry of kite)
        $\angle PTQ = 90°$ (diagonals cross at right angles)
        $a = 180 - 90 - 30$ (angle sum of triangle)

$b = 65°$ $∠SRQ = 50°$ (angles on a straight line)

$∠TRQ = 25°$ (symmetry of kite)
$b = 180 - 90 - 25$ (angle sum of triangle)

(c) $a = 40°$ (angle between tangent and radius $= 90°$)
$b = 90°$ (angle in a semicircle $= 90°$)
$c = 50°$ (angle sum of a triangle)

(d) $a = 70°$ $∠YZX = 110°$ (alternate angles)
$a = 70°$ (angles on a straight line)
$b = 40°$ (angle sum of triangle)

(e) $a = 90°$ (angle in a semicircle)
$b = 40°$ (angle sum of triangle)
$c = 60°$ (isosceles triangle)
$d = 60°$ (angle sum of triangle)
$e = 120°$ (angles on a straight line)

$f = \dfrac{1}{2}(180 - 120) = 30°$ (isosceles triangle)

$g = 30°$ (isosceles triangle)

4. (a) obtuse   (b) reflex
5. (a) 72°   (b) 36°
6.

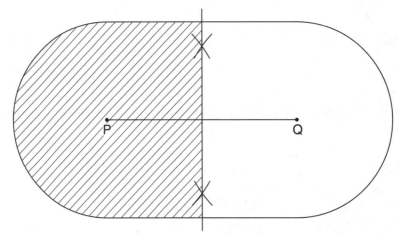

7. (a) (ii) 54 to 58°
   (b)

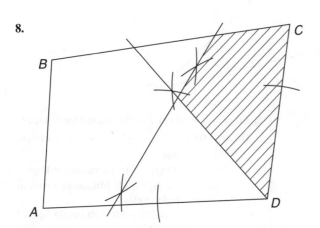

8.

9.

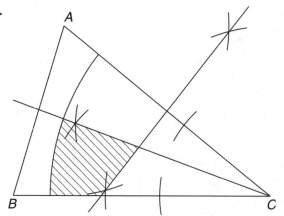

10. 110°
11. (a) 90°   **(b)** 65°   **(c)** 25°
12. (a) equilateral   **(b)** (triangular) prism
13. $x = 120$ $y = 150$
14. (a) 120   **(b)** 70
   (c) (i) 130   **(ii)** 100   **(iii)** 70, 30
15. (a) (i) 84 to 86   **(ii)** 70 to 74°
   (b) (ii) 82 to 84 m

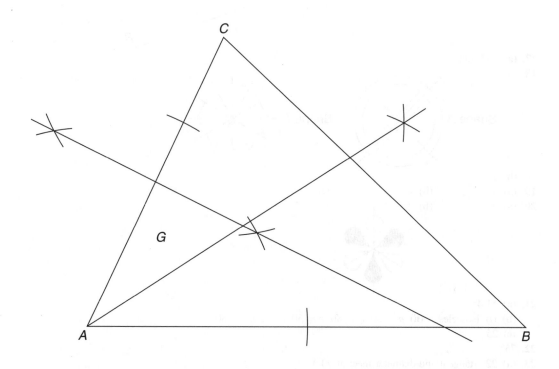

16. (a) (ii) 90°   **(iv)** angle in a semicircle

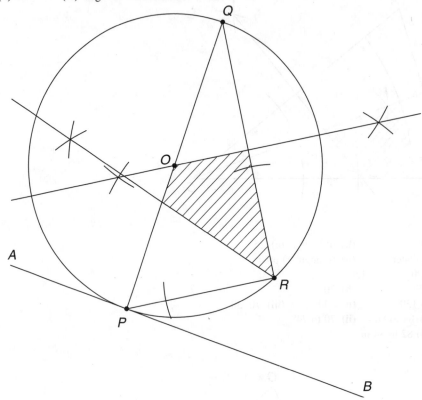

17. (a) 263 mm
18. (a)

Shape *A*   Shape *B*

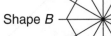

   (b) 2
19. (a) 4   **(b)** 4
20. (a) 3   **(b)**

21. (a) 51.4°
   (b) (i) isosceles   **(ii)** $p = 50$, $q = 80$, $r = 50$, $s = 50$, $t = 80$
   (c) 25
22. 75°
23. (a) 22° (tangent and diameter meet at 90°)
   (b) 90° (angle in a semicircle)
   (c) 68° (angle sum of triangle)
   (d) 68° (alternate angles)

Chapter ⑦

Exercise 7.1

2.

1.

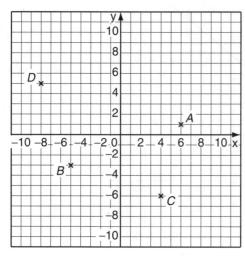

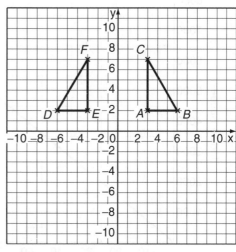

(e) The triangles are the same shape and size
(f) The triangles are facing in different directions

3. *A*: (8, 9)   *D*: (−3, 3)   *G*: (9, 0)   *K*: (9, −3)
   B: (0, 8)   *E*: (3, 3)   *H*: (0, 0)   *L*: (−6, −5)
   C: (−6, 8)   *F*: (−7, 0)   *J*: (0, −3)   *M*: (5, −7)

Exercise 7.2

1.

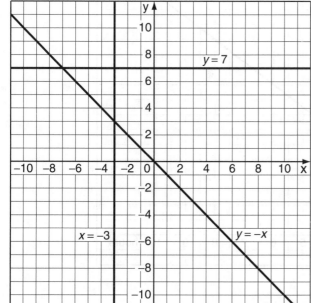

2. *l*: $x = 10$ *m*: $y = -5$

Exercise 7.3

1. (a) July   **(b)** 9

2. (a)

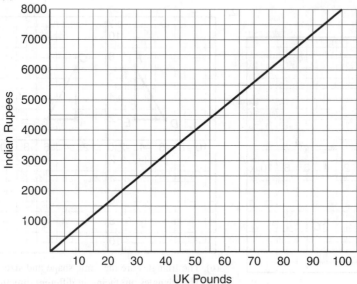

(b) (i) 6000 rupees   **(ii)** £37.50

3.

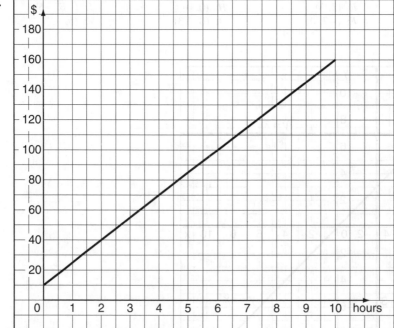

(c) $92.50   **(d)** 7.5 hours

4. (a) 1 hour   **(b)** 1.5 hours   **(c)** 1.5 km
   (d) 1 kilometre per hour   **(e)** Anton

Exercise 7.4

1.

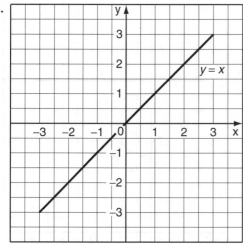

2.

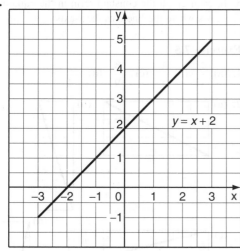

3.

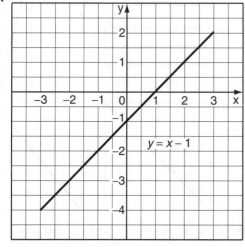

4.

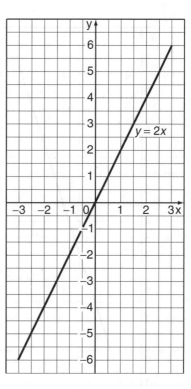

5.

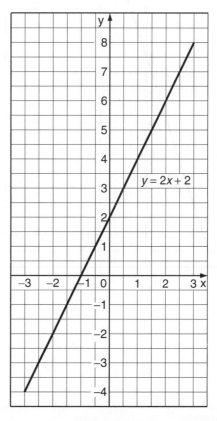

6.

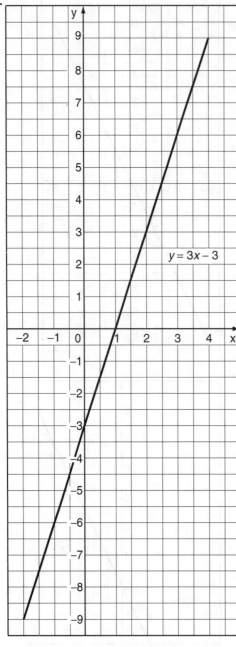

$y = 3x - 3$

7.

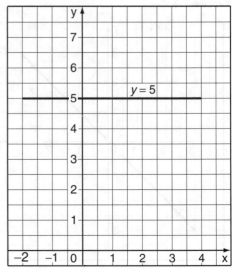

$y = 5$

8. In each graph the constant term in the equation shows where the line cuts the *y*-axis

Exercise 7.5

1. (a) $\dfrac{4}{5}$ **(b)** $\dfrac{4}{3}$ **(c)** $-\dfrac{8}{9}$ **(d)** $-\dfrac{5}{2}$

2. (a) 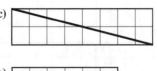 **(b)**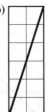

(c) 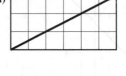 **(d)**

(e)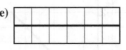

(f) 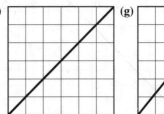 **(g)**

3. (a) (ii)   **(b)** (iv)   **(c)** (i)   **(d)** (iii)

4.

angle made with *x*-axis	less than 45°	exactly 45°	between 45° and 90°
gradient	$\dfrac{1}{2}, \dfrac{1}{5}, \dfrac{3}{5}$	$\dfrac{2}{2}$	$5, \dfrac{4}{3}$

Exercise 7.6

1.

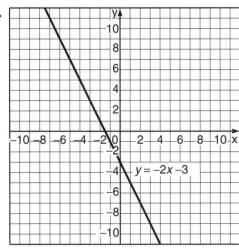

3.

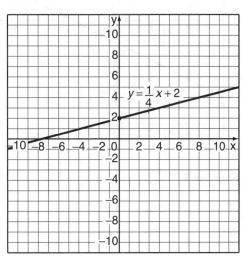

2.

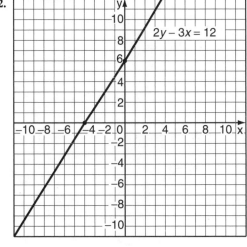

4.

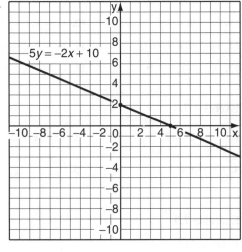

5.

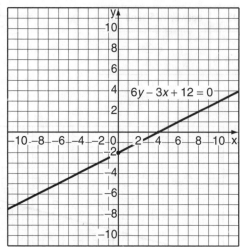

6.

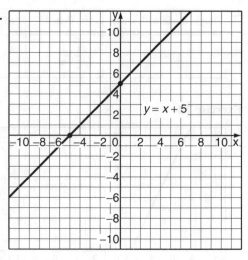

Exercise 7.7

1. $m = \dfrac{1}{2}, c = -5$

2. $m = -\dfrac{1}{2}, c = \dfrac{3}{2}$

3. $m = -\dfrac{2}{5}, c = \dfrac{4}{5}$

4. $m = -1, c = -1$

5. $m = \dfrac{5}{4}, c = -5$

6. $m = \dfrac{1}{2}, c = \dfrac{1}{2}$

7. $m = -\dfrac{2}{3}, c = \dfrac{4}{3}$

8. $m = \dfrac{1}{2}, c = 3$

9. $m = 0, c = 6$

10. $m = 1, c = 0$

11. $m = 0, c = -10$

12. $m = 2, c = 0$

13. $m = -1, c = 0$

14. $m = -1, c = 4$

Exercise 7.8

1. $y = -2x + 2$

2. $y = 2x - 2$

3. $y = \dfrac{1}{2}x + 1$

4. $y = -x + 3$

5. $\dfrac{1}{8}$

6. -3

Exercise 7.9

1. (a) and (c)

2. for example, $y = \dfrac{1}{4}x + 5$

3. (a) $\dfrac{1}{2}$ (b) $\dfrac{2}{1} = 2$ (c) $\dfrac{2}{1} = 2$

(b) and (c) are parallel

Exercise 7.10

1. (a)

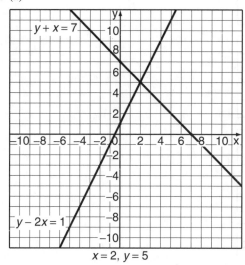

$x = 2, y = 5$

(b) $x = 2, y = 5$

2. (a)

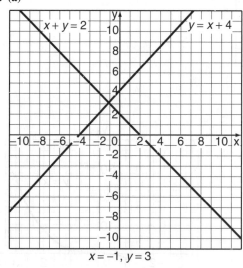

$x = -1, y = 3$

(b)

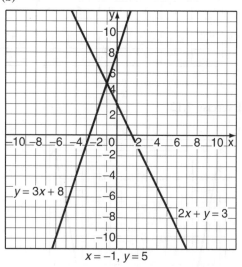

$x = -1, y = 5$

Exercise 7.11

1.

x	-3	-2	-1	0	1	2	3
y	9	4	1	0	1	4	9

2.

x	-3	-2	-1	0	1	2	3
y	-27	-8	-1	0	1	8	27

3.

x	−3	−2	−1	0	1	2	3
y	−1	$-\dfrac{3}{2}$	−3	...	3	$\dfrac{3}{2}$	1

4.

x	−3	−2	−1	0	1	2	3
y	6	2	0	0	2	6	12

5.

x	−3	−2	−1	0	1	2	3
y	12	6	2	0	0	2	6

6.

x	−3	−2	−1	0	1	2	3
y	−25	−6	1	2	3	10	29

7.

x	−3	−2	−1	0	1	2	3
y	−9	−4	−1	0	−1	−4	−9

8.

x	−3	−2	−1	0	1	2	3
y	27	8	1	0	−1	−8	−27

9.

x	−3	−2	−1	0	1	2	3
y	7	1	−3	−5	−5	−3	1

10.

x	−3	−2	−1	0	1	2	3
y	−2	1	2	1	−2	−7	−14

Exercise 7.12

1.

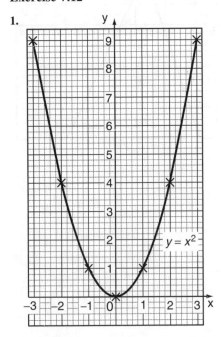

2.

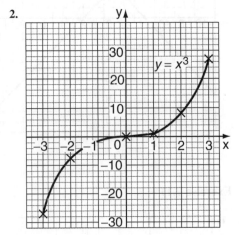

3.

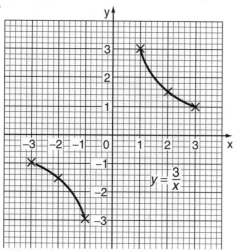

$y = \dfrac{3}{x}$

5.

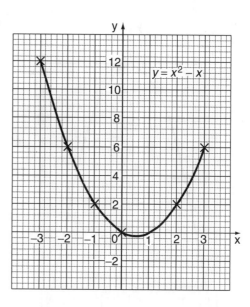

$y = x^2 - x$

4.

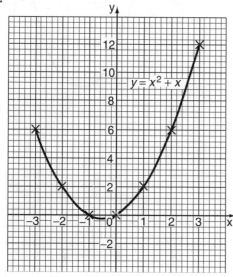

$y = x^2 + x$

6.

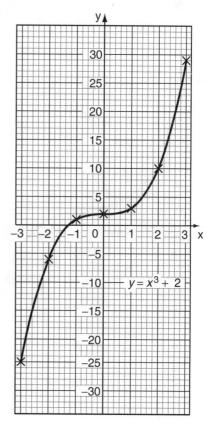

$y = x^3 + 2$

7.

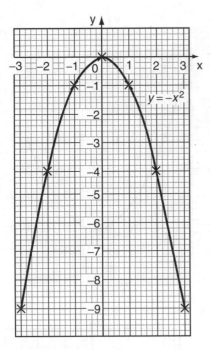

$y = -x^2$

8.

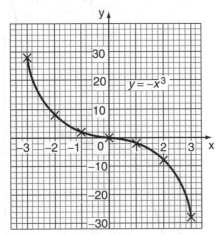

$y = -x^3$

9.

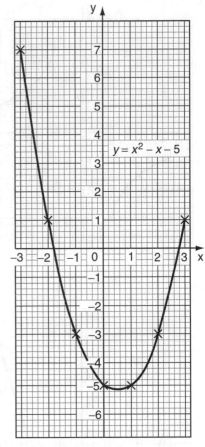

$y = x^2 - x - 5$

10.

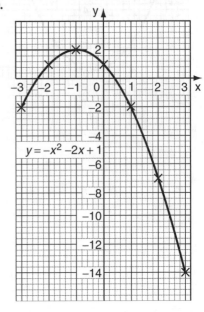

$y = -x^2 - 2x + 1$

Exercise 7.13

1. (a) 9m   **(b)** 2.5m   **(c)** no

2. (a) $y = x$ **(b)** $y = -x$ **(c)** $x = 5$
   (d) $y = -2$

3. (a)

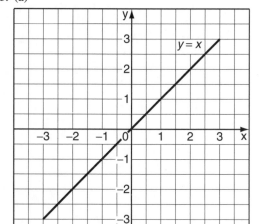

(b)

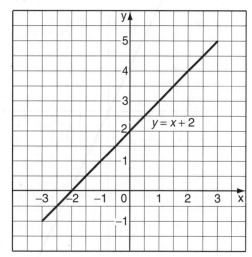

(c)

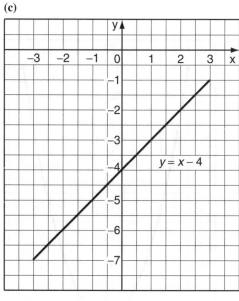

(d)

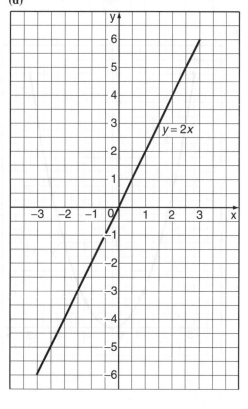

(e)

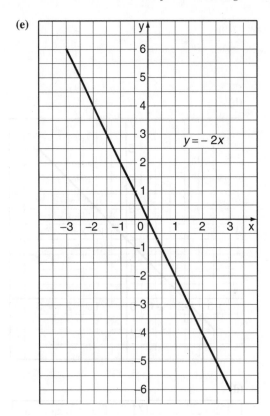

$y = -2x$

4. (a)

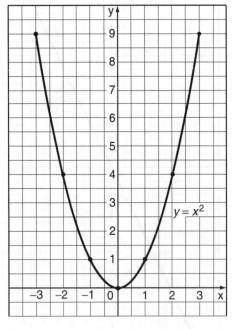

$y = x^2$

(b)

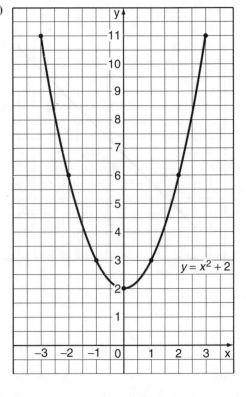

$y = x^2 + 2$

(c)

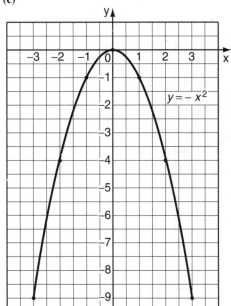

(d)

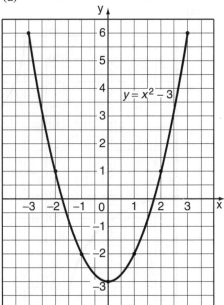

5. (a) This should be a continuous smooth curve, not flat at the base

    (b) This should be a smooth curve, not made up of straight line segments

    (c) One point is out of line because it has been incorrectly plotted

6. (a) (ii)   **(b)** (iv)   **(c)** (i)

    (d) (vi)   **(e)** (v)   **(f)** (iii)

7. (a) 1   **(b)** $\dfrac{1}{2}$ **(c)** 2

8. (a)

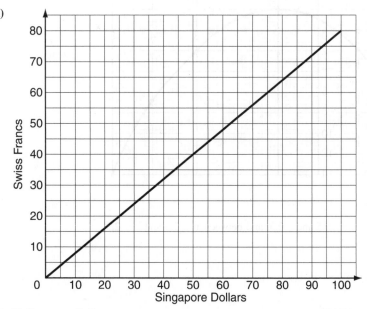

    (b) 55 Singapore Dollars

9. **(a)** **(i)** −10, −20, −60, 30, 20, 15

   (ii)

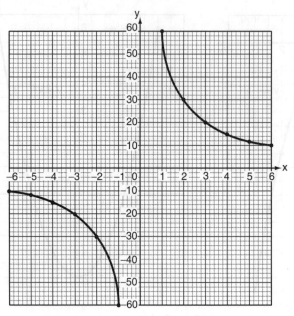

   (b) 2

10. **(a)** −2, 1, 2, −7

   (b)

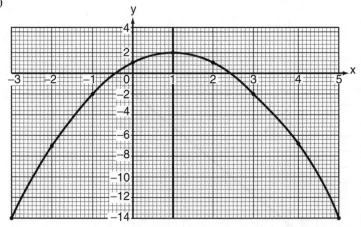

   (c) answers between $x = -0.5$ and -0.3, and between $x = 2.3$ and $x = 2.5$

   (d) **(ii)** $x = 1$

11. **(a)** 90

   (b) between D and E

   (c) **(i)** 1300 m

   　　(ii) 3.25 metres per second

12.

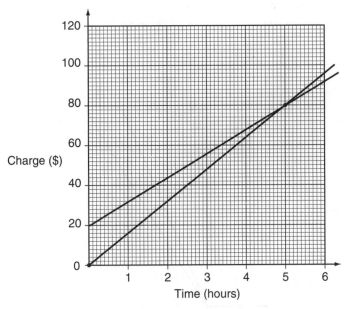

(a) $62
(b) 2.5 hours
(c) (ii) 5 hours
13. $y = 2x - 3$
14. (a) (i) Minimum temp on Sunday $= -3$, Maximum temp on Sunday $= 9$
   (ii) 9°C
(b)

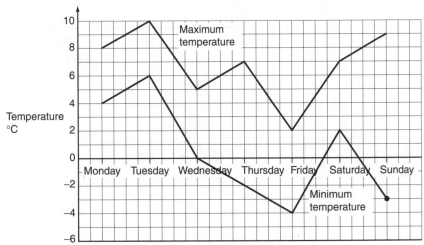

(c) (i) 3 days
   (ii) Sunday
(d) 42.8
15. (a) 10 kilometres per hour
(b) 20 minutes

(c)

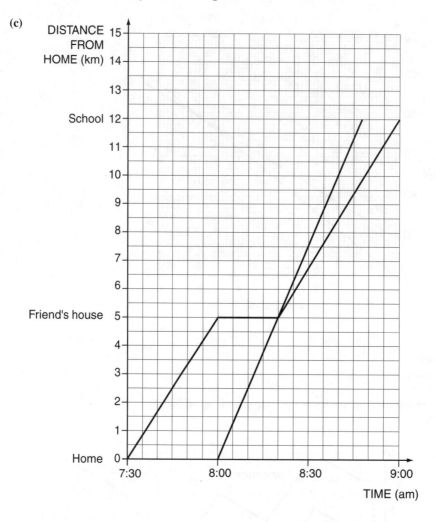

DISTANCE FROM HOME (km)

School 12

Friend's house 5

Home 0

TIME (am)

(d) 12

16. (a)

x	−4	−3	−2	−1	0	1	2	3
y	9	3	−1	−3	−3	−1	3	9

(b)

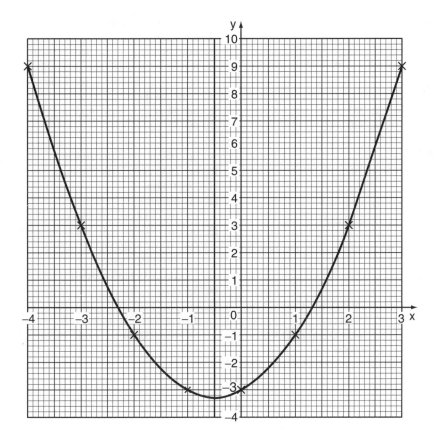

(c) $x = -0.5$, $y = -3.2$ or -3.3

(d) (ii) $x = -0.5$

Chapter 8

Exercise 8.1

1. (a) 28 cm   **(b)** 22 cm
   (c) 24 cm   **(d)** 29 cm
   (e) 32 cm

2. (a) 31.4 cm   **(b)** 15.7 cm
   (c) 44.9 m   **(d)** 330 cm

3. (a) 20.6 cm   **(b)** 26.7 cm
   (c) 41.1 cm   **(d)** 23.7 cm
   (e) 30.3 cm

4. (a) $x = 2.6$ **(b)** $x = 4.77$
   (c) $x = 1.75$ **(d)** $x = 3$

5. (a) 2 cm   **(b)** 9 cm

Exercise 8.2

1. (a) 12 cm²   **(b)** 15 cm²   **(c)** 25 cm²
   (d) 12 cm²   **(e)** 15 cm²   **(f)** 21 cm²

2. (a) $x = 3.16$ cm   **(b)** $x = 1.75$ cm

Exercise 8.3

1. (a) 113 cm²   **(b)** 32.2 m²   **(c)** 50.3 cm²
   (d) 56.5 cm²   **(e)** 25.8 cm²

2. (a) 25.1 cm²   **(b)** 41.9 cm²   **(c)** 113 cm²
   (d) 26.5 cm²   **(e)** 14.1 cm²

3. (a) $x = 2.33$ **(b)** $x = 3.91$
   (c) $x = 1.78$ **(d)** $x = 3.91$

4. 37.1 cm²

Exercise 8.4

1. 15 cm²   2. 54 cm²   3. 31.5 cm²
4. 163 cm²   5. 33 cm²   6. 36 cm²

Exercise 8.5

1. 56 cm²   2. 283 cm²
3. 96 cm²   4. 540 cm³

Exercise 8.6

1. 32 ml   2. 339 ml
3. 42 cm³   4. 648 cm³
5. (a) 240 cm³   (b) 860 cm³
6. 2 cm   7. 2.25 cm²
8. 1.20 cm
9. (a) 1180 cm²   (b) 118000 cm³   (c) 7850 ml

Exercise 8.7

1. (a) Surface area: A: 28000 cm²   B: 39500 cm²
   C: 35000 cm²
   (b) Volume: A: 442000 cm³ B: 518000 cm³
   C: 500000 cm³
   Volume ÷ surface area: A : 15.8   B : 13.1
   C : 14.3

A is the best value
2. h = 12.7
3. (a) 1.26 cm   (b) 5.09 cm
4. (b) 2a + 2b = 16 or a + b = 8
   (c) a = 2   b = 6   (d) 8 cm, 2 cm
5. 7 cm and 4 cm
6. (a) 54 m   (b) 9.15 m
7. (b) 52 cm²
8. 2.71
9. (a) 160 m   (b) 50.9 m
10. (a) 2830 cm²   (b) 226 litres
11. 24500 litres
12. (a) diameter
    (b) (i) 30.8 cm (ii) 56.5 cm²
13. (a) (i) 10.8 m   (ii) 32400 litres (iii) 36 litres
    (b) (i) 61 hours and 30 minutes
        (ii) 13500 gallons
        (iii) 3.38 litres   (iv) 4
14. 6.5 cm
15. (a) (i) 43.0 cm²   (ii) 10.0
    (b) (i) 22.2 cm, 14.8 cm, 20 cm
        (ii) 6570cm³   (iii) 78.5%

Chapter **9**

Exercise 9.1

1. (a) 1.4826   (b) 3.7321   (c) 0.5122
   (d) 0.5774   (e) 1.7321   (f) 1.1667
2. (a) 9.4°   (b) 38.7°   (c) 58.9°
   (d) 80.5°   (e) 86.1°   (f) 26.6°

Exercise 9.2

1. 20.6°   2. 71.6°
3. 27.4°   4. 45°
5. 71.0°   6. 35.0°
7. 66.1°

Exercise 9.3

1. (a) 3.57 cm   (b) 4.11 cm   (c) 2.29 cm
   (d) 2.84 cm   (e) 17.3 cm   (f) 9.18 cm
2. (a) 27.5°   (b) 9.60 cm   (c) 4.43 cm
   (d) 13.5 cm   (e) 65.3°

Exercise 9.4

1. 52.4°   2. 2.96 metres   3. 66.4°
4. 5.44 cm   5. 20.6 cm   6. 57.2°
7. 11.7 cm   8. 42.8 metres   9. 62.8°
10. 9.05 m

Exercise 9.5

1. 5.50 cm   2. 5.71 cm   3. 12.5 m   4. 3.85 cm

Exercise 9.6

1. 110 cm   2. 88.8 cm   3. 49.6 m
4. 9.05 cm   5. 128 mm   6. 68.3 mm

Exercise 9.7

1. 10.2 cm   2. 12.8 cm   3. 9.40 cm
4. 4.68 m   5. 4.46 cm   6. 5.80 m
7. 41.8°   8. 31.7°   9. 53.0°

Exercise 9.8

1. **(a)** **(i)** 12   **(ii)** 10   **(iii)** 7
   (b) Cut the 7 m length into two pieces, 3 m and 4 m long. Join all three lengths to form a triangle with sides 3 m, 4 m and 5 m long
2. 5.66 cm   **3.** 8.66 cm   **4.** 8.06 m
5. **(a)** 5.32 cm   **(b)** 48.8°   **(c)** 131°
6. 15.1 cm or 15.2 cm
7. **(a)** 12 cm   **(b)** 9 cm   **(c)** 36.9°

Exercise 9.9

1. **(a)** 240°   **(b)** 280°   **(c)** 50°
   (d) 120°
2. 240°   **3.** 015°
4. **(a)** answer between 18.5 and 19 km
   (b) 282° to 284°
5. 16.2 nautical miles, 158°
6. 16.4 km
7. **(a)** **(i)** SE   **(ii)** NW
   (b) **(i)** 090°   **(ii)** 225°

Exercise 9.10

1. **(a)** 32.6°, 6.11 cm   **(b)** 3.36 cm, 42.1°
   (c) 53.1°, 36.9°   **(d)** 7.28 cm, 74.1°

2. 2.90 m
3. 6.54 cm
4. **(a)** Bearing of D from W is 300° so the angle between the North line and the line DW is 60°. 60° + 30° is 90°
   (b) $a = 110°$, $b = 30°$, $c = 40°$, $d = 50°$
   (c) 6.43 km   **(d)** 7.66 km
5. $a = 11.3°$
6. **(i)** $x = 70$ (alternate angles), $y = 20$ (angles on a straight line), 70 + 20 = 90
   (ii) 50.2°   **(iii)** 120°   **(iv)** 300°
7. 325°
8. **(a)** 1500 m   **(b)** 36.9°
9. **(a)** 270°   **(b)** 045°
10. **(a)** 5.66 cm   **(b)** 32.0 cm²
11. **(a)** **(i)** $<COB = \dfrac{1}{2}(180 - 56) = 62$
    (ii) 2.82 m   **(iii)** 5.63 or 5.64 m
    (iv) 5.30 m
    (b) **(i)** 29.8 or 29.9 m² **(ii)** 12.5 m²
    (iii) 42.3 or 42.4 m²
    (c) **(i)** 21100 or 21200 m³   **(ii)** 30
12. **(a)** 208 cm²   **(b)** 192 cm³
    (c) **(ii)** 12.8 cm   **(iii)** 51.3 or 51.4°
13. 430 m

Chapter 10

Exercise 10.1

1.
(a)
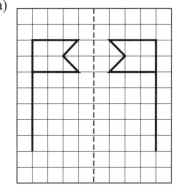
(b)
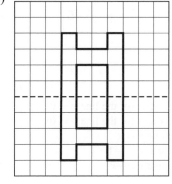

(c)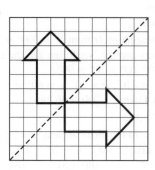

(d)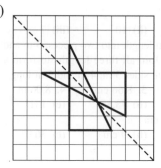

2.

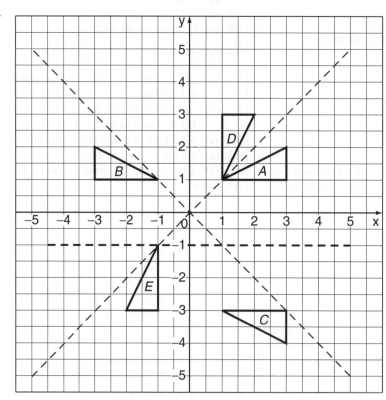

Exercise 10.2

1. Reflection in the *y*-axis (or the line *x* = 0)
2. Reflection in the *x*-axis (or the line *y* = 0)
3. Reflection in the line *y* = *x*
4. Reflection in the line *x* = −1
5. Reflection in the line *y* = −*x*

Exercise 10.3

1. (a)

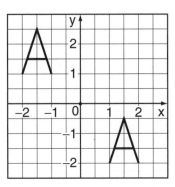

(b)

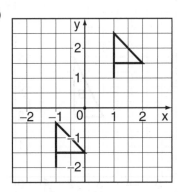

(c)

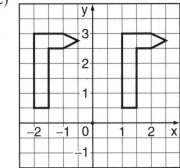

2. For example,

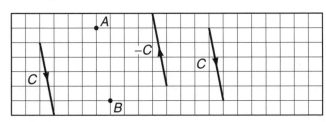

3.

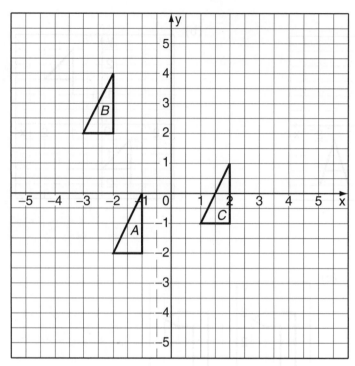

(c) Translation $\begin{pmatrix} 4 \\ -3 \end{pmatrix}$

4. (a) (i) Translation $\begin{pmatrix} 5 \\ 0 \end{pmatrix}$

   (ii) Translation $\begin{pmatrix} 0 \\ -4.5 \end{pmatrix}$

   (iii) Translation $\begin{pmatrix} -5 \\ -4.5 \end{pmatrix}$

(b) (i) Translation $\begin{pmatrix} 4 \\ -0.5 \end{pmatrix}$

   (ii) Translation $\begin{pmatrix} 2 \\ 3.5 \end{pmatrix}$

   (iii) Translation $\begin{pmatrix} -2 \\ -3.5 \end{pmatrix}$

Exercise 10.4

1. (a)

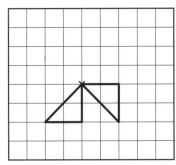

(b)

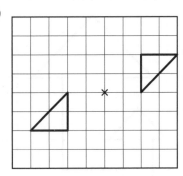

(c)

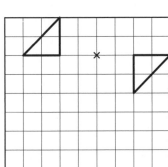

(d)

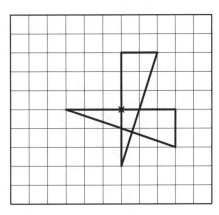

(e)

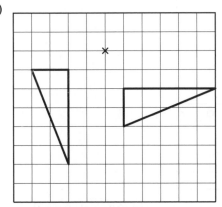

2. (a) Rotation 90° clockwise about (0,0)
   (b) Rotation 90° anticlockwise about (0, −1)
   (c) Rotation 90° clockwise about (1, 2)
   (d) Rotation 180° about (0, 0)
   (e) Rotation 90° clockwise about (2, 1)
   (f) Rotation 180° about (1.5, 1.5)
   (g) Rotation 90° anticlockwise about (0, 0)

Exercise 10.5

1. (a)

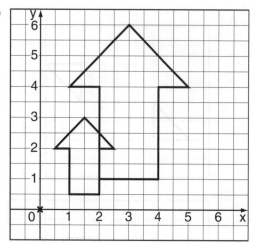

   (b)

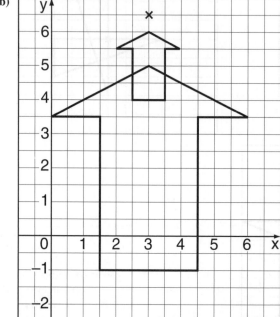

2. (a) Enlargement, centre (0,0), scale factor 2   (b) Enlargement, centre (0,0), scale factor $\frac{1}{3}$

Exercise 10.6

1. (a) $\begin{pmatrix} 3 \\ 2 \\ 1 \end{pmatrix}$ (b) $\begin{pmatrix} -10 \\ 0 \end{pmatrix}$ (c) $\begin{pmatrix} 2 \\ 9 \end{pmatrix}$ (d) $\begin{pmatrix} 60 \\ 28 \end{pmatrix}$

2. $a = 4$

3. $x = 2, \quad y = 3$

4. a, b and f are parallel

5. $\begin{pmatrix} 3 \\ 12 \end{pmatrix} = 3\begin{pmatrix} 1 \\ 4 \end{pmatrix}$

6. They are parallel and the second is twice as long as the first

Exercise 10.7

1. **(a)**

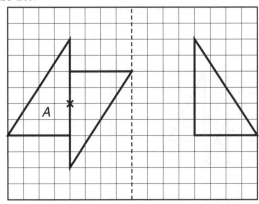

(b)

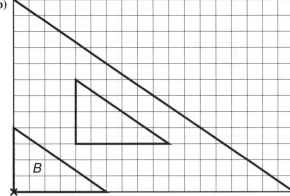

2. **(a)** **(i)** Reflector in the line $y = x$ **(ii)** Reflection in the line $y = -1$

(b)

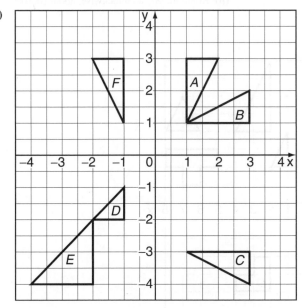

3. (a) Translation $\begin{pmatrix} -6 \\ -7 \end{pmatrix}$

(b) Rotation 90° Clockwise about (0, 0)

(c) Centre (0, 0), scale factor 1.5

(d)

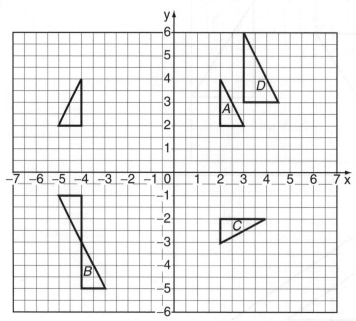

4. (a) $x = 6$, $y = -4$

(b) (i) Rotation, 180° about (2.5, 6)

(ii) Enlargement, centre (1, 7), scale factor 3

5. (a) (i) Translation $\begin{pmatrix} 10 \\ -2 \end{pmatrix}$

(ii) Rotation, 90° anticlockwise, about (0, 0)

(b)

6. (a)

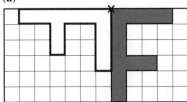

Rotation, 90° Clockwise, about point marked X

(b)

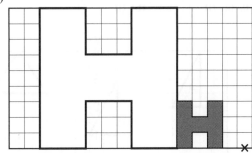

Enlargement, scale factor 3, centre the point marked X

7. (a) $\begin{pmatrix} -3 \\ 12 \end{pmatrix}$ **(b)** *AB* and *CD* are parallel.
CD is three times the length of *AB*

8. (a) $\begin{pmatrix} -1 \\ 3 \end{pmatrix}$ **(b)** $(-2, -1)$

9. (a) Trapezium

(b)

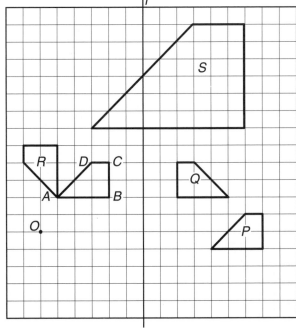

10. (a) (i) $\begin{pmatrix} 3 \\ -1 \end{pmatrix}$ **(b) (i)** $\begin{pmatrix} -2 \\ 2 \end{pmatrix}$

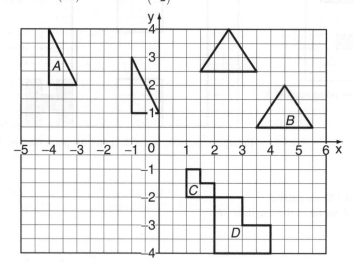

(c) Enlargement, centre (0, 0), scale factor 2
(d) (i) 1   **(ii)** 1
(iii)

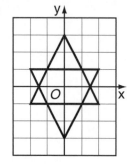

(iv) Reflection in the *x*-axis.

11. (a) $\begin{pmatrix} 5 \\ -2 \end{pmatrix}$

(b)

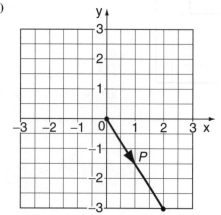

12.

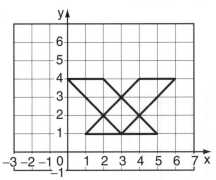

13. (a) (i) $(-3, -2)$

(ii) $\overrightarrow{AB} = \begin{pmatrix} 4 \\ 2 \end{pmatrix}$ $\overrightarrow{BC} = \begin{pmatrix} -3 \\ 2 \end{pmatrix}$

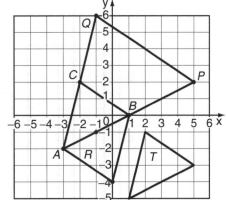

(c) (ii) Enlargement, centre $(-3, -2)$, scale factor 2

Chapter 11

Exercise 11.1

1. (a)

Score	1	2	3	4	5	6
Frequency	6	8	10	6	10	4

(b)

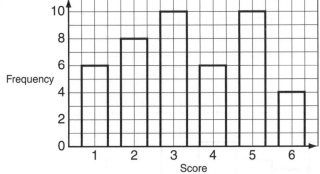

2. (a)

Class	1–10	11–20	21–30	31–40	41–50
Frequency	15	10	7	12	11

(b)

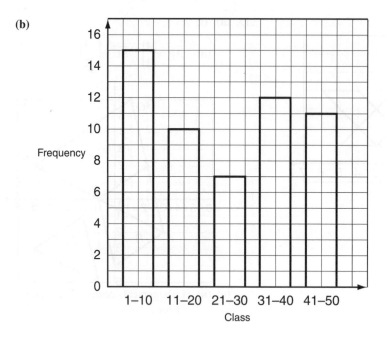

3. (a)

Class	Frequency
$0 \leqslant x < 10$	15
$10 \leqslant x < 20$	5
$20 \leqslant x < 30$	2
$30 \leqslant x < 40$	3
$40 \leqslant x < 50$	5
$50 \leqslant x < 60$	13
$60 \leqslant x < 70$	1

(b)

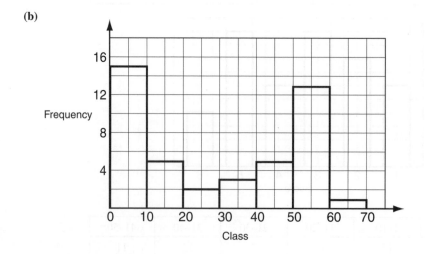

4. (a)

Height (h)	Frequency
$120 \leqslant h < 130$	3
$130 \leqslant h < 140$	4
$140 \leqslant h < 150$	6
$150 \leqslant h < 160$	8
$160 \leqslant h < 170$	4

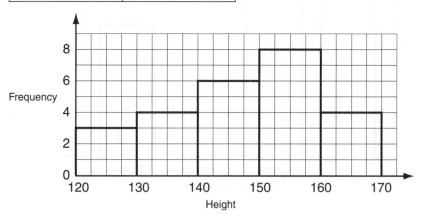

5. (a) and (b)

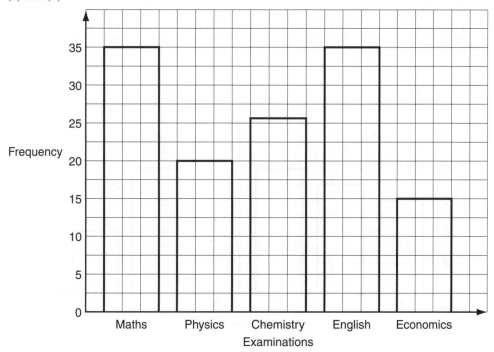

(c) 131

6.

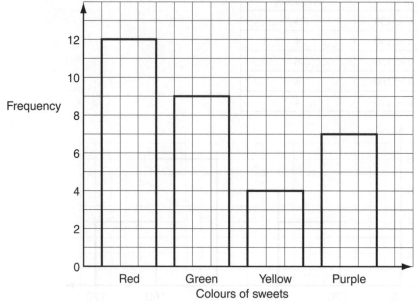

7. (a)

Box ticked	0	1	2	3	4	5
Frequency	6	7	8	11	8	4

(b)

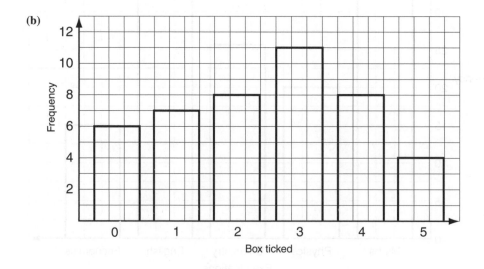

Exercise 11.2

1.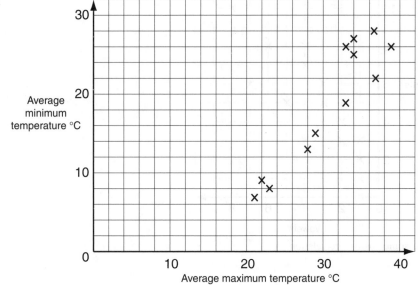

She is right. There is a positive correlation. Warmer days have warmer nights.

2. (a)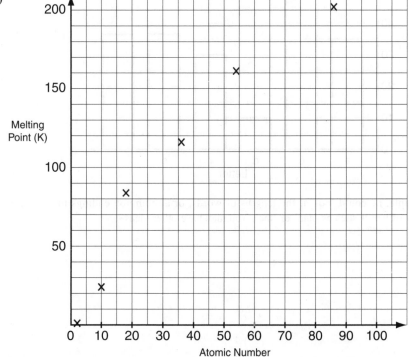

(b) There is a positive correlation. The higher the atomic number, the higher the melting point

3. (a) (i)

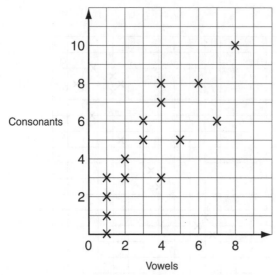

(ii) There is some positive correlation

(b)

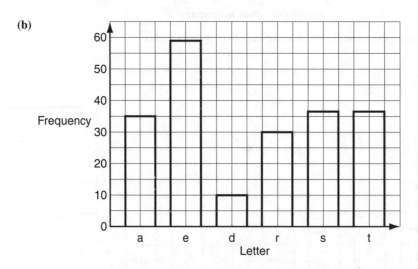

(c) For example, he could look at the lengths of words, or at common endings to words, or at the frequency of words such as 'and' or 'the'. You may have other ideas

4. (a) and **(b)**

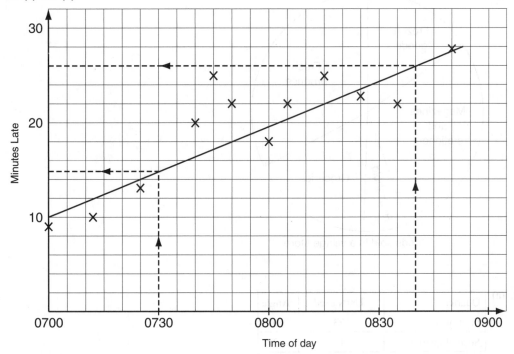

NOTE: The line of best fit is drawn by eye. Yours may not be the same. Your answers to parts **(c)** and **(d)** should be read from *your* graph.

  (c) 15 minutes
  (d) Journey will take **26** minutes.
    They will be **6** minutes late.

Exercise 11.3

1. (a) and **(c)**

Category	Fresh produce	Groceries	Household products	Magazines and stationery	Frozen goods	TOTAL
Number of items	351	183	66	315	165	1080
Angle	117°	61°	22°	105°	55°	360°
Floor space (m²)	32.5	16.9	6.1	29.2	15.3	100

(b)

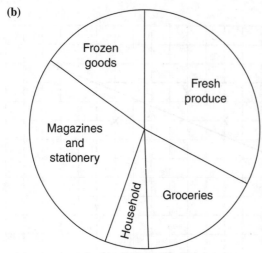

Goods sold in a village store

2. (a)

Quarter	Units used	Angle
First quarter	23	69°
Second quarter	11	33°
Third quarter	21	63°
Fourth quarter	65	195°
TOTAL	120	360°

(b)

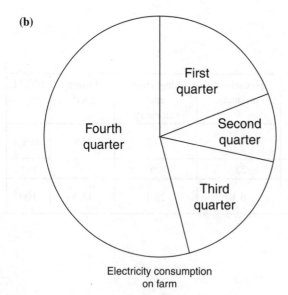

Electricity consumption
on farm

3. (a)

	Percentage of total stock	Angle on pie chart
Newspapers	35	126°
Magazines	50	180°
Snacks	15	54°
TOTAL	100	360°

(b)

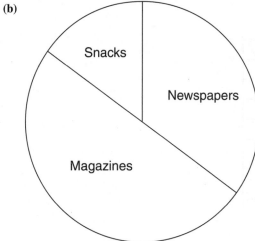

Items stocked by a newsagent

4.

	Number of students	Angle on pie chart
Psychology	14	70°
Sociology	20	100°
Economics	22	110°
History	16	80°
TOTAL	72	360°

Subjects studied by a group of students

Exercise 11.4

1. (a) (i) 5.875   (ii) 5.5   (iii) 9   (iv) 9
    (b) (i) 4.93   (ii) 6   (iii) 7   (iv) 8
    (c) (i) 3.98   (ii) 4.2   (iii) 2.6   (iv) 3.1
2. (a) 800 cm   (b) 159 cm
3. (a) 28.5   (b) 5 new classrooms   (c) 24.5
4. (a) 1500   (b) 100 bags

Exercise 11.5

1.

Data value	Frequency	Value × frequency
100	7	700
110	10	1100
120	15	1800
130	2	260
140	6	840
150	3	450
160	7	1120
TOTALS	50	6270

(a) 125.4   (b) 120   (c) 120

2.

Data value	25	26	27	28	29	30	31	TOTALS
Frequency	51	70	69	32	15	43	15	295
Value × frequency	1275	1820	1863	896	435	1290	465	8044

(a) 27.3   (b) 27   (c) 26

3.

Data value	Frequency	Value × frequency
12.4	3	37.2
12.5	5	62.5
12.6	2	25.2
12.7	1	12.7
12.8	0	0
12.9	5	64.5
13.0	0	0
13.1	2	26.2
TOTAL	18	228.3

(a) 12.7   (b) 12.6   (c) 12.5 and 12.9

Exercise 11.6

1. (a) mode $= 8$, range $= 25$

(b)

Days absent	Frequency
0 to 4	10
5 to 9	10
10 to 14	7
15 to 19	3
20 to 24	2
25 to 30	1
Total frequency	33

(c)

2. (a) 840

(b) The mode might give an indication of the possible results for the whole school, but the median only refers to the data for the 100 students.

3. $a = 3$, $\qquad b = 6$, $\qquad c = 7$

4.

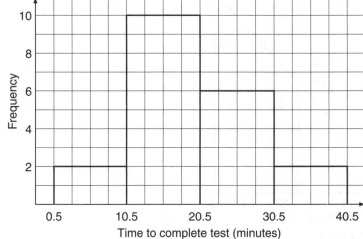

5. Mean = 3.41. Median = 3. There are two modes: 3 and 5
6. Mean = 4.9. Median = 4.5. Mode = 4
7. 1.38
8. **(a)** and **(c)**

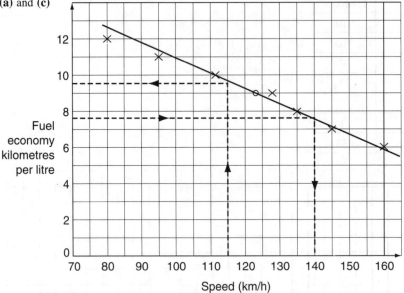

Speed (km/h)

(b) **(i)** 122 km/h **(ii)** 9 kilometres per litre
(d) Within the range of speeds shown there is a negative correlation, so that the faster the car travels the fewer kilometres it will do per litre, so the less economical it is to run. However, this does not mean that this is the case for speeds outside this range.
(e) 9.5 Kilometres per litre   **(f)** 140 km/h
9. **(a)** **(i)** 163 cm   **(ii)** 24 cm
   (b)

(iii) 163 cm   **(iv)** positive
(v) That taller students tend to have greater handspans

10. **(a)** 126   **(b)** 40%

11. negative

12.

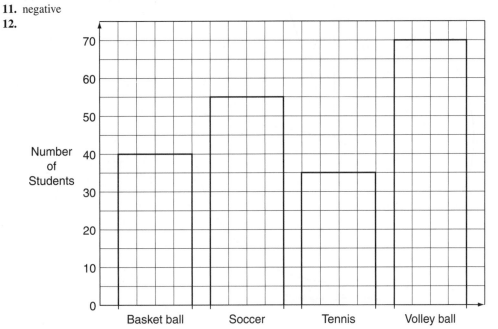

13.

(i)

Grade	Number of students	Angle on pie chart
A	5	20°
B	15	60°
C	40	160°
D	20	80°
E	10	40°
TOTALS	90	360°

(ii)

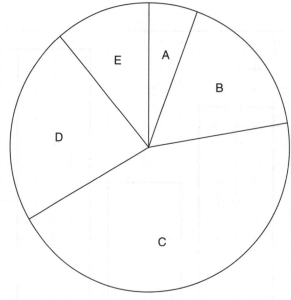

Examination grades

14. negative

15. **(a)** 590°C   **(b)** Neptune

16. **(a)** 22   **(b)** 77   **(c)** 89

17. **(a)** **(i)** 50   **(ii)** 43.9   **(iii)** 47

   (b) Two of the estimates (20 and 24 cm) were much lower that the others and have too great an influence
      on the mean

18. **(a)**

Sport	Volleyball	Football	Hockey	Cricket
Number of students	6	9	7	2
Angle on pie chart	90°	135°	105°	30°

(b)

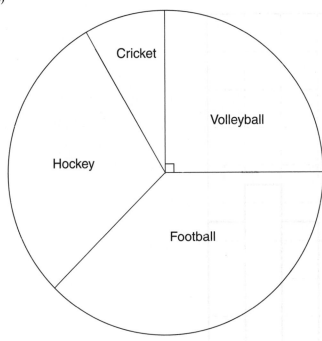

(c) football

19. **(a)** 8.36   **(b)** 8   **(c)** 6   **(d)** Frequencies: 3, 4, 4, 3
20. **(a)** Frequencies: 4, 7, 6, 4, 4, 2, 3   **(b)** 1   **(c)** 2   **(d)** 2.5   **(e)** 40
21. **(a)** 7   **(b)** 42   **(c) (i)** 9   **(ii)** 8   **(iii)** 8.3
    (d) 5 cm   **(e)** 36°   **(f)** $7.50   **(g)** 22%
22. **(a) (i)**

Number of people in a car	Tally	Number of cars
1	𝐽𝐻𝑇 𝐼	6
2	𝐽𝐻𝑇 𝐽𝐻𝑇 𝐽𝐻𝑇 𝐼𝐼	17
3	𝐽𝐻𝑇 𝐼𝐼𝐼	8
4	𝐽𝐻𝑇 𝐼𝐼𝐼𝐼	9
5	𝐽𝐻𝑇 𝐽𝐻𝑇 𝐼	11
6	𝐽𝐻𝑇 𝐼𝐼𝐼𝐼	9

(ii)

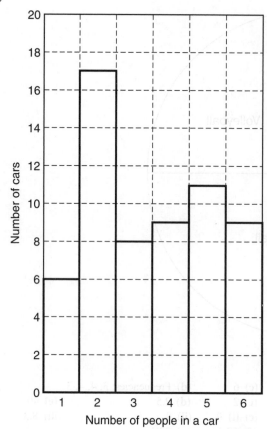

(iii) 2   **(iv)** 3   **(v)** 3.48

(b) 66°

Chapter 12

NOTE: Unless specified in the question answers to probability questions may be given as fractions, or, if exact, as decimals or percentages. The fractions do not have to be simplified. Some, but not all, of the alternatives have been given in these answers.

Exercise 12.1

1. (a) (iii)   **(b)** (i)   **(c)** (iv)
   (d) (v)   **(e)** (ii)

2. (a) biased
   (b) mutually exclusive
   (c) mutually exclusive
   (d) random, outcome

3. (a) and (d)

4. 60% or 0.6

5. (a) $\dfrac{5}{12}$ **(b)** $\dfrac{4}{12}$ or $\dfrac{1}{3}$

   (c) 0   **(d)** $\dfrac{8}{12}$ or $\dfrac{2}{3}$

6. $\dfrac{2}{5}$

7. $\dfrac{1}{5}$

8. (a) $\dfrac{3}{10}$ **(b)** $\dfrac{6}{10}$ or $\dfrac{3}{5}$

    (c) $\dfrac{3}{10}$ **(d)** 0

9. (a) $\dfrac{1}{6}$ **(b)** $\dfrac{1}{5}$

10. $\dfrac{5}{18}$

11. Because the areas are not all equal in size. The larger the area the higher the probability that the counter will land on it.

12. (a) Because the numbers of each car sold and awaiting collection are different so the outcomes are not all equally likely. The probability that the next car to be collected will be blue is $\dfrac{140}{360} = \dfrac{7}{18}$.

    (b) $\dfrac{60}{360}$ or $\dfrac{1}{6}$

    (c) 0

    (d) $\dfrac{200}{360}$ or $\dfrac{5}{9}$

Exercise 12.2

1. (a) 0.95

    (b) (i) 50   **(ii)** No

2. 49% or 0.49

3. (a) $\dfrac{1}{5}$ **(b)** 5   **(c)** 20%

4. 3880

5. (a) $\dfrac{20}{47}$ **(b)** $\dfrac{35}{47}$

6. (a) $\dfrac{3}{7}$ **(b)** $\dfrac{6}{7}$

7. $\dfrac{14}{30}$

8. (a) 10   **(b)** $\dfrac{4}{35}$

9. (a) $\dfrac{2}{5}$ **(b)** 0

10. (a) 20, 60, 160, 80, 40

    (c) (i) $\dfrac{4}{9}$ **(ii)** $\dfrac{3}{9}$ or $\dfrac{1}{3}$

11. (a) (i) $\dfrac{7}{30}$ **(ii)** $\dfrac{9}{30}$ **(b)** 40

12. (a) 15% or 0.15

    (b) (i) $\dfrac{4}{15}$ **(ii)** $\dfrac{10}{15}$ **(iii)** 0

13. (b) (i) $\dfrac{10}{24}$ **(ii)** $\dfrac{15}{24}$ **(iii)** $\dfrac{19}{24}$

    (c)

Probability Scale

Impossible   Certain

(i) 0   A   B   C   (ii) 1

14. (a) $\dfrac{12}{23}$ **(b)** $\dfrac{11}{20}$

15. (a) (i) $\dfrac{31}{36}$ **(ii)** 0   **(iii)** 1

    (b) $\dfrac{17}{99}$ **(c)** Piero's

Index

NOTE: The page numbers given are for the principal pages on which the items occur.